Frozen Assets:
British mining, exploration, and geopolitics on
Spitsbergen, 1904-53

Dear Dmitry

Thank you for your contribution to the Adaptation Futures 2016 symposium.

Annette
Scheepstra

Circumpolar Studies

Volume 9

Circumpolar Studies is a series on Dutch research in the Polar Regions published by the Arctic Centre of the University of Groningen in the Netherlands.

For other titles in the series, see the website of the publisher at www.barkhuis.nl.

This research was funded by the Groningen Institute of Archaeology (GIA) of the University of Groningen as part of the University's Ubbo Emmius Scholarship Programme.

NWO AWL financially supported the LASHIPA expeditions to Spitsbergen (Svalbard) as part of the International Polar Year (IPY) 2007-2008.

Cover photograph: D. Avango.
Cover design: Nynke Tiekstra, ColtsfootMedia, Rotterdam.
Book design: Barkhuis Publishing, Eelde.

ISBN: 9789491431333

RIJKSUNIVERSITEIT GRONINGEN

FROZEN ASSETS

British mining, exploration, and geopolitics on Spitsbergen, 1904-53

Proefschrift

ter verkrijging van het doctoraat in de
Faculteit der Letteren
aan de Rijksuniversiteit Groningen
op gezag van de
Rector Magnificus, dr. E. Sterken,
in het openbaar te verdedigen op
19 december 2013
om 9.00 uur

door

Frigga Kruse

geboren op 19 januari 1978
te Rendsburg, Duitsland.

1 Introduction and contextualisation

1.1 Introduction

British mining enterprise reached Spitsbergen by the beginning of the twentieth century. Spitsbergen lies in the European High Arctic. Its industrial development was subject to two defining factors. Firstly, the uninhabited archipelago comprised an unknown quantity of natural resources. Secondly, it was a no man's land. Although English adventurers[1] had already reported the presence of coal some 300 years earlier, a Norwegian sealer undertook the first commercial shipment to Norway in 1899. Subsequently, Spitsbergen experienced a coal rush.[2] Norwegian, British, and American entrepreneurs were the first to compete for coal and other economic minerals. Swedish, Russian, and Dutch companies soon followed. Simultaneously, an international conflict over the sovereignty of Spitsbergen evolved, which is commonly referred to as the Spitsbergen Question.[3] It was only resolved with the signing of the Treaty concerning Spitsbergen in 1920.[4] The treaty placed the administration of the islands in Norwegian hands, but it allowed for the continued commercial exploitation by other nations. Despite the settled legal status, the British companies nonetheless fell victim to unfavourable market forces during the ensuing decades. At present, only a Norwegian and a Russian company continue to mine coal here.

The focus of this book is the role of Great Britain in the industrial development of Spitsbergen. Following the initial documentation by the Dutch in 1596, English whalers were among the first people to frequent the archipelago in the early seventeenth century. These were followed by a surge of naval and scientific expeditions during the second half of the nineteenth century. At the beginning of the twentieth century, a number of British mining and exploration

---

[1] An early definition of adventurer, which makes a comeback in this book, is a business investor who 'ventures' capital, commonly in risky commercial enterprises that hold the promise of financial gain. A case in point is the Company of Merchant Adventurers (in full: *The Mystery, Company, and Fellowship of Merchant Adventurers for the Discovery of Regions, Dominions, Islands, and Places Unknown*) founded in London in the mid-sixteenth century. It was the precursor to the Muscovy Company, whose activities centred around Spitsbergen in the seventeenth century.

[2] The scale of this coal rush was small compared to, say, the gold rushes in California (1848), Witwatersrand/SA (1886), or Klondike/Yukon (1896). Yet, it entailed a flurry of human activity hitherto unknown in this part of the world.

[3] In British diplomacy of the late nineteenth and the early twentieth century, issues pertaining to individual balances of power were commonly referred to as 'questions'.

[4] The full title is 'Treaty between Norway, The United States of America, Denmark, France, Italy, Japan, the Netherlands, Great Britain and Ireland and the British overseas Dominions and Sweden concerning Spitsbergen signed in Paris 9th February 1920'. For the remainder of this book, it will be called the Spitsbergen Treaty. Any references to a Svalbard treaty are anachronous.

companies was registered in an attempt to make their fortune in this Arctic region. The four companies of interest to this study were:

> the Spitzbergen Coal & Trading Company, Ltd. (1904-18);
> the Spitzbergen Mining & Exploration Syndicate, Ltd. (1906-11);
> the Northern Exploration Company, Ltd. (1910-34);[5]
> and the Scottish Spitsbergen Syndicate, Ltd. (1909-53).[6]

At the time of their arrival, they knew little of this unpeopled land located in a potentially hostile environment. Furthermore, the companies operated against a backdrop of political upheaval and economic uncertainty. The prosperous Victorian era had ended with the monarch's death in 1901; the spirit of Edwardian England had already been shaken by the Second Boer War in South Africa (1899-1902); and the British Empire found itself en route to the First World War (1914-8). The post-war economic boom did not last. Instead, the European markets of the interwar years were marred by economic depression. After the Spitsbergen Treaty was ratified in 1925, Spitsbergen and Bear Island to the south were joined under the current name of Svalbard, a Norwegian territory administered by a governor or *sysselmannen*. Only the Scottish syndicate prevailed in Svalbard until after the Second World War (1939-45). It was sold to a Scottish development firm in 1950, which has not been treated in this study.

1.2 The central research question and hypothesis

The primary aim of this study is to explain the British operations on Spitsbergen from a historical international comparative perspective. Hence, the central research question is:

> *What were the driving forces behind the development of the British mining industry on Spitsbergen between 1904 and 1953?*

---

[5] The Northern Exploration Co. was instrumental in the registration of the Spitzbergen Syndicate, Ltd. (1924-7), not to be confused with the Scottish Spitsbergen Syndicate. It was formed to underwrite the company's shares as will be shown in Chapter 7.

[6] The other registrations were Spitzbergen United, Ltd. (1906-12), Spitzbergen Mineral, Ltd. (1919-24), the Anglo-Russian Grumant Co., Ltd. (1920-32), and Scottish Spitsbergen (Development) Ltd. (1950-53). The first two were excluded from this study because they never came into operation. The Anglo-Russian firm was left out because its operations were under Russian management. Although the Scottish Spitsbergen Syndicate had transferred its assets and rights to the Scottish development company, the latter was the product of a new era: the international discussions surrounding its sale to Norway in 1953 echoed early Cold War sentiments and in part created the Russo-Norwegian relations of today. Its inclusion was outside the scope of this work.

Driving forces are understood to be those factors that created and drove the whole historical process. Generally, they include economy, technology, politics, and culture. Previous work on the Spitsbergen coal rush has only shown, however, that all companies of the represented nationalities had broad economic and political motives.[7] Yet, the research did not discriminate between different companies harbouring these or other motives in different proportions against the backdrop of rapidly and dramatically changing global contexts. In Britain especially, there were times when anti-German sentiments in particular had a profound effect on all walks of life. The related hypothesis, therefore, is:

> *While economic driving forces initially determined the British presence on Spitsbergen, political motivations amplified at specific moments in time and the repercussions of these motivations sustained the companies beyond economic feasibility.*

The central research question can be divided into four sub-questions. Firstly, why were the British mining and exploration companies started? This question deals predominantly with the original economic incentives and lends itself to analysis along the lines of a core-periphery model, in which a European industrial core invested heavily in an Arctic margin. The underlying sub-hypothesis is that British stakeholders financed the exploitation of Spitsbergen for as long as the archipelago could promise marketable raw materials.

Secondly, how did the British companies choose to operate, and why? Besides economic factors, this question addresses the other aforementioned driving forces of technology, politics, and culture. In addition, knowledge, competence, management, and labour relations also played an important role. Thus, British actions entailed complex interactions in a specific environment, which lends itself to analysis according to an actor-network approach, in which success of an undertaking hinges on the ability to enlist human and non-human agents alike into a network designed to assume power and maintain control. The corresponding sub-hypothesis is that British operations purposefully enrolled meaningful actors into their networks to create a powerful rhetoric and strengthen their hold on Spitsbergen.

Thirdly, why did the British companies discontinue? This query takes into consideration both Spitsbergen's marginality and the actor-networks and looks at external and internal factors that contributed to the companies' dissolution. The supposition is that since Spitsbergen could deliver neither the raw materials,

[7] For references, refer to section 1.5 on previous work.

international prestige, nor strategic importance, backers withdrew and networks crumbled against a backdrop of unfavourable markets.

Lastly, what were the consequences of the companies' participation on Spitsbergen? This question quantifies the British role in terms of profitability, political influence, and environmental impact. By and large, the companies were small, short-lived, unsuccessful, and their claims were usually not worked again. Spitsbergen never became a British colony, although British attitudes were woven into the Spitsbergen Treaty and the associated Mining Code.[8] Today, the companies' archaeological remains on the archipelago are extensive and increasingly popular with tourists. Thus, the last sub-hypothesis is that the British legacy on Spitsbergen is one of minimal economic exploitation, negligible environmental impact but inclusive Arctic mining regulations and enduring industrial heritage. These four sub-questions move away firstly from national narratives to include the international dimension and secondly from purely economic reasoning to include an array of external and internal driving forces specific to this Arctic region at the time in question.

The ambition of this book is to accentuate the little known chapter of British mining and exploration on Spitsbergen within the greater story of the historical exploitation of the Polar Regions. Until now, the exploitation of the poles has almost exclusively been considered from a national if not nationalistic approach. Research on this topic has been based primarily on written sources. So the objective is to move away from these limitations by introducing complimentary dimensions as well as archaeological fieldwork on Spitsbergen. This study is therefore suitable for a wide readership of scholars of polar history, mining history, environmental history, industrial archaeology, cultural heritage, geopolitics, sustainable development, and any combination thereof.

1.3 The research context

This study has a place within a wider international research context. It is a sub-project of LASHIPA (in full: *Large-scale Historical Exploitation of Polar Areas*), which investigates the history of three exploitative industries at both poles over the last four centuries.[9] The LASHIPA project is based at the Arctic Centre of the

---

[8] The full title of the consulted document is 'The Mining Code (the Mining Regulations) for Spitsbergen (Svalbard), laid down by Royal Decree of 7 August 1925 as amended by Royal Decree of 11 June 1975.'

[9] Avango, D. and Hacquebord, L. (2008) 'The history and heritage of natural resource exploitation in the Arctic and Antarctic: the LASHIPA project', *Industrial Patrimony*, 19, pp. 7-16; Hacquebord, L. and Avango, D. (2009) 'Settlements in an Arctic Resource Frontier Region', *Arctic Anthropology*, 46 (1-2), pp. 25-39; Avango, D., Hacquebord, L., Aalders, Y., Haas, H. R. de, Gustafsson, U., and Kruse, F. (2011) 'Between markets and geo-politics: natural resource exploitation on Spitsbergen from 1600 to

University of Groningen in the Netherlands and involves researchers from universities and institutes in Sweden, the United States of America, Russia, Norway, and Great Britain. LASHIPA was part of the International Polar Year (IPY) 2007-2008.[10] The IPY was organised through the International Council for Science (ICSU) and the World Meteorological Organization (WMO) and involved over 200 projects with thousands of scientists from 60 nations.[11] Earlier IPYs focussed on meteorological and geophysical phenomena; this fourth IPY also included projects in the hitherto underrepresented disciplines of social science and humanities.

LASHIPA was one of these projects, primarily rooted in the distinct yet complementary fields of polar history and industrial archaeology. It concerns the story and legacy of natural resource exploitation at the poles and provides an illustrative example of the way in which people have utilised and are utilising raw materials around the world. As such, it aims to explain the development of large-scale historical exploitation at the poles since the seventeenth century and to reveal its geopolitical and environmental consequences. LASHIPA is aware that the polar past has previously been viewed almost exclusively from regional or national perspectives, using archival inquiries only.[12] Its objectives are therefore to move away from any patriotic bias and to broaden the scope of research in polar history to include a transnational angle as well as standardised archaeological recording.

This book contributes to answering the main questions of the parent project: why, how, and under what economic and geopolitical preconditions have the natural resources of the poles been explored and exploited? What was the function of the settlements and stations in this development? What were the economic results and the effects on the natural environment and the geopolitical situation? To meet the objectives and address the research questions, LASHIPA pinpoints three distinct industrial periods at the poles and formulates five interrelated research themes.[13] The three periods are: a) whaling and hunting in the Arctic and Antarctic; b) coal and mineral mining in the Arctic; and c) oil extraction in the Arctic. The chosen case studies are Spitsbergen in the Arctic, and South Georgia and South Shetland in the Antarctic. The five themes are: 1) the driving forces behind polar industrial development; 2) the design of polar industrial

---

the present day', *Polar Record*, 47 (240), pp. 29-39; Hacquebord. L. (ed.) (2012) *LASHIPA. History of large scale exploitation in polar areas*, Groningen: Barkhuis.

[10] *International Polar Year 2007-2008* (2010) Available at: www.ipy.org (Accessed: 25 May 2009).

[11] Krupnik, I. *et al.* (eds) (2011) *Understanding earth's polar challenges: International Polar Year 2007-2008,* Edmonton: CCI Press in collaboration with the University of the Arctic and ICSU/WMO Joint Committee for International Polar Year 2007-2008.

[12] Leading descriptive reference works are: Hoel, A. (1966-7) *Svalbard: Svalbards historie 1596-1965.* (3 vols.) Oslo: Sverre Kildahls Boktrykkeri; Arlov, T. B. (1989) *A short history of Svalbard*, Oslo: Norwegian Polar Institute.

[13] Avango and Hacquebord (2008).

technology; 3) the design of polar industrial settlements and the organisation of production; 4) the political influence and control over polar natural resources; and 5) the consequence of polar industries on the local environment. Within LASHIPA, the core-periphery model and the actor-network theory (ANT) provide the theoretical tools and analytical framework. Hence, industrial sub-projects such as this can be compared with different branches, different times, different national origins, and across both poles. In addition, historical and archaeological sources can be streamlined. Ultimately, there is room for synthesis and overarching conclusions.

1.4 The theoretical context

The research question enquires into the reasons behind British industry on Spitsbergen between 1904 and 1953, a key difference between Britain and Spitsbergen being their location and associated physical environment. Britain in North-West Europe has a temperate maritime climate; Spitsbergen above the Arctic Circle has an arctic climate. Here, striking annual variations occur regarding light conditions, sea ice extent, and ocean currents.[14] While these annual occurrences invariably affected the historical actors in later chapters, they at this point serve to emphasise the importance of space and how to operate therein. It links up with debates in economic history and geography, science and technology studies, and any multi-disciplinary variants thereof. While the LASHIPA project on the whole adheres to the *core-periphery model* as a theoretical tool with which to explain the traditional economic relationship between the two regions, the idea of *marginality* is presented here as a feasible, inclusive alternative. Secondly, the *actor-network theory* (ANT) moves away from, for instance, business history or technological systems to also include non-economic determinants. Both concepts are borrowed from to create a study-specific analytical frame.

In the late 1960s, Friedmann inaugurated the core-periphery debate.[15] He constructed different types of global economic regions and described the relationship between them. Relevant to this work are core regions and resource

[14] The warm and relatively salty West Spitsbergen Current keeps the area west of the shelf essentially ice-free, even in winter; on the shelf, cold and comparatively fresh Arctic Water arrives from the east, rounds the southern tip, and flows northwards. The common wind direction is along the valleys or fjords from inland to sea, the prevailing winds are from the north-east to the south-east sectors, except in summer. The break-up of sea ice occurs between April and July. An outer fjord can then be either ice-free or covered with drift ice. Ice conditions in the middle of a fjord are subject to highly variable winds, waves, and tides. Fast ice remains in an inner fjord until the onset of the melt. (Svendsen, H., *et al.* (2002) 'The physical environment of Kongsfjorden-Krossfjorden, an Arctic fjord system in Svalbard', *Polar Research,* 21 (1), pp. 133-66.)

[15] Friedmann, J. (1966) *Regional development policy: a case study of Venezuela*, Cambridge/MA: MIT Press.

frontier regions. The core has a high potential of innovation and growth; the resource frontier is often remote and sparsely peopled with limited development and diversification. For the core to expand, it needs to engulf the periphery and exploit its resources. Sugden adopted the notion in his research on the Polar Regions in the early 1980s.[16] He observed that the poles comprise similar natural resources to other parts of the world, with the exception of plant-based products. They have witnessed intrusive waves of economic development, which was occasionally superimposed on indigenous systems. By the 1980s, the polar peripheries were fast becoming extensions of their respective national cores. This was aided by the improvement of transport links between the regions. Sugden's assessment of Spitsbergen's capital, however, has been overtaken by subsequent events. Nowadays, Longyearbyen has outgrown the limited expectations of a specialised mining town and transformed into a diverse cultural centre. More recently, Bone applied the core-periphery model to perceptions of the Canadian North. He maintained that 'much of the economic destiny of the resource hinterlands is controlled by external forces.'[17] Development is highly dependent on global demand and often only government intervention may assure continuation, let alone success.

The core-periphery model has been criticised. The uneven development of the core and the periphery, for instance, is not a consequence of development *per sé* but rather a result of the chosen modes of production. Of communist, socialist, and capitalist modes, to name a few, Friedmann's model only represents the effects of the latter. Hayter, Barnes, and Bradshaw question the usefulness of the model in light of increasing globalisation.[18] They claim that industrial cores have dominated the discourse for too long, defining peripheries as 'remote; elsewhere, foreign, uncomfortable, expensive to reach and sometimes dangerous.'[19] In fact, 'studying resource peripheries can provide new insights into the global economy that cannot be derived from the experience of cores, and which then act as a catalyst for new forms of economic geography theorizing.'[20] The authors move away from Friedmann's purely economic driving forces and recognise that 'for resource peripheries around the globe, environmental, cultural and geopolitical factors are intersecting with industrial dynamics in unique ways.'[21]

---

[16] Sugden, D. E. (1982) *Arctic and Antarctic: a modern geographical synthesis*, Oxford: Basil Blackwell.
[17] Bone, R. M. (2003) *The geography of the Canadian North*, Oxford: Oxford University Press, p. 11.
[18] Hayter, R., Barnes, T. J., and Bradshaw, M. J. (2003) 'Relocating resource peripheries to the core of economic geography's theorizing: rationale and agenda', *Area*, 35 (1), pp. 15-23.
[19] Hayter, Barnes, and Bradshaw (2003) p. 17.
[20] Hayter, Barnes, and Bradshaw (2003) p. 17.
[21] Hayter, Barnes, and Bradshaw (2003) p. 21. Similar conclusions have also been reached by Hacquebord, L. (1996) 'Whaling stations as bridgeheads for exploration of the Arctic regions in the sixteenth and seventeenth century', *International Conference on Shipping, Factories and Colonization*

For the purposes of this book, a near-linear relationship between a core and a periphery does not apply to Britain and Spitsbergen at the turn of the twentieth century. In addition to economic relationships, there is a strong human determinant. The assertion of power dictated whether Spitsbergen was far away and foreign – or close by and familiar.[22] If beneficial, a heroic pioneer might push the frontier further into the remote and untamed Arctic wild (which would have appealed to Romanticists and New Imperialists); similarly, company promoters placed territorial claims within easy reach of Europe (Map 1) and in a summer climate comparable to that of Scotland in autumn to pacify stakeholders. Spitsbergen's location therefore seemed perfectly fluid. As such, this work is better served with the emerging and interlinked notion of marginality as both an economic concept as well as a social construct. In turn, social, cultural, and political facets and uses of marginality facilitate the leap to ANT.

From the literature, the basic building blocks and processes of the actor-network emerge.[23] An *actor* is a person, group, or organisation that acts consciously, while an *actant* can be anyone or anything that acquires meaning and performs only through an actor. A prospector on Spitsbergen, for example, might

---

1994. Brussels 24-26 November, Brussels: Koninklijke Academie van Beligië, pp. 289-97; Hacquebord and Avango (2009).

[22] To remedy the short-comings of the core-periphery model, Cullen and Pretes pursue the meaning of marginality. Every society views itself as central; foreigners and strangers are on the outside. The contrast between the centre and the periphery can be seen everywhere and is perhaps not a model at all. 'Ethnocentrism appears to be a universal human trait.' On one hand, marginality can be understood as an economic concept and the core-periphery model can be applied to analyse marginal or peripheral regions as discussed above. On the other hand, marginality is a social construction. '[...] *Power* becomes the central determinant of marginality. The social constructivist view perceives marginality as a power relationship between a group viewing itself as a "center," and consequently viewing all minorities and non-members as marginal or "other."' The authors conclude that no clear meaning of marginality exists as of yet. While survey respondents understood it to be a fixed concept or real condition that can be observed, emerging literature increasingly perceives marginality as a social construction: it is relevant to one's point of reference, fluid, and can thus be projected onto a landscape. (Cullen, B. T. and Pretes, M. (2000) 'The meaning of marginality: interpretations and perceptions in social science', *Social Science Journal*, 37 (2), pp. 215-29.)

[23] Callon, M. (1986) 'Some elements of a sociology of translation: domestication of the scallops and the fishermen of St Brieuc Bay', in Law, J. (ed.) *Power, action and belief: a new sociology of knowledge*, London: Routledge & Kegan Paul; Latour, B. (1987) *Science in action: how to follow scientists and engineers through society*, Milton Keynes: Open University Press; Law, J. (1987) 'Technology and heterogeneous engineering: the case of Portuguese expansion', in Bijker, W. E., Hughes, T. P., and Pinch, T. J. (eds.) *The social construction of technological systems: new directions in the sociology and history of technology,* Cambridge/MA: MIT Press; Law, J. and Hassard, J. (eds.) (1987) *Actor network theory and after*, Oxford & Keele: Blackwell & Sociological Review; Law, J. and Callon, M. (1992) 'The life and death of an aircraft: a network analysis of technical change', in Bijker, W. E. and Law, J. (eds) *Shaping technology/building society, studies in sociotechnical change,* London: MIT Press, pp. 21-52; Avango, D. (2003) 'Aktanter i ingenmanslandet', in Avango, D. and Lundström, B. (eds.) *Industrins avtryck*, Stockholm: Brutus Östlings Förlag Symposium, pp. 173-206 ; Avango, D. (2005) *Sveagruvan – Svensk gruvhantering mellan industri, diplomati och geovetenskap*, Stockholm: Jernkontoret; Latour, B. (2005) *Reassembling the social: an introduction to actor-network-theory*, Oxford: Oxford University Press.

deliberately claim an area with the use of place-names and claim markers. In his absence, the area may be protected either by an appointed Norwegian hunter or by the symbolic function of the toponym and the marker. A *network builder*, in turn, is a person, group, or organisation wanting to start and maintain a project of sorts. With regards to this book, the network builder is invariably the company, represented by a head office with varying degrees of in-house expertise (Fig. 1.1), and its goal is financial gain through the exploitation of Spitsbergen.

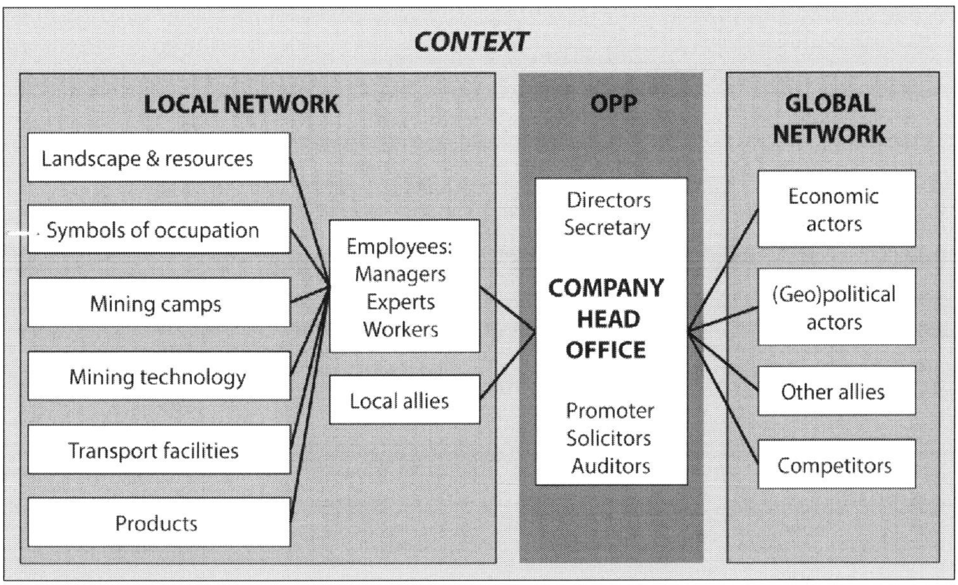

1.1 Graphic representation of a likely actor-network for Spitsbergen. (After Avango (2003) p. 198.)

Building the network hinges on four specific phases. During *problematisation*, the company defines the problems that need to be solved to get the project off the ground as well as the actors who can solve them. It assumes control over all contacts between different actors, establishing itself as the obligatory point of passage (OPP). During *interessement*, the company gets actors interested in the project and discusses the terms of their involvement. During *enrollment*, the actors agree that the project is worth pursuing and accept their roles. During *mobilisation*, the actors begin to actively support the network. Managing the project depends on the directors' ability to maintain and strengthen the global and the local networks and assure the flow of resources between them.

The *global network* provides resources in the form of capital, equipment, legal advice, political influence, and more. The main actors in the global network tend to be entrepreneurs, politicians and governmental departments, scientists and academic institutions, the media, and the public. The *local network* provides natural resources, which may be accompanied by prestige and strategic importance. Its actors are commonly individuals working on the project such as managers, engineers, scientists, and workers. On Spitsbergen, its actants included the landscape and resources, symbols of occupation, mining camps and technology, transport facilities, and, of course, products. The actor-network can only survive if the global network continuously supplies the local network with the means it needs to function and deliver and if the local network in turn delivers the products that the actors in the global network desire. The company must establish itself as the obligatory point of passage between the networks to assure this flow. It must make its services indispensable and if necessary, it will try to re-define the overall context in which the components are placed and downplay any detrimental influences. Relationships in a network cannot only be started, maintained, and strengthened; if unavoidable, they can also be broken, used, and abused.

Politics – particularly *geopolitics* – play an important role in both marginality studies and ANT. The political history of Spitsbergen has been described in some detail.[24] Geopolitical processes, their reality being an underlying assumption here, have on the whole been overlooked. Dodds deduces that 'geopolitics, precisely because it is preoccupied with borders, resources, flows, territories, and identities, can provide a pathway for critical analysis and understanding – albeit a controversial one.'[25] It is a way of looking at the world and dividing it into particular zones, thereby generating distinct attitudes towards political actions along presumed geographical templates. These templates are constructed using a threefold division into formal, practical, and popular geopolitics (Fig. 1.2) and presented to target audiences.

What transpires is that almost every geographical and political reality has a geopolitical construct, which was intentionally formed and performed by one party in order to be perceived by another. Geopolitics, therefore, lends itself to the invention and reinvention of national identity and purpose as '[...] everyday life is replete with practices and symbols indicative of national identities and territories such as flags, currency, 'national news', and references to territory as either the 'fatherland' or in the case of the United States the 'homeland'.'[26] On Spitsbergen during the first quarter of the twentieth century, mining companies of all nationalities made conscious but not always conscientious use of geopolitical

[24] For references, refer to section 1.5 on previous work and to Ulfstein, G. (1995) *The Svalbard treaty,* Oslo: Scandinavian University Press.
[25] Dodds, K. (2007) *Geopolitics – A very short introduction,* Oxford: Oxford University Press, p. 3.
[26] Dodds (2007) p. 85.

practices and representations as they staked out their claims and lobbied for support from their respective core regions.

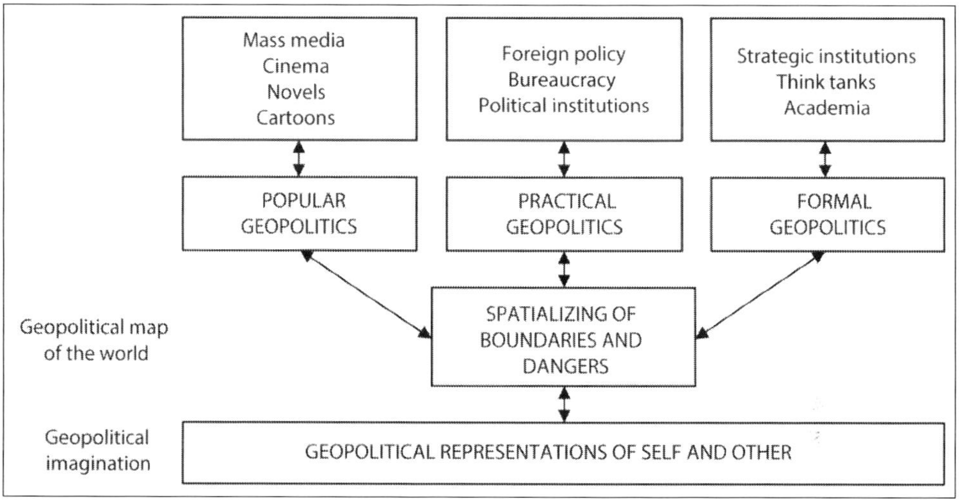

1.2 The threefold division of geopolitics: 'The formal is concerned with [how] academics and commentators self-consciously invoke an intellectual tradition associated with geopolitics. Practical geopolitics refers to the policy-orientated geographical templates used by political leaders such as President Bush as they represent global politics. Finally, popular geopolitics includes the role of the media and other forms of popular culture, which citizens use to make sense of events in their own locale, county, region, and the wider world.' (Source: Dodds (2007) p. 45-6.)

In view of the above theoretical tools, a particular analytical frame develops to assist the meaningful assessment of British companies on Spitsbergen during the twentieth century. This analytical frame is characterised by the merging of the core-periphery model into the non-model of marginality and by both marginal fluidity and geopolitical symbolism being deliberate and powerful actants in the emerging actor-networks.

1.5 Previous work

In addition to the previous work already cited, there is a sizable corpus of literature on the early exploitation and later industrialisation of Spitsbergen.[27] Although highly

[27] Selected reading: Arlov, T. B. (1991) *Store Norske 75 år*, Longyearbyen: Store Norske Spitsbergen Kulkompani A/S; Arlov, T. B. (1996) *Svalbards historie 1596-1996*, Oslo: Aschehoug; Arlov, T. B. (2005) 'The discovery and early exploitation of Svalbard. Some historical notes', *Acta Borealia*, 22 (1), pp. 3-20; Barr, S. (1985) 'Ernest Mansfield, drømmer, svindler, gentleman og eventyrer' in Amundsen, B. (ed.)

relevant to the LASHIPA project as a whole, these publications usually concern a different industry, period, or national origin than lie at the heart of this book. The British involvement is commonly sidelined or inaccurately portrayed. Nonetheless, four examples have been dealt with in more detail below.[28]

Hoel's monumental three-volume effort to capture Spitsbergen's human past is unavoidable for any scholar or enthusiast with an interest in the history of the archipelago.[29] Its accessibility is, however, severely restricted by it only being available in Norwegian. Painstaking translation is of limited use because the sources of the vast amount of detail are barely referenced, and although the author will undoubtedly have witnessed many of the events himself, his descriptions may be subject to imperfect memory or bias. Nonetheless, he commonly remains the only source obtainable, and where this is the case, it has duly been stated in the footnotes. The value of his version, which lacks any attempt at interpretation, lies in its comparison with British evidence and, if they differ, the question as to why they differ. Some answers are provided in the following chapters.

Mathisen offers a descriptive and empirical multi-national overview in the English language.[30] The British sources he consulted for a general perspective – referring to the British Government rather than any other actors – included state papers; documents on the origin of the First World War, 1898-1914; documents on foreign policy, 1919-39; and parliamentary debates. They did not include company archives. Mathisen notes that a growing number of companies on Spitsbergen caused escalating disputes in need of an authority to settle these and to register

---

Svalbardboka 1985-1986, Tromsø: Ursus Forlag, pp. 159-73; Evjen, B. (1996) *Longyearbyen 1916-1975. Fra arktisk arbeidsplass til etablert industrisamfunn?,* PhD thesis, Tromsø; Hacquebord, L. (1984) *Smeerenburg. Het verblijf van Nederlandse walvisvaarders op de westkust van Spitsbergen in the 17e eeuw*, PhD thesis, University of Amsterdam; Hacquebord (1996); Hacquebord, L., Steenhuisen, F., and Waterbolk, H. J. (2003) 'English and Dutch whaling trade and whaling stations in Spitsbergen (Svalbard) before 1660', *International Journal of Maritime History*, 2 (15), pp. 117-34; Hanoa, R. (1993) *Kings Bay Kull Comp. A/S 1917-1992 – fra gruvedrift til forskningsservice på Svalbard*, Oslo: Schibsted; Lajus, J. (2004) 'From fishing to mining: the change of priorities in the development of the North and Russian expeditions to Spitsbergen in the early 20th century', in Wråkberg, U. (ed.) *Arktisk gruvdrift II. Teknik, vetenskap och historia i nor*, Stockholm, Jernkantoret, pp. 93-106; Lund-Mathiesen, I. (1974) *Grubesamfunnene på Spitsbergen 1905-1917 – syndikaliststreiken og militærintervensjonen sommeren* 1917, Oslo; Mikalsen, H. (1995) *The Scottish Spitsbergen Syndicate og the Northern Exploration Company på Svalbard 1909-1920*, unpublished thesis, Oslo; Reymert, P. K. (2004) 'Den förste kulldrift på Svalbard. Søren Zakariassen og Kulkompagniet Isefjord Spitsbergen', in Wråkberg, U. (ed.) *Arktisk gruvdrift II. Teknik, vetenskap och historia i norr*, Stockholm, Jernkantoret, pp. 107-18; Sörlin, S. (2002) 'Rituals and resources of natural history: the North and the Arctic in Swedish scientific nationalism', in Sörlin, S. and Bravo, M. (eds.), *Narrating the Arctic. A cultural History of Nordic scientific practices*, Canton/MA: Science History Publications, pp. 73-122; Vattens, O. (1980) *Longyearbyen 1905-1935: et norsk samfunns oppbygging og utvikling*, unpublished thesis, Trondheim.

[28] For previous work concerning the industrial archaeology of Spitsbergen and the LASHIPA publications, refer to section 1.6.1 on the archaeological record.

[29] Hoel (1966-7).

[30] Mathisen, T. (1954) *Svalbard in international politics*, Oslo: Norwegian Polar Institute.

claims appropriately. He uses the early strikes of Scandinavian workers at the British and American mines to expose the potential for international complications, adding that '[it] was not surprising that those circles in Great Britain which were interested in coal mining in Svalbard should prefer a Norwegian occupation.'[31] With regards to the strike at the British mine in 1907, Mathisen infers that the British Government refused to act on behalf of its citizens in fear of provoking Russia.[32] He primarily concludes that the Spitsbergen Question arose for political reasons at a time 'marked by the race between powers for new territories and spheres of interests.'[33] Coal mining was therefore not an actor in its own right; it merely provided the impetus needed to settle the question of sovereignty, which (the Governments of) Britain and Russia were apparently willing to concede to Norway, at least in 1907. Ultimately, 'the decisive change in relative strength of the powers brought about by the war [...] were to influence the nature of the Svalbard question to a large degree.'[34] In critique of Mathisen, however, it was actually very surprising that Britain, let alone the British mining companies, should put aside strong Victorian and imperial traditions and agree to a foreign overlord. If anything, the shift in strength of the powers after the war was in Britain's favour and could have swayed the Spitsbergen Treaty accordingly. Mathisen fails to fully explain why this did not happen.

Østreng adopts yet a more aggressive nationalistic view of the events leading up to the signing of the Spitsbergen Treaty in 1920.[35] He reiterates that Norwegians visited the islands in the twelfth century, which has not been proven archaeologically, and maintains that while seventeenth-century English and Dutch whalers soon lost interest, Denmark-Norway upheld its claims until the nineteenth century, after which the notion of a no man's land was admittedly also accepted by Norway. Completely bypassing the era of the Russian Pomor hunters, Norwegian trappers then dominated natural-resource exploitation in the late nineteenth century, while ensuing mining was merely a cause for chaotic conditions and disputes in need of international settlement. Prior to the ratification of the Spitsbergen Treaty, Norway endeavoured '(1) to draft a mining code for the archipelago; (2) to obtain recognition of the Svalbard treaty by the governments of Germany and the Soviet Union; (3) to introduce measures to secure Norwegian

[31] Mathisen (1954) p. 45.
[32] This statement is controversial in view of Russia having lost the Russo-Japanese War in 1905 and being a much weakened contender in the European balance of power thereafter.
[33] Mathisen (1954) p. 174.
[34] Mathisen (1954) p. 175.
[35] Østreng, W. (1978) *Politics in high latitudes - The Svalbard archipelago,* Montreal: McGill-Queen's University.

interests and claims for the Svalbard Commissary; and (4) to decide on the national status of the archipelago.'[36]

The mining regulations were to be drafted in cooperation with Sweden and Britain, whose subjects also held substantial mining interests. Østreng argues that the Northern Exploration Co. was behind the many fundamental objections raised by the British representatives because the company was almost bankrupt and envisaged that any improvements would be to its advantage. This simplification, however, is unfounded. Eventually, all established proprietors of the represented nations received sole operating rights for a period of ten years. Nonetheless, the Northern Exploration Co.'s only remaining option was to sell its rights in 1932. The Scottish Spitsbergen Syndicate followed suit in 1950. According to Østreng, the trend of company failure was due to two sets of factors. Firstly, the archipelago's economic potential had been overestimated while the problems of industrial development in the Arctic had been underestimated. Secondly, the uncertain economic climate of Europe in the 1920s made it difficult for companies to find a market for their products.

Despite these local and global difficulties, Østreng maintains that Britain harboured belligerent political and military intentions for the islands. As such, '[the] possibility of the British exploiting their position in Svalbard [...] could not be discounted: the old British theme of an 'Arctic Gibraltar' in Svalbard had been publically mooted in Great Britain as far back as 1912.'[37] Additionally, Winston Churchill's anti-communism elevated suspicions. As with Mathisen, Østreng makes several sweeping statements and does not explain the apparent shift from initial British disinterest to firstly a seemingly greater presence and influence of the mining companies and secondly a more assertive foreign policy regarding Spitsbergen. Ultimately, he does not clarify why British enterprises failed where Norwegian and Russian companies evidently did not.

Singh follows up the Norwegian views with a strictly American focus.[38] America's policies were developed between 1907 and 1916 in conjunction with the Arctic Coal Co. from Boston, Massachusetts, developing a claim on Spitsbergen. Singh assumes that the State Department entertained the suggestions of the Arctic Coal Co., albeit hesitantly, because of the company's successful lobbying attempts. Although the company's activities ceased in 1916 with the profitable sale of its property in Advent Bay to Norway's Store Norske Spitsbergen Kulkompani, the State Department extended its interest in the islands until the Paris Peace Conference in 1919 and the resultant Spitsbergen Treaty. According to Singh, this

---

[36] Østreng (1978) p. 16.
[37] Østreng (1978) p. 53.
[38] Singh, E. (1980) *The Spitsbergen (Svalbard) Question: United States Foreign Policy, 1907-1935*, Oslo: Universitetsforlaget.

interest in a localised affair was anchored in the greater international sphere. She reasons,

> Though the Spitsbergen Question was not an issue in international politics over which nations were willing to go to war, it was a question which reflected facets of larger issues in international politics, namely the definition of territory together with claim and occupation, the politics of conference diplomacy, the question of living and mineral resource management including access and utilization, the politics of recognition, and the question of administration of non-contiguous territory having no indigenous population.[39]

In her assessment of the Arctic Coal Co., Singh lays the foundation for similar research on British companies and, in fact, companies of any nationality on Spitsbergen. In its comparative entirety, such research would enable the re-definition of mining companies from passively providing a point of discussion to actors consciously participating in if not occasionally leading the debate. It is a mistake to undervalue their impact, direct and indirect, on the proceedings on the global stage. Last but not least, the above quote rings true for the current 'Arctic Question'.

1.6 Sources and methods

Spitsbergen was chosen as one of the case studies for LASHIPA for three reasons. Firstly, it had a long history of natural resource exploitation. Secondly, it had an equally long history of political disputes over ownership and sovereignty. Lastly, both written and archaeological sources are available to discern these histories. Spitsbergen was one of the last remaining no man's lands in the world, and its industrial development occurred relatively late, that is to say quite recently. As such, Spitsbergen offers a rare glimpse into the processes at work in a country that belonged to no one, and chances were that the survival of relevant written sources had been assured in a number of national, local, and personal archives. In addition, the fragile archaeological remains of early industry and exploration have not been destroyed or masked by later agricultural or industrial expansion as is the case in most European countries.

Throughout this book, the reconstruction of the British companies is provided in the way in which the historical actors themselves experienced history. The reconstruction is, however, laced with contextual references that may have escaped them at the time. Hence, the Arctic archipelago is predominately referred to as Spitsbergen, according to the original Dutch spelling of 1596. The rendition of

[39] Singh (1980) p. 5.

Spitzbergen with a central 'z' is kept in company names and quotations to indicate an unawareness of or indifference towards increasingly anti-German sentiments. Following the ratification of the Spitsbergen Treaty in 1925, the subsequent Norwegian name of Svalbard is used where contemporaries themselves applied it or where the author refers to the present-day situation. It is a personal choice to write *on* Spitsbergen to describe the island and no man's land, admittedly supporting a peripheral connotation, and *in* Svalbard, where the territory now administered by Norway is meant.

As for other place-names, many English versions originated during the whaling period in the seventeenth century, but the book adheres to later British renderings, which were observed in contemporary documentation and doubled-checked against *Place-names of Svalbard*.[40] The issue of naming in the Arctic was and continues to be complicated with profound geopolitical implications.[41] Where no historical English term was known to exist, fieldwork benefitted from current Norwegian designations. A list of former place-names used throughout the text and their modern counterparts has been provided in Appendix 1.

1.6.1 The archaeological record

Whereas historians generally rely on written sources to understand the human past, archaeologists aim to reconstruct it using its physical remains. The physical remains typically include artefacts, ecofacts, structures, and whole landscapes that together form the archaeological record. The archaeological record is an integral part of our cultural heritage. It is vulnerable and non-renewable and – if we intend to keep our cultural heritage for posterity – in need of conservation. In Svalbard, its safekeeping falls under the Environmental Protection Act, which automatically embraces all signs of early human activity, be it fixed or moveable, up to and including the year 1945.[42] There are problems with the act in that archaeological sites are notoriously difficult to define and the conservation of portable finds *in situ* is challenging, to say the least. Furthermore, there are potentially valuable sites which date from after the Second World War that are not routinely protected and therefore at great risk.

---

[40] Hoel, A. (1942) *The place-names of Svalbard*, Oslo: Norges Svalbard- og Ishavs-Undersøkelser.

[41] Debenham, F. (1942) 'Place-names in the Polar Regions', *Polar Record*, 3, pp. 541-52; Brown, R. N. R. (1943) 'Review: Place-names of Svalbard', *Geographical Journal*, 102 (4), pp. 180-4; Wråkberg, U. (2002) 'The politics of naming: contested observations and the shaping of geographical knowledge', in Bravo, M. and Sörlin, S. (eds.) *Narrating the Arctic. A cultural history of Nordic scientific practices*, Canton/MA: Watson Publishing International.

[42] The Governor of Svalbard (2011) *Environmental Act*. Available at: http://www.sysselmannen.no/hovedEnkel.aspx?m=45282 (Accessed: 2 November 2011).

This book focuses on industrial archaeology. The term has been coined during the spurt of indiscriminate post-war redevelopment in Britain fifty years ago and denotes the study and conservation of industrial monuments.[43] It has since generated allies such as the International Council of Monuments and Sites (ICOMOS) and The International Committee for the Conservation of the Industrial Heritage (TICCIH) and has matured with the inclusion of industrial monuments in the World Heritage List.[44] In fact,

> It is not exaggerated to say that public consciousness in relation to industrial society has been profoundly changed over the last three decades, and that attention has been shifted from specific monuments to groups or whole landscapes of industrial monuments. The point about landscapes rather than isolated monuments is the value of contexts in conservation.[45]

The justification for the study and conservation of industrial archaeological landscapes is threefold: firstly, industrial monuments may actually aesthetically enhance their surroundings; secondly, they have a second lease on life as tourist attractions and are legitimate sources of income; and thirdly, they communicate the experience of previous industrialisation to our post-industrial society, thereby creating roots and identity as well as informing, for instance, planning and management decisions. Future efforts must not only be directed towards the greater designation and protection of conservation areas but also towards improving the underdeveloped educational aspect of industrial archaeology, particularly in collaboration with historical research and other academic disciplines.

Although the phenomenon of company towns as distinctive archaeological manifestations in economic pioneering has been explored extensively in the literature, the move away from isolated monuments to portray the former industrial landscape has intensified with the advent of historical archaeology in recent decades.[46] More recently still, there have been attempts to characterise the

---

[43] Buchanan, A. (2005) 'Industrial archaeology: past, present and prospective', *Industrial Archaeology Review,* 27 (1), pp. 19-21; Oglethorpe, M. (2005) 'Industrial heritage and national identity – sharing data, the importance of context and strategic priorities', *Industrial Archaeology Review,* 27 (1), pp. 27-31.

[44] British inscriptions are Ironbridge Gorge, Blaenavon Industrial Landscape, Derwent Valley Mills, New Lanark, Saltaire, Cornwall and West Devon Mining Landscape, and Pontcysyllte Aqueduct and Canal. In Norway, Røros Mining Town and the Circumference is as of yet the only listed industrial site.

[45] Buchanan (2005) pp. 19-20.

[46] Selected reading: Allen, J. B. (1966) *The company town in the American West*, Norman: University of Oklahoma Press; Porteous, J. D. (1970) 'The nature of the company town', *Transactions of the Institute of British Geographers,* 51, pp. 127-42; Cassell, M. S. (2005) 'The landscape of Iñupiat Eskimo industrial labor', *Historical Archaeology*, 39 (3), pp. 132-51; Given, M. (2005) 'Mining landscapes and colonial rule in early-twentieth-century Cyprus', *Historical Archaeology*, 39 (3), pp. 49-60; Hardesty, D. L. (2010) *Mining archaeology in the American West: a view from the Silver State*, Lincoln: University of Nebraska Press and the Society for Historical Archaeology; Dinius, O. J. and Vergara, A. (eds.) (2011)

industrial landscape of Spitsbergen.[47] A concentrated effort of documentation[48] and publication[49] is occurring under the auspices of the LASHIPA project. While these works have come a long way, they cannot as of yet completely fathom the complexities of the early mining industry on the islands. Studies such as this potentially provide the missing detail for an overall synthesis, but their integral

---

Company towns in the Americas: landscape, power, and working-class communities, Athens: University of Georgia Press.

[47] Selected reading: Arisholm, T., Devold, E. M., Hoem, S., and Rossnes, G. (2000) *Gruve 2b, Longyearbyen, Svalbard. Rapport – dokumentasjon 2000*, Oslo: DAR-gruppen; Catford, K. (2002) 'The industrial archaeology of Spitsbergen', *Industrial Archaeology Review,* 24 (1), pp. 23-36; Catford, K. (2004) 'Svalbard coal – a hundred years'. [Online]. Available at: http://website.lineone.net/~polar.publishing/svalbardcoal.htm (Accessed: 7 November 2011); Planke, T. and Wammen, M. (2008) *Dokumentasjon av fangsthytter på Prins Karls Forland.* [Online]. Available at: http://www.sysselmannen.no/hoved.aspx?m=44267&amid=2380979 (Accessed: 8 November 2011).

[48] Martin, P. *et al.* (2006) *[LASHIPA 1] Industrial heritage in the Arctic: research and training on Svalbard, August 2004,* Michigan: MTU; Avango, D. *et al.* (2006) *LASHIPA 2, Archaeological expedition on Svalbard, August 8-22, 2005,* Groningen: Arctic Centre; Avango, D. *et al.* (2008a) *LASHIPA 3, Archaeological expedition on Spitsbergen, August 7-24, 2006,* Groningen: Arctic Centre; Avango, D. *et al.* (2008b) *LASHIPA 4, Archaeological expedition on Spitsbergen, August 2-25, 2007,* Groningen: Arctic Centre; Avango, D. *et al.* (2009) *LASHIPA 5, Archaeological expedition on Spitsbergen, 27 July – 17 August 2008*, Groningen: Arctic Centre; Avango, D. *et al.* (2011) *LASHIPA 6, Archaeological expedition on South Georgia, 3 March – 12 April 2009,* Groningen: Arctic Centre; Haas, H. R. de *et al.* (2009) *LASHIPA 7, Archaeological expedition on Spitsbergen, 2009,* Groningen: Arctic Centre; Avango, D. *et al.* (2012) *LASHIPA 8, Archaeological expedition to Deception Island and South Shetland Islands, 2010,* Groningen: Arctic Centre; Avango, D. *et al.* (2010) *LASHIPA 9, Archaeological expedition to Spitsbergen, 31 July – 15 August 2010,* Groningen: Arctic Centre.

[49] Oglethorpe, M. and Nisser, M. (2004) 'Industrial archaeology field course', *TICCIH Bulletin,* 27, p. 5; Avango (2005); Michigan Technological University (2005) *Svalbard archaeology.* [Online]. Available at: http://www.svalbardarchaeology.org/ (Accessed: 8 November 2011); Aalders, Y. I. and Hacquebord, L. (2008) 'Europese walvisvaarders en Russische jagers in Green Harbour, Spitsbergen', *Paleoaktueel,* 19, pp. 210-5; Haas, H. R. de (2008a) 'Spitsbergen's resources, Dutch entrepreneurs and geo-politics', *Industrial Patrimony,* 19, pp. 25-36; Haas, H. R. de (2008b) 'Nederlandse ondernemers op poolavontuur. De Nederlandsche Spitsbergen Compagnie, 1920-1932', in Veerman, D. (ed.) *Tegenpolen. Op expeditie naar contrastrijke poolgebieden*, Zutphen: Walburg Pers, p. 10; Gustafsson, U. I. (2008) 'Modern whaling industry in Spitsbergen as a tool for territorial claiming and national sovereignty strives', *Industrial Patrimony,* 19, pp. 17-24; Hacquebord and Avango (2008); Haas, H. R. de (2009) 'NV Nederlandsche Spitsbergen Compagnie. The forgotten history of a Dutch coal mining company in the High Arctic', *Natural Resource*, 0 (3), pp. 26-8; DePasqual, S. (2009) *Winning coal at 78° north: mining, contingency and the Chaîne Opératoire in Old Longyear City,* unpublished MSc thesis, Michigan Technological University; Hacquebord and Avango (2009); Hartnell, C. C. (2009) *Arctic network builders: the Arctic Coal Company's operations on Spitsbergen and its relationship with the environment*, PhD thesis, Michigan Technological University; Gustafsson, U. I. (2010) 'Early whaling in the north. Industrialising the Arctic: settlement design of modern whaling stations in Spitsbergen and Bear Island', *The 3rd Symposium on Whaling and History.* Commander Chr. Christensen Whaling Museum, Sandefjord, June 2009. Sandefjord: Commander Chr. Christensen Museum, pp. 45-58; Starkov, V. F., Hacquebord, L., and Avango, D. (2010) 'Archaeological studies at Svalbard archipelago according to IPY Programme', in Kotlyąkov, V. M. (ed.) *Changes of natural environment and climate; natural and possible consequent human-induced catastrophes. Vol. 3 Part 2: Natural processes in Polar Regions*, Moscow: Institute of Geography RAS, Institute of Physics of the Earth RAS, pp. 301-7; Avango *et al.* (2011); Kruse, F. (2011) 'Four former British mining settlements on Spitsbergen', *Mining Perspective: the 8th International Mining History Congress 2009.* Penventon Park Hotel, Redruth 12-15 June. Truro: Cornwall and West Devon Mining Landscape World Heritage Site, Cornwall Council, pp. 117-24; Hacquebord (ed.) (2012).

fieldwork component must be undertaken within appropriate research frameworks and along applicable standards.

Sysselmannen does not insist on Norwegian standards for international fieldwork carried out on the islands. Consequently, the LASHIPA expeditions profited from teams of international researchers, who diligently applied their national ethics and commonly offered complementary perspectives and alternative interpretations. The LASHIPA 1 and 3 expeditions in 2004 and 2006, respectively, visited a significant number of British archaeological sites prior to the author's arrival on the project. Thereafter, the author participated in the LASHIPA 5 and 9 surveys in 2008 and 2010. Known sites were generally chosen first, followed by sites identified in written sources. There were no means to undertake purely speculative surveys in the hope of rediscovering sites lost from living memory.

The archaeological potential of the sites was assessed according to a four-stage strategy. During *pre-fieldwork* preparation, previous work was evaluated and a project design created. The *field surveys* benefitted from the recording of environmental factors on the one hand and of material remains collectively referred to as features on the other.[50] The features were documented on two levels. The first was a basic visual survey, the second a descriptive record supplemented by measured drawing and digital photography. A total station was available, but the surveys were usually conducted over large distances and under time pressure, which was best served by a downsized, mobile kit including a handheld GPS unit. During the *post-fieldwork* processes, all written, drawn, and photographic records were cross-correlated and analysed to provide a synthesis of the results. The GPS data was compiled into ArcGIS maps. To date, the results of the LASHIPA expeditions have formed the basis of several field reports.[51] Where possible, the author undertook her own *archiving* to enable indexing, ordering, quantification, and checking for consistency of all original records. Although these site archives stay in her possession, they feed into the LASHIPA database, which the project is obliged to create as part of the IPY requirements. The metadata thereby enters the public domain, foreseeably via the archiving system of the Data Archiving and Networked Services (DANS) of the Netherlands.[52]

[50] Hardesty proposes the "features system" to be the best analytical tool for the history and archaeology of mining and defines this 'as a group of archaeologically visible features and objects that is the product of a specific human activity.' (Hardesty, 2010, p. 5) The feature system is made up of associated structural features and may include some that are widely dispersed geographically. While the notion of the feature system finds mention here, it is only very loosely adhered to throughout the text.

[51] See footnote 48. The field reports of LASHIPA 1, 3, 5, and 9 include British material remains on Spitsbergen.

[52] Data Archiving and Networked Services (2013) *EASY*. Available at: https://easy.dans.knaw.nl/ui/home (Accessed: 1 January 2013).

The wealth of information generated over the course of this project contributed to the topical explosion of archaeological activity at historical mining sites.[53] The question arises how this information must be evaluated and enquires not into the above theoretical framework but asks specifically how to 'read' archaeological evidence. Cassell and Stachiw pinpoint only physical, material, and social factors, and even in his highly relevant study of the landscape of Eskimo industrial labour, Cassell bypasses an interpretation of the Arctic environment.[54] For Spitsbergen, it is useful to revisit the unpeopled and undefined polar space in order to discern what effect this natural and cultural wilderness may have had on newly arriving actors. Attention has therefore been paid to both the accessibility of a site at different times of year and the attractiveness of all its natural resources prior to any material or social construction occurring.

Several researchers have identified the capacity of archaeology for the fine-grained resolution of mining sites.[55] Considering this capacity, the prevailing consensus is that industrial landscapes are best served by the means of historical archaeology. Addressing the limitations of history, Cassell states,

> Having such a plethora of documentary and ethnohistoric data concerning the western Arctic commercial whaling industry and the Eskimos who labored in its

[53] Hardesty (2010) p. ix.

[54] Cassell, M. S. and Stachiw, M. O. (2005) 'Perspectives on landscapes of industrial labour', *Historical Archaeology,* 39 (3), pp. 1-7; Cassell (2005).

[55] Despite the remoteness of the mining frontier, Hardesty identifies the company networks that enabled far-reaching material and social adaptations. Mining sites mushroomed in accordance with Victorian ideology. 'That the Victorian ideology was carried onto the mining frontier is quite clear in both the documentary and archaeological records. The archaeological record of Victorian Culture is most visible in the layout of settlements and in trash dumps.' (Hardesty, 2010, p. 5) Yet, most of his characteristics are too advanced for the simple British manifestations on Spitsbergen. It is barely possible to apply divisions of mining technology, residential settlement, and household. The former no man's land must also be understood in terms of spatial control. Cassell's evaluation of a short-lived commercial whaling station offers additional information. It was panoptic and allowed for close surveillance of the stores and the employees. (The authoritative text on panoptics is Foucault, M. (1977) *Discipline and punishment: the birth of the prison*, New York, NY: Pantheon Books.) A distant, differently built house demonstrated the presence of 'others', in this case Eskimos, who intended to participate in whaling while at the same time escaping observation and maintaining their immigrant status. Surveillance features in Given's work on Cypriot mining, too. He recognises naming, agriculture, architecture, and community as agents of colonial control. 'A mining landscape, then, was a colonial creation with its boundaries, name, and identity imposed by the colonizers of the land. In this respect it was a model *of* the colony, as it followed the same system, and also a model *for* the colony because it illustrated the principle clearly in a limited and manageable area.' (Given, 2005, p. 52). Where the company failed to win the loyalty of its workers, it was met with resistance. Given finds resistance in the conflicting traditions of foodways, archaeologically expressed in pottery. Food for the large Cypriot workforce was imported and the company sold overpriced imperial goods back to the impoverished population as another means of control. While Porteous observes that in remote areas – such as Spitsbergen – the companies were obliged to provide facilities and sustain their workers, Given sees ensuing smuggling, pilfering, and theft as practical and ideological acts against a foreign overlord. This serves as a reminder to search the physical remains on Spitsbergen for expressions of control and resistance, keeping in mind that they may reveal hitherto unknown facets.

> support is fortunate. Archaeological understanding has been lacking, however. With its emphasis on material culture, archaeology alone can offer a database capable of providing important patterns and details of the everyday lives of Eskimos as workers in the whaling industry.[56]

Correspondingly, Given asserts that 'the history of mining [...] has been discussed from perspectives that are technical, descriptive, or political. Historical archaeology adds a perspective that does not rely on the rhetoric of government reports or political pamphlets, or on the impersonal statistics of colonial bureaucracies.'[57] He maintains that it is only through historical archaeology that artefacts are able to convey the fullest meaning. It is clear that 'both documentary and archaeological "images" of the mining frontier can be constructed, each with its own strengths and weaknesses. The most accurate image, however, comes from overlapping documentary and archaeological accounts.'[58] For this reason, the approach of combining archaeological fieldwork on Spitsbergen with the archival research outlined below has been most beneficial to this study and to the LASHIPA project as a whole.

1.6.2 Contemporary documentation

In view of British companies having explored and exploited a polar region under the watchful eye of Americans and other competitors before being bought by Norwegians, it made sense to access the national and appropriate regional archives of England, Scotland, Norway, and the United States as well as a number of British institutions focused on mining history and polar geography. As such, this research benefitted primarily, but not exclusively, from the collections at the following depositories:

- British Geological Service Library, Nottingham
- The National Archives of the United Kingdom, Kew
- National Coal Mining Museum, Wakefield
- North of England Institute of Mining and Mechanical Engineers Library, Newcastle upon Tyne
- Royal Geographical Society (with IBG) Library, London
- Scott Polar Research Institute Library & Archives, Cambridge
- Edinburgh University Library, Centre for Research Collections, Edinburgh

[56] Cassell (2005) p. 148.
[57] Given (2005) p. 50.
[58] Hardesty (2010) p. ix.

- Falkirk Council Archives, Falkirk
- Glasgow University Library, Special Collections, Glasgow
- National Records of Scotland, Edinburgh
- National Library of Scotland, Edinburgh
- National Museums of Scotland Library, Edinburgh
- Royal Scottish Geographical Society Collections, Perth
- National Archives of Norway, Oslo
- Norwegian Polar Institute Library & Archives, Tromsø
- Polar Museum Collections, Tromsø
- Regional State Archives, Tromsø
- Governor of Svalbard Collections, Longyearbyen
- Svalbard Museum Archives, Longyearbyen
- Michigan Technological University Archives & Copper Country Historical Collections, Houghton/MI

In addition, online access to *The Times*, *The Geographical Journal*, *Polar Record*, and other newspapers and journals was granted by the University of Groningen, while *The Gazettes* and *Hansard* were freely available on the internet.[59] The amount and content of information that would be found in these collections was not entirely clear before the research began in earnest. Written sources that were eventually consulted included so-called dissolved company files, British Foreign Office correspondence, Norwegian Foreign Ministry correspondence, company correspondence, company field reports, company circulars, individual testimonies, maps, photographs, and newspaper articles. Occasionally, the author was able to access a personal collection or interview a living relative of a Spitsbergen veteran.

The varied nature of the documentary evidence encompasses both strengths and weaknesses. It is possible to build up an impression of each company using the dissolved company files. These comprise 'the data generated in the legal process of incorporation [and] are particularly important, because of the information contained in these sources concerning the occupations or business interests of the founders of companies and of shareholders, and their geographical location.'[60] The financial facts supplied enable an analysis of share capital as well as the accumulation of assets by other means such as debentures, mortgages, loans, or retained profits. Yet, the information may be ambiguous. Descriptions of

---

[59] *The Gazettes* are available for London, Edinburgh, and Belfast. They are the Official Newspaper of Record for the UK and disseminate and record official, regulatory, and legal information. Available at: http://www.london-gazette.co.uk/ (Accessed: 24 October 2011); *Hansard* is the Official Report of debates in Parliament. Available at: http://hansard.millbanksystems.com/ (Accessed: 24 October 2011).

[60] Church, R. (1986) *The history of the British coal industry: Vol. 3 1830-1913 Victorian pre-eminence*, Oxford: Clarendon Press, p. 133.

occupations and business interests may be vague. The terms 'coalowner' or 'colliery proprietor' may say little about the true source of a person's wealth. The 'founders' may not always have been the prime movers behind company formation, while dominant shareholders may have had greater influences on policy-making than is obvious. As a whole, the dissolved company files provide an insight into the boardroom and the global network, but they reveal little about the companies' expeditions to Spitsbergen and operations locally.

The Foreign Office files, *Hansard*, and the collections of the Norwegian Foreign Ministry give substance to the issue of practical geopolitics. The Foreign Office files contain both internal memos as well as authorised correspondence. Hence, it is commonly possible to trace the development of an argument before the department arrived at an official position. Unfortunately, the files are often incomplete with whole years missing at a time. In addition, they rarely offer sensitive information, and the true reason for a particular foreign policy may forever remain unknown. Besides textual information, the foreign departments had collected several maps and plans of relevance to Spitsbergen. These maps were not only tools for exploration and territorial possession, they were weapons.[61] Yves Lacoste asserts,

> The map, perhaps the central referent of geography, is, and has been, fundamentally an instrument of power. A map is an abstraction from concrete reality, which was designed and motivated by practical (political and military) concerns; it is a way of representing space, which facilities its domination and control. To map ... serves the practical interests of the State machine.[62]

Maps must therefore be interpreted with caution and ideally in conjunction with other sources to gain knowledge about their accuracy, purpose, method of production, uniqueness, and map details.[63] Spitsbergen's claim maps, for example, can be verified by corresponding legal documentation. Underlying questions include when the map was made, how it was made, and most importantly why. The question of why enquires after the purpose of the map as well as the criteria selected to make it, the consistency in the selection, any bias on behalf of the map maker, any deliberate omission, the intended market, and the use of conventional symbols. Last but not least, it queries whether what is shown actually existed – a query that puts the role of archaeology into perspective. While the Foreign Office

---

[61] Burnett, D. G. (2000) *Masters of all they surveyed. Exploration, geography, and a British El Dorado*, Chicago: The University of Chicago Press.

[62] Lacoste, Y. in: Dodds (2007) p. 110.

[63] National Library of Scotland (no date) *Using maps as historical sources*. Available at: http://www.nls.uk/collections/maps/subject-info/historical (Accessed: 24 August 2011).

afforded a wealth of information, Spitsbergen found comparatively little mention in Parliament. The Norwegian files offer a complementary political view.

The formal geopolitical processes at work are evident in institutions such as the Royal Geographical Society, the Royal Scottish Geographical Society, the Scott Polar Research Institute, and the Norwegian Polar Institute. Besides maps, their collections comprise numerous historical photographs, which are also potent geopolitical tools requiring a cautious approach. It is important to know who took the photograph as well as why and for whom. It may also be instructive to know how it was taken and whether companion images exist. Such a sequence of images may provide additional information on why they were taken. Lastly, some attention must be paid to how images are presented. The question of presentation does not necessarily arise when negatives or original prints can be consulted, but in this research, the photographs commonly appeared in company circulars or other printed material, where cropping, for instance, may have occurred for promotion purposes. Promotion through texts, maps, and photographs, especially in the media, was a cornerstone of company strategies. Although historically incorrect, biased sources in particular generate an understanding of how network builders attempted to enrol actors, thereby constructing their actor-networks.

Newspapers disclose the contemporary popular geopolitical climate. Traditionally, newspapers have been regarded as 'mere chroniclers of the passing scene'.[64] Yet, studies have shown that they play an important role in instigating or hindering social change. In keeping with the approach of reconstructing historical experiences, therefore,

> [...] history is concerned – or should be concerned – not only with what actually happened in any given time or place, but also with what people *thought* was happening, as revealed to them through the means of mass communication, which may have conditioned their subsequent actions. Thus, perception of events as filtered through the press may have changed the historical outcome. According to this concept, it does not matter if the news is false or distorted as long as readers believed it and acted on their belief. To the historian trying to understand public opinion then, newspapers become primary rather than secondary sources.[65]

Newspaper articles, of course, are restricted by the pressure of deadlines, limited access to information, and available space. Nonetheless, *The Times* demonstrated the prevailing public mood as did several Scottish papers. Publications with narrower readerships such as *The Colliery Guardian* and *The Mining Journal* may be less suitable to expose the opinions of British society. Yet, they indicate the views of a specific audience, in this case the mining community, and show how the

---

[64] Knudson, J. W. (1993) 'Late to the feast: newspapers as historical sources', *Perspectives*, 31 (7).
[65] Knudson (1993).

Spitsbergen coal rush was perceived by professionals, experts, and potential investors in the mining sector.

The volume of contemporary documents encountered was by no means infinite and most material could easily be processed within the scope of this study. Merely the Foreign Office correspondence with the British legation at the Paris Peace Conference in 1919 proved too much to deal with in its entirety, so only those documents referring to any of the four British companies were selected. Generally, it was not so much a question of choice but of necessity to use these rather than other sources. The amount to choose from was simply too limited, at least where sources traceable on the internet and available in the English language were concerned. More material may be hidden in foreign archives and personal collections, their owners unaware of the mounting research interest in them. The author was fortunate enough to trace a few of these collections and speak to relatives, but that was the exception, not the norm.

1.7 The composition of the study

This book is divided into four parts comprising nine chapters. Part I includes this introductory chapter. Chapter 2 accentuates *Britishness* as distinct from other national identities active on Spitsbergen and provides selective yet instructive background knowledge of Britain's coal industry, its empire, and its contributions to polar exploration. The chapter is intended to hint at, but not to exhaust, historical, cultural, and ideological reasons as to why British subjects may have been interested in a potentially resource-bearing polar no man's land.

Part II pays deserved attention to the immense body of empirical archaeological data gathered during four LASHIPA expeditions to Spitsbergen over a period of six years. Chapter 3 focuses on the physical remains of two mining companies, Chapter 4 on those of two exploration companies. Both chapters adhere to a simple structure of *site narratives*, *site interpretations*, and a short conclusion for better readability. The lengthy site narratives are justified in this book as the primary scholarly outlet for this information and are even necessary to generate a comprehensive archaeological image not only of the monuments themselves but also of the Arctic industrial landscape as a whole.

The archaeological image is enhanced by the historical image produced in Part III. Correspondingly, Chapters 5 and 6 deal with the empirical information relating to the Spitzbergen Coal & Trading Co. (1904-18) and the Spitzbergen Mining & Exploration Syndicate (1906-11), both of which were primarily involved in mining. The Northern Exploration Co. (1910-34) and the Scottish Spitsbergen Syndicate (1909-53) of Chapters 7 and 8 were, in fact, exploration companies, and their output must be measured in something other than tonnage. Structurally, details of the companies' formation are followed by a characterisation of their global and local networks and an assessment of their achievements.

Both the archaeological image and the historical image feed into Part IV – the combined image. The discussion in Chapter 9 is structured according to the study's four sub-questions. Thus, it looks at the reasons why the British companies were started, their operational choices, factors at the heart of their discontinuation, as well as their economic, political, and environmental impact. The actor-network theory is used to generate answers and provide plausible explanations for the rise and fall of this British industry. Where appropriate, international examples serve to compare and contrast. The conclusion revisits the central research question and tests the related hypothesis.

2 *Britishness*[1]

2.1 Introduction

There are several basic facts about the entwined histories of Britain and Spitsbergen one needs to know to appreciate the international historical context in which the research question is rooted. These facts are that the British core region had a long tradition in mining[2]; that it had acquired a vast empire; and that it played a leading role in early polar exploration. The latter included a history of expeditions to Spitsbergen. The coal mining industry, the British Empire, and polar exploration have been extensively written about elsewhere.[3] However, simple summaries of these works or broad generalisations would be inadequate to emphasise their connection with this research. The author has therefore identified persons who capture the essence of *Britishness*: of coal mining, Empire, and Arctic exploration, and who can in turn be linked to the events on Spitsbergen in a number of different ways. Their actions hint at historical, cultural, and ideological reasons as to why British subjects may have been interested in a potentially resource-bearing polar no man's land. Their achievements provide a yardstick against which to measure the conduct of British companies on the archipelago.

2.2 The coal industry

Emerson Muschamp Bainbridge (1845-1911) personified the British coal industry at the end of the nineteenth century. His link with the Arctic is immediately obvious in that he was also the founding director of the Spitzbergen Coal & Trading Co. in

---

[1] For further reading on British national identity, see 'Themed section of varieties of Britishness' (2006) *Nations and Nationalism,* 12 (3).

[2] Britain's industrial minerals include building stone, coal, gypsum, metals, and onshore oil and gas among others. By 1830, the coal industry was one of Britain's major industries and growing. Unless otherwise stated, coal mining is representative of all mining for the remainder of this study.

[3] Selected reading: Stoker, D. (1984) *The history of the British coal industry Vol. 2 1700-1830 The Industrial Revolution*, Oxford: Clarendon Press; Church (1986); Supple, B. (1987) *The history of the British coal industry: Vol. 4 1913-1946 The political economy of decline*, Oxford: Clarendon Press.; Mitchell, B. R. (1984) *Economic development of the British coal industry 1800-1914*, Cambridge: Cambridge University Press; Morris, J. (1979) *Pax Britannica: the climax of an empire*, London: Penguin Books; Coleman, E. C. (2006) *The Royal Navy and Polar exploration from Frobischer to Ross*, Stroud: Tempus; Coleman, E. C. (2007) *The Royal Navy and Polar exploration Vol 2: from Franklin to Scott*, Stroud: Tempus; Rice, T. (1986) *British oceanographic vessels 1800-1950,* London: The Ray Society; Spufford, F. (1996) *I may be some time,* London: Faber and Faber; Speak, P. (2003) *William Speirs Bruce: Polar explorer and Scottish nationalist*, Edinburgh: National Museums of Scotland; Dodds, K. (2002) *Pink Ice: Britain and the South Atlantic empire*, London: I. B. Tauris.

1904. Bainbridge was born in Newcastle upon Tyne, which lay in the heart of the large Northumberland and Durham Coalfield in North East England (Fig. 2.1).[4]

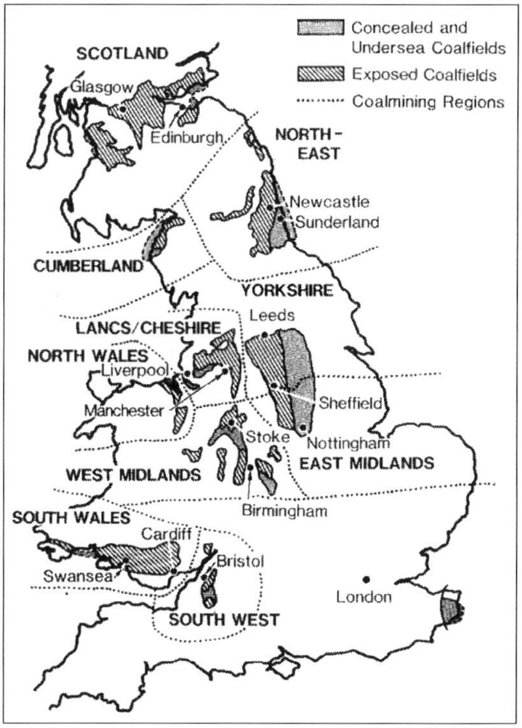

2.1 British coalfields in 1913. (Source: Church (1986) p. 8.)

The North East is one of the major coal regions in Britain.[5] Notably, Britain's capital and financial hub London is not one of them. At the time, the regions' capacity to meet the increasing industrial demands depended mainly on geological factors such as the extent of outcrops or near-surface seams that did not require deep shafts. However, 'so rapid was the growth in UK output that in one after another region coalowners found themselves under pressure to sink below the Permian foundation, a decision which involved commitment of greater amounts of capital to increasingly high-risk investment in the uncharted areas where the concealed coalfields were believed to be.'[6] Even by 1900, mapping of the coalfields was incomplete or already outdated. The speculative character of the coal industry notwithstanding, Britain dominated the European markets due to superior coalfields, excellent steam coal[7], and low-cost shipping. The North East and Scotland in particular controlled the trade with northern Europe, while Wales

[4] Unless otherwise stated, biographical details from: Wilmot, D. (2005) 'Emerson Bainbridge of Newcastle & Sheffield, an overlooked entrepreneur', *Yorkshire Archaeological Journal*, 77, pp. 241-52; Pimlott Baker, A. (2004) 'Bainbridge, Emerson Muschamp (1845-1911)', *Oxford Dictionary of National Biography*, Oxford: Oxford University Press.

[5] Unless otherwise stated, coal mining details from: Church (1986), the third volume in the authoritative *History of British coal industry*. The author felt it unnecessary to recreate his colossal research efforts within the limited scope of this book.

[6] Church (1986) p. 8.

[7] Coal had three distinct uses: burning; distillation to coke, gas, and other by-products; and smelting. Steam coal did not exist as a category until 1830 and then mainly in South Wales. Steam coal was high-quality bituminous coal that combined rapid ignition, intense burning (no caking), minimal clinker, and hot, smokeless fire (no raking), thereby economising labour.

tended southward. This dominance peaked in 1900, but it was still considerable before 1913. Much to the dismay of British steam coalowners, the Admiralty then switched to oil-burning vessels.[8]

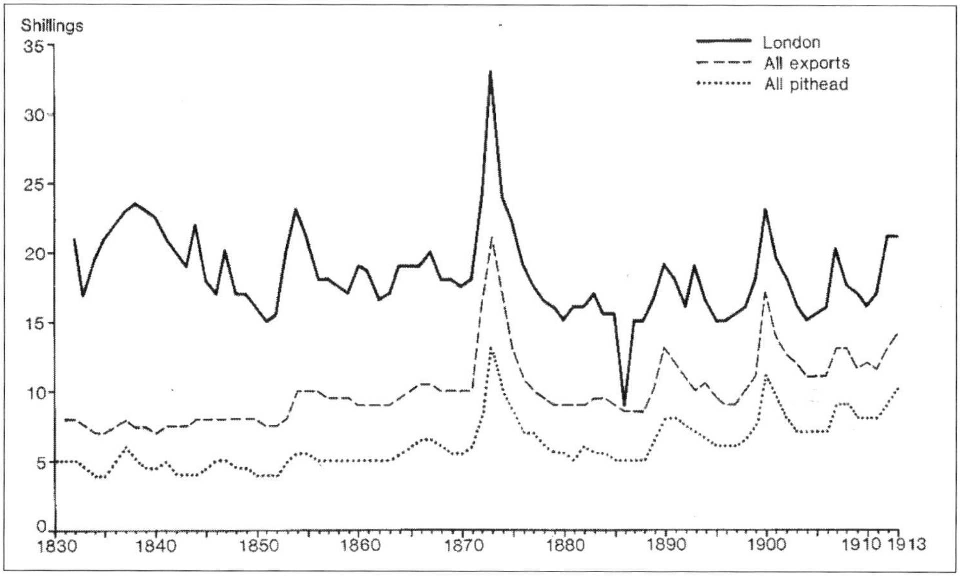

2.2 Comparison of coal price movements, 1830-1913. (Source: Church (1986) p. 53.) Of particular interest to this study is the period after 1904.

Bainbridge's father established Britain's first department store and also invested heavily in collieries. Collieries were traditionally partnerships, but during the depression of the 1860s (Fig. 2.2), many were converted into limited joint-stock companies, which benefitted from transferable shares, the right to sue, and limited liability. While the occupations and business interests of company founders and shareholders are often vague, it is clear that Bainbridge senior had attained his wealth in retail, not in coal. Yet, he and others like him were often the companies' chief sources of capital. Theoretically, private shares were harder to come by than public ones, but private companies booked the greater growth because the public was by and large not interested in colliery shares, except in peak years such as the 1870s and the 1900s. Preference shares were an attempt to counteract this unpopularity, although they were commonly issued as part of a company's financial reconstruction. Over a third of colliery investors originated from outside the coal regions, usually London. The distance between them and their investments raises

[8] Engdahl, F. W. (1993) *A century of war. Anglo-American oil politics and the new world order*, Wiesbaden: Dr Bottinger Verlags-GmbH.

questions about their knowledge of the industry and its profitability. Other sources of capital included debentures and mortgages, bank loans, and inter-company loans. Bainbridge 'stressed the difficulties in securing external finance, which as a result of the need to sink deeper mines had greatly increased capital needs, and which by 1900 might require £300,000. In general, he regarded the credit standing of colliery proprietors as "very bad".'[9] If a colliery were to survive, it needed a conservative financial policy. The average lifespan of a colliery was a mere 16.4 years, and non-landed coalowners were uncommon among the half-millionaires before the First World War. Bainbridge was one of the wealthiest colliery proprietors, leaving £461,769 upon his death in 1911.

2.3 Emerson Muschamp Bainbridge (1845-1911). (Source: http://www.victoriacountyhistory.ac.uk/explore/items/bolsover-colliery-company.)

Bainbridge (Fig. 2.3) received a private education, the choice of schools being influenced by the family's strong faith in Methodism.[10] Subsequently, he studied mathematics and mining engineering at Durham University and concurrently undertook an apprenticeship at the collieries of the Marquis of Londonderry in County Durham. Besides geological factors, the supply of coal depended on landowners such as the Marquis, who in Britain, unlike in the rest of Europe, owned the minerals of their land and the right to exploit them. Yet, landed colliery proprietors were a dying breed. Firstly, capital requirements were becoming too great. Secondly, colliery organisation was becoming too complex. Thirdly, conceding too much power and influence to growing numbers of viewers, agents, and managers was undesirable. Instead, landowners acted as lessors to private individuals or colliery companies. They negotiated leases and fixed royalties, the terms of which could either promote or obstruct the pace and efficiency of the industry, depending on whether or not the landowners chose to exercise monopoly powers over their coal reserves. Further hindrances to mining were taxation and inconsistent government regulations.

[9] Church (1986) p. 157.

[10] Methodism was one of the leading non-Conformist religions in Britain. It was strongly connected to social issues and injustices and reacted to social ills such as gambling and drinking with the temperance movement and teetotalism. Nonetheless, Bainbridge later married in the Church of England.

Viewers, agents, and managers had the technical and legal knowledge to reduce risks and maximise gains. Viewers were mining engineers who assessed the workability and profitability of the coal to be opened up. They had the best information and most experience as to all mining matters. Viewers could have several employers and because they usually had no personal interest in the collieries, they were thought to be more professional than colliery managers. Whereas the supervision and control of the mines had at first been undertaken by former miners who depended on a profit, the sinking of deeper pits witnessed the switch to direct salaried management. From 1863 till 1867, Bainbridge was a viewer at the Marquis' properties. In 1868, he was promoted to assistant manager. In 1870, he became manager at the Nunnery & Tinsley Park Collieries near Sheffield in the South Yorkshire Coalfield. These collieries belonged to the Duke of Norfolk. In 1874, they were transformed into separate limited companies, and Bainbridge assumed the position of managing director at Nunnery Colliery Co., Ltd.

Bainbridge's impact on the policies and strategies of the company has not been evaluated. It is, in any case, difficult to assess whether a company was controlled by its owners or its directors. If directors were also major shareholders, the firm could indeed have been run by its owners. If the directors were salaried and held only few shares as part of their payment, shareholders could have had a major influence on policy. Salaried directors may have been employed on the basis of knowledge and skill; or there could have been a degree of patronage. Church assumes that shareholder control is evident if no more than eight individuals hold less than half of the shares.[11] '[...] smaller firms would have been private, limited-shareholding companies in which owner-directors would have considerable impact on policy and management. Professional managers, an expensive commodity, might be entirely eschewed in favour of advice on contract from viewers and consultant engineers. Where they were employed managers would be less able, less knowledgeable, less well paid, and probably less powerful.'[12] The overall trend, however, was one from ownership control to professional leadership.

The development of a new coal mine required boring to find the coal, laying out of surface plant to sink the shaft, and the actual sinking operations. Drilling boreholes to retrieve core samples was not cheap, and the samples needed specialist interpretation. By the late nineteenth century, shafts in the North East had reached depths of 850 metres, often in extreme watery conditions. Ventilation was poor, and the build-up of methane, carbon monoxide, and carbon dioxide were a constant threat. Following a serious fire at one of his pits in 1871, Bainbridge was nonetheless able to put the men back to work within 48 hours. The

[11] Church (1986) p. 435.
[12] Church (1986) p. 447.

presence of water posed an even greater problem, and pumping was costly. In 1883, one of Bainbridge's pits flooded due to a railway station building interfering with the course of a river. Subsequently, he acted as an expert witness in the ensuing court proceedings against the railway company. At depth, ventilation and haulage routes were established, and where it existed, a sound roof occasionally reduced the need for supportive timber.

The seams were worked either by traditional room-and-pillar or by longwall methods. In room-and-pillar mining, the coal was essentially extracted by leaving pillars of coal to support the roof. In longwalling, the seam was worked along one long face, using wooden props to control the fall of the roof behind the face. 'The conditions under which longwall methods were most suitable [...] included thin, hard coal seams capable of bearing pressure, and abundant supply of land, rubbish or ironstone mixed with coals or debris for filling the hollows, water-free roofs, and the workings free from buildings or rivers.'[13] Where applicable, longwall methods were cheaper and yielded larger coals. Where sufficient capital existed, mechanisation was introduced to reduce labour costs and increase productivity, but it was only slowly adopted. Regarding labour costs, the Nunnery Co. repeatedly reduced the workers' wages, which caused strikes in 1878 and again in 1880/1. Bainbridge thwarted the first by bringing in non-union workers. The second resulted in a lawsuit against the workers, which was rejected by the magistrate. Despite this drawback, Bainbridge enjoyed a growing reputation as a mining engineer.

In 1889, Bainbridge's entrepreneurial skills became evident, when he leased land in Derbyshire and founded the Bolsover Colliery Co., Ltd. Bolsover lies roughly halfway between Sheffield and Nottingham. Coal was reached here in 1891, at nearby Creswell in 1896. The 1890s saw periods of high demand and scarce labour. Hewers, whose earnings had irreversibly risen above the cost of comfortable living (Fig. 2.4), were unwilling to work the difficult thin seams except at increased wages. This prompted Bainbridge, who knew the conditions in Yorkshire and in the East Midlands, to adopt coal-cutting machinery at those collieries with which he was connected. Further nationwide developments in mechanisation entailed electric coal-cutters, 1904 marking the turning point in colliery electrification. That year, Bainbridge founded the Spitzbergen Coal & Trading Co. and simultaneously attended to the preparation of coal. Preparation removed impurities and graded the coal according to size. Different types of coal had different uses and markets. Small coals, for instance, were used in the manufacture of patent fuels and some collieries were heavily dependent on coking and briquetting. In an attempt to make unsaleable coal saleable, Bainbridge was driven to washing his coal against his will.

[13] Church (1986) p. 335.

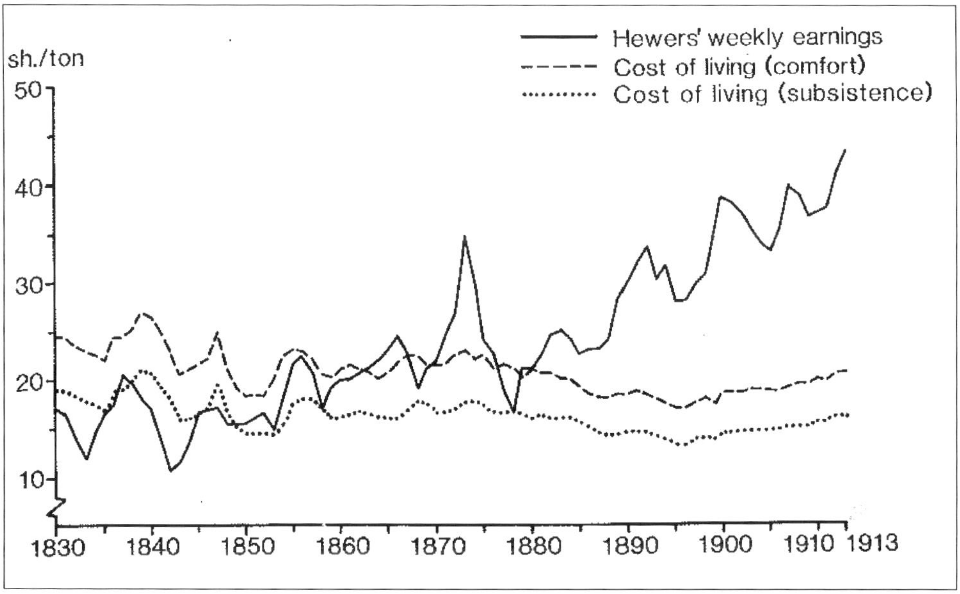

2.4 Estimates of hewers' earnings and the cost of living, 1830-1913. (Source: Church (1986) p. 573.)

Labour intensity and physical exertion had long been a trait of the British coal industry. Underground production workers cut, loaded, and moved the coal to the surface; maintenance staff did the deadwork, ensured continuous production, advanced the face, and drove the road; and above-ground workers saw to the movement and preparation of coal, the maintenance and repair of colliery equipment, and the care of the horses. The supply of mining labour had traditionally been dictated by customs as well as poverty. There was no long-term unemployment during the nineteenth century, and the option to be absent from work, i.e. the number of days worked per week, was actually the choice of the worker. This voluntary absenteeism commonly hindered productivity. Working hours, however, were subject to legislation and constantly caused disputes.

To control the workers, managers could apply a number of strategies. As such, contracts specified duties, rates of payment, and penalties for transgressions or neglect and were enforceable by law. Managers could also make use of long pay, whereby the miners had to purchase from company stores or truck shops. While the miners could end up indebted to the company, long pay was a means of controlling alcohol consumption, too. Yet, free beer had also acted as an incentive during the early 1870s boom. Weekly pays were in use after 1870. Depending on the colliery and the company, there were other approaches.

Managers used colliery housing as a strategy to recruit and keep labour.[14] In the North East, it was free, which was not the case everywhere in England. Generally, miners and their families faced eviction if they did not perform well or were unruly.[15] Colliery housing occasionally evolved into entire model villages. Although Britain's oldest model village was already constructed in Norfolk in 1805, as many as seven were erected in the newly developing coalfields in Yorkshire and the Midlands in the second half of the nineteenth century. Notably, Bainbridge completed Creswell Village in 1895 (Fig. 2.5) and Bolsover Village in 1896.

2.5 Creswell Model Village, refurbished in 2000 and 2004. (Source: http://www.mvmp.co.uk/settlement/index.php?settlement_id=57 (Accessed: 20 June 2011.)

Here, labour management strategies not only extended to housing but influenced entire institutional structures such as schools, church and chapel, institute, and club. In the first instance, this allowed for greater social control over education, religion (Methodism being the preferred faith), and the sanction of public opinion, but Bainbridge's strict paternalism and philanthropy also played an important role.[16] As will be seen in Chapter 3, Advent City gives the impression of a model village, but its mere presence does not confirm that it actually worked.

[14] While the economics of colliery housing were less compelling, many coalowners were persuaded that considerations of regularity and quality of the labour force justified such investment. Rent-free housing was the single most powerful lever on labour [...] Yet, it is clear that while financial return was one consideration influencing coalowners' decisions on house building and housing policy, much more important were the labour management strategies of individual firms, affected powerfully by the geographical location of collieries in relation to accessible supplies of labour and the retention of experienced and steady working miners. (Church (1986) pp. 279-81.)

[15] Unruliness included trade-union activities or strikes.

[16] 'Studies of paternalism [...] have overlooked the extent to which, certainly in mining, the need to socialize a migrant or inexperienced work-force continued to present a problem throughout the nineteenth century. Thus, when the first of several villages established by Emerson Bainbridge's Bolsover Colliery was begun in Derbyshire in 1892, the *Colliery Guardian* reporter noted that 'the village will be governed, as far as possible, by a council of men themselves; and drunkenness and swearing and gambling will render offenders liable to instant dismissal'. (Church (1986) p. 288.)

It was not unusual for men like Bainbridge to perform a range of concurrent professional, social, and political functions. Hence, the coalowner also headed an engineering consultancy firm in London, sat on the boards of several railway, hardware, and mining companies, and held memberships to a number of institutions, for which he occasionally prepared learned papers. If Church notices a public dissociation from the coal industry as a symptom of social aspirations, Bainbridge's intricate connection with coal mining was all the more obvious. Yet, he harboured only limited political ambitions, being elected Liberal Member of Parliament for Gainsborough in Lincolnshire in 1895 before losing his seat again in 1900, albeit by a small margin. Wilmot assumes that his political ambition was 'more likely to have been related in both timing and purpose to the quest for funding and support for the construction of the [Lancashire, Derbyshire & East Coast Railway] than with any local connections or desire for public service.'[17] This railway would have serviced Bainbridge's Bolsover collieries, direct access to markets being very advantageous. Railways had opened up the British inland trade, whereas shipping was instrumental to Britain's coal exports. In shipping, sailing vessels were dependent on the weather, while steamships could make more trips. Both were hindered by fog, gales, and ice. In turn, the shipping industry witnessed important technological advances in the type and speed of unloading machinery. As for Bainbridge's political convictions, he only spoke of light railways twice before the House of Commons.[18] The remainder of his 58 contributions concerned the rights of coal miners and the Boer War.

The above synopsis by no means completes the picture of the British coal industry in the nineteenth century and more specifically at the time when British companies became involved in the undeveloped Arctic. However, it conveys an idea of the traditions, knowledge, experience, and competence, which British entrepreneurs, managers, engineers, and miners may have brought with them to Spitsbergen. Coal had been found on the islands in 1610. In 1891, a German also observed seams on Bear Island. In fact, 'Herr Cremer was able to ascertain that coal-mining was quite possible. Although, perhaps, owing to the climate, not always in a regular manner.'[19] This discovery was duly reported in Britain as was the first commercial shipment of coal by the Norwegian sealing captain Søren Zakariassen in 1899. The newspaper predicted,

> Until recently no one has ever found reindeer or anything else on Spitzbergen that could give value to this great Arctic region. Perhaps this is the reason why none of

[17] Wilmot (2005) p. 250.

[18] *Hansard* is the official record of debates in Parliament. Available at: http://hansard.millbanksystems.com/ (Accessed: 7 June 2011).

[19] 'Coal in Spitzbergen' (1892) *Colliery Guardian*, 63, p. 254.

> the European Governments has ever thought it worthwhile to assert ownership over Spitzbergen. But a cargo of coal has now arrived in Norway from Spitzbergen and this may give these bleak northern islands greater value in the opinion of Europe.[20]

The article further consented to a Norwegian take-over of the Arctic no man's land but opposed Russian or German control. Despite the unsettled question of ownership, several British actors recognised an opportunity. In light of aspects that often hindered the development of collieries at home, Bainbridge may have appreciated Spitsbergen's advantages. These were said to include its geographical position on the seaboard; high-grade coal and metal ores; low working expenses; low extraction costs; good natural facilities for shipping, power, coal, electricity, etc.; good natural facilities for loading and shipping coal and ore; absence of taxes and duties; and absence of government interference. In the following chapters, it will be seen how British proprietors attempted to use these advantages for the furtherance of their newly formed companies.

2.3 The Empire

Cecil John Rhodes (1853-1902) embodied the British Empire at its zenith in the late nineteenth century. The archetypal imperialist, mining entrepreneur, and colonial politician explored and occupied a vast African territory that became the British colony known as Rhodesia (now Zimbabwe and Zambia) and founded the British South Africa Co., Ltd. to exploit it. Although Rhodes never set foot on Spitsbergen, among his pioneers in Rhodesia was one Frederick William Salisbury-Jones (1870?-1939). The connection was made when Salisbury-Jones became the managing director of the Northern Exploration Co. in 1918.[21]

Rhodes was born in Hertfordshire in 1853.[22] He was the fifth son of a Church of England clergyman. Educated at grammar school, he made no clear career choices. So at the age of 17, his father sent him to grow cotton in the British colony of Natal in south-eastern Africa. The Kimberley diamond fields had just been discovered 800 kilometres to the west and were quickly annexed by the British Crown. This instigated a new phase of aggressive imperial expansion in southern Africa. At the end of 1871, Rhodes had arrived at the diamond diggings and applied himself so diligently that within two years he could afford land and a higher education in England. At Oxford University, he consolidated his vehement

[20] 'A cargo of it taken to Norway from the land that nobody owns' (1899) RGS/SSC/23, RGS/IBG Archives, London.
[21] *Copy of paper read by Mr. Herbert G. Ponting* (1918) RGS/CB8/Conway, RGS/IBG Archives, London.
[22] Unless otherwise stated, biographical details from: Marks, S. and Trapido, S. (2004) 'Rhodes, Cecil John (1853-1902)', *Oxford Dictionary of National Biography*, Oxford: Oxford University Press.

imperial principles. Already in 1872, he had penned the first of eight wills, leaving his possessions to the British Secretary of State for the Colonies for the extension of the British Empire. Furthermore, during his induction as a freemason in 1877, he presented his *Confession of faith*, which supported the formation of a secret society 'with but one object the furtherance of the British Empire and the bringing of the whole uncivilized world under British rule for the recovery of the United States for the making the Anglo-Saxon race but one Empire.'[23] Rhodes aspired to spread ideas of imperial race pride and world domination and gathered around him men who shared his vision. Salisbury-Jones evidently was one of these men.

Generally, Victorian Britain at the end of the nineteenth century was 'a dynamic State in an age of excitement: capital looking for markets, vitality looking for opportunity, success looking for new fields.'[24] The British Empire was accordingly expansionist and sensational. 'Supremacy, dominion, authority, size, were the watchwords of the time.'[25] Yet, empire meant different things to different people. For Rhodes, military power was initially not a determinant, commercial prospects and prestige appealing to him instead. Like Rhodes, the Brits who sought to make their fortune in Britain's colonies were 'unmistakably British.'[26] According to Morris,

> They were the children of a unique culture, that of the English public schools, with its celibate discipline, its classical loyalties, its emphasis on self-reliance, team spirit, delegated responsibility, Christian duty and stoic control. [...] It produced men of high spirits, courage and assurance, ready to rough it and unafraid of responsibility. If it was intellectually narrowing and chauvinist, well, this was an Empire that survived by the separateness of its rulers, their conviction that what they did was right, and that all else was second best. [...] At his worst the public school man was a snobbish hearty: at his best he combined authority with Christian kindness and what he would have called *grit*: the rarest of his virtues was human sympathy, the rarest of his vices cowardice. And the most irritating of his traits, at least in the imperial context, seems to have been smugness.'[27]

Yet, besides military power, commercial opportunity, and prestige, Morris recognises a deeper, more fundamental desire behind the Empire. Hence,

> Much of the driving force of imperialism, as of Victorian progress in general, was the energy sparked by man's struggle with his own environment, and to many of the imperialists the struggle was an end in itself. The notion of perpetual striving was essential to the morality of the day. [...] A puritanical pleasure in hardship was

[23] Marks and Trapido (2004).
[24] Morris (1979) p. 23.
[25] Morris (1979) p. 22.
[26] Morris (1979) p. 217.
[27] Morris (1979) pp. 220-1.

> often allied with a boyish delight in rip-roar, the two formidably combining to produce a breed of stoic adventurers, for whom the imperial mission was a larger embodiment of a personal challenge. [...] most of those who went out gold-digging for the fun of it could afford to do it without the gold anyway.[28]

Against this background of Britishness and Empire, Rhodes conducted his diverse and often controversial activities in Africa. For the purpose of this book, the fact that the territorial claims of one man's mining company could develop into a British colony and eventually into a country is most significant. This development can be illustrated using the following cornerstones: From the 1860s, the existence of gold between the Limpopo and Zambezi rivers had been known. Gold was undoubtedly among the reasons for rapid imperial expansion in Africa, which led to the international Berlin Conference to be called in 1884.[29] In 1888, Rhodes was able to obtain mineral rights from the dominant tribal leaders and to buy out his rivals. In 1889, the British Government granted his British South Africa Co. a royal charter to exploit and control a vast, ill-defined region between the Limpopo and Lake Tanganyika. In 1890, Rhodes used the charter to send 200 white and 150 black settlers to establish Fort Salisbury (named after the then British Prime Minister, now Harare). This Pioneer Column was protected by the company's own 500-strong British South Africa Police. Following several wars and revolts, which were decided in favour of the British settlers, the Legislative Council was formed in 1899. It governed the civil affairs of the company and through it the company could pass government measures. The company's rule survived Cecil Rhodes, who died in 1902. The charter was renewed in 1914 but became impractical in 1920. Rhodesian self-government began in 1923.

The death of Queen Victoria in 1901 marked the end of the Victorian era and the shift to an Edwardian culture. Edwardian England (1901-10) displayed, for instance, a greater interest in socialism. Although the British Empire had passed its peak, there continued to be 'something attractive about a country that belongs to nobody'[30] even until after the First World War. The country in question was, of course, Spitsbergen. A former pioneer in Rhodesia, Frederick William Salisbury-Jones had assumed the role of managing director of the Northern Exploration Co.

---

[28] Morris (1979) p. 305.

[29] The outcome of the Berlin Conference in 1884-5 was the General Act. Besides regulating African colonisation and trade, the act would be relevant to the occupation of Spitsbergen because of its Principle of Effectivity. According to the principle, a power could only hold colonies if it had adhered to certain standards. Effective occupation was proven by treaties with local chiefs, flying the country's flag, establishing administration and a police force, and using the territory economically. If a nation failed to meet the requirements, another power could take possession of the territory. The act foresaw that each new act of occupation should be accompanied by a formal notification by the power making the claim to the other signatory powers in Berlin.

[30] Cecil Harmsworth, at the time Under-Secretary of State for Foreign Affairs, HC Deb 15 May 1919 vol 115 c1927.

just before the conclusion of the war. His incentive may have been the opportunity to recreate the former glory of the British colonies and duplicate or even surpass Rhodes' successes. His conduct in the Arctic was in any case greatly politicised. Failing to attain absolute company control and ultimately British sovereignty over the islands, Salisbury-Jones terminated his employment soon after the signing of the Spitsbergen Treaty in Norway's favour.

2.4 Contributions to polar exploration

Spufford envisages the history of polar exploration as a history in two parts.[31] Firstly, he refers to the *technical history*, which factually recounts the sequence of expeditions. Of relevance to this study were the expeditions of Poole; of Parry, Sabine, and Scoresby; and of William Speirs Bruce. Poole reported the discovery of coal a few years prior to the first English claim on Spitsbergen in 1614. Parry, Sabine, and Scoresby were posthumously embedded into the rhetoric of the Northern Exploration Co. Bruce was instrumental in the formation of the Scottish Spitsbergen Syndicate. Secondly, Spufford's *imaginative history* relies heavily on the notion of the sublime. It offers an alternative explanation as to why British explorers chose the Polar Regions before other destinations. The language of the sublime is unmistakable in the writings of Ernest Mansfield, instigator behind as many as three companies on Spitsbergen.

Jonas Poole (baptised 1566, died 1612) personified Arctic exploration in the early seventeenth century and was among the first Englishmen to come ashore on Spitsbergen. Poole was the son of a naval captain.[32] Having volunteered to learn about trade and exploration, he accompanied the Muscovy Trading Co. to Archangel in 1603. This expedition confirmed the Dutch sighting of Spitsbergen. Subsequently, but with the exception of 1607, Poole made yearly northward journeys to hunt walrus. In 1610, he christened several prominent bays, notably Horn Sound, Bell Sound, and Icefiord, the names of which persist to this day and are used throughout this work. He also reported both the abundance of whales in these bays and the existence of coal that burnt well.[33] Although Poole's career ended prematurely with his being murdered in 1612, his observations marked the beginning of English whaling in Spitsbergen waters. English whalers were strongly rivalled by the Dutch. In turn, each made an effort to defend themselves and to protect their rights on the islands. Not only did the English and the Dutch sign an

---

[31] Unless otherwise stated, general statements of polar exploration and the sublime from: Spufford (1996).

[32] Baldwin, R. C. D. (2004) 'Poole, Jonas (bap. 1566, d. 1612)', *Oxford Dictionary of National Biography*, Oxford: Oxford University Press.

[33] Conway, M. (1906) *No man's land*. Cambridge: Cambridge University Press.

agreement regarding the location of their operations in the season of 1614; the English also ceremoniously erected the arms of King James I., thereby taking what in their home country passed for lawful possession of their claims.[34] The story of early claiming and counterclaiming is indeed much more complex than illustrated here. Suffice to say that it was this British annexation of Spitsbergen, albeit of only part of the islands, that was stressed in political debate in the first quarter of the twentieth century.

Almost 200 years later, Parry, Sabine, and Scoresby entered the polar stage. On the one hand, they had much in common with other Arctic explorers in the mid-nineteenth century; on the other hand, each made individual contributions to discovery and science. Besides adding to the knowledge of Spitsbergen, they are linked to its history because the Northern Exploration Co. saw fit to name claim huts after them. It was an attempt to tap into the men's international renown in order to appeal to potential backers and to display British astuteness in the face of adversaries. Whether or not the intended symbolism matched the reality will be seen in the archaeological site narratives in Chapter 4.

William Edward Parry (1790-1855) was a naval officer and Arctic explorer.[35] From 1810, he spent three years protecting British whalers in Spitsbergen waters. In 1818, he joined John Ross's search for a north-westerly passage.[36] In the following year, he commanded his own search, his party becoming the first British expedition to deliberately winter in the Arctic. On his journeys, Parry amassed much scientific material and was afterwards elected a Fellow of the Royal Society (FRS). It was his personal conduct, care of his men, ability to solve problems, and meticulous scientific work that set the methodological and moral standard for future Arctic exploration. His proposal to reach the North Pole via Spitsbergen was approved by the Royal Society in 1826. Although he did not attain it, his furthest north was an unbroken record for over 50 years. He came into conflict with the First Secretary of the Admiralty[37] and after being knighted in 1829, he resigned from the Navy to take up a position in Australia. The remainder of his life is inconsequential to this book.

[34] King James VI. of Scotland, crowned in 1567, became King James I. of England and Ireland, when he inherited the English and Irish Crown in 1603. Thus, one can arguably speak of a British annexation of Spitsbergen.

[35] Laughton, J. K. (2004) 'Parry, Sir (William) Edward (1790-1855), *Oxford Dictionary of National Biography*, Oxford: Oxford University Press.

[36] In 1815, the Anglo-American War and the Napoleonic wars had ended. Therefore, the Admiralty could employ surplus ships elsewhere. Ross's was the first in a series of Admiralty expeditions in search for a north-west passage. It also included Edward Sabine as scientific officer.

[37] The First Secretary of the Admiralty was the political secretary in charge of the government of the Royal Navy. From 1809 till 1830, this was John Wilson Croker.

Edward Sabine (1788-1883) was an army officer and physicist.[38] After 1815, he turned to exploration and science.[39] With the help of his brother Joseph Sabine, he penetrated London's scientific circles, and his brother-in-law Henry Browne provided him with his first magnetic instruments. In 1818, Sabine served as astronomer on John Ross's expedition. Thereafter, he was among Parry's wintering party. From 1821 till 1823, he sailed as far north as Spitsbergen and gathered evidence of the Gulf Stream and the length of the arc of the meridian. One of his most influential patrons was the Second Secretary of the Admiralty.[40] In fact, his ambitious project to understand the earth's magnetism was supported by the Royal Society, the British Association, the Admiralty and Ordnance, and Parliament. He subsequently produced the first global magnetic data set.

William Scoresby, junior (1789-1857) was an Arctic whaler, scientist, and Church of England clergyman.[41] Aboard his father's vessel, Scoresby first sailed northward at the age of ten. In all but one summer between 1803 and 1823, he took part in the Greenland whale fishery, 1823 being his last Arctic voyage. In 1806, he enrolled at Edinburgh University to study chemistry and natural science in the off-season. In 1817/8, he instigated a Government-sponsored expedition to the Arctic, which was nonetheless placed under the command of a naval officer. Whereas Parry and Sabine had enjoyed the support of a wide network, Scoresby was snubbed by the naval and scientific establishments, which greatly hindered his research. Moved by the death of his first wife, Scoresby entered the church in 1822. During the course of his life, he published widely on Arctic geography, meteorology, and oceanography, particularly on Arctic currents, waves, and the Gulf Stream.

William Speirs Bruce (1867-1921; Fig. 2.6) formed the link between the Heroic Age of Antarctic Exploration, generally agreed to have occurred between 1897 and 1922, and the emergence of modern oceanography. His scientific work and Antarctic expeditions are well-known.[42] What is more, his surveys on Spitsbergen led directly to the formation of the Scottish Spitsbergen Syndicate.

---

[38] Good, G. A. (2004) 'Sabine, Sir Edward (1788-1883)', *Oxford Dictionary of National Biography*, Oxford: Oxford University Press.

[39] After the Anglo-American War and the Napoleonic wars, many British officers turned to exploration and science.

[40] The Second Secretary of the Admiralty was an assistant to the first and usually a civil servant. From 1807 till 1845, this was John Barrow. He was a great promoter of Arctic expeditions, including those of John Ross, William Edward Parry, James Clarke Ross, and John Franklin.

[41] Baigent, E. (2004) 'Scoresby, William, junior (1789-1857)', *Oxford Dictionary of National Biography*, Oxford: Oxford University Press.

[42] Barr, W. (2008) 'The Arctic cruises of Prince Albert I of Monaco', *Polar Record*, 44 (228), pp. 1-14; Bernstein, R. E. (1985) 'The Scottish National Antarctic Expedition 1902-04', *Polar Record*, 22 (139), pp. 379-92; Speak, P. (1992) 'William Speirs Bruce: Scottish nationalist and polar explorer', *Polar Record*, 28 (167), pp. 285-92; Speak, P. (2003) *William Speirs Bruce. Polar explorer and Scottish nationalist,* Edinburgh: National Museums of Scotland Publishing; Speak, P. (2004) 'Bruce, William

2.6 William Speirs Bruce (1867-1921). (Source: Speak, 2003.)

The London-born polar explorer and scientist longed for the Scotland of his forefathers, a lifelong trait being his fervent Scottish nationalism.[43] At Edinburgh University, Bruce studied medicine and natural science and gained invaluable experience in early oceanography as a volunteer to the *Challenger* laboratories.[44] He also met influential mentors and friends, one being the imminent librarian of the Royal Geographical Society.[45] He recommended Bruce to the Dundee Whaling Expedition to the Antarctic in 1892-3, after which the latter quit medicine and focused instead on polar science. After a short interlude at the Ben Nevis meteorological station, Bruce accompanied the Jackson-Harmsworth expedition to Franz Josef Land in 1896, during which he first met the Norwegian Fridtjof Nansen, returning from his famed *Fram* expedition.[46]

In 1898, Bruce had been in the Barents Sea, when the Prince of Monaco invited him to partake in an oceanographic survey in Spitsbergen waters. The invitation was extended to a second survey in 1899. That year, he also applied to the Royal Geographical Society to accompany Scott's *Discovery* expedition to the

---

Speirs (1867-1921)', *Oxford Dictionary of National Biography*, Oxford: Oxford University Press; Swinney, G. N. (2001) 'Some new perspective on the life of William Speirs Bruce (1867-1921), with a preliminary catalogue of the Bruce collection of manuscripts in the University of Edinburgh', *Archives of Natural History*, 28 (3), pp. 285-311; Swinney, G. N. (2002a) 'The training of a polar scientist: Patrick Geddes and the student career of William Speirs Bruce', *Archives of Natural History*, 29 (3), pp. 287-301; Swinney, G. N. (2002b) 'William Speirs Bruce, the Ben Nevis Observatory, and Antarctic Meteorology', *Scottish Geographical Journal*, 118 (4), pp. 263-82; Swinney, G. N. (2003) 'From the Arctic and Antarctic to 'the back parts of Mull': The life and career of William Gordon Burn Murdoch (1862-1939)', *Scottish Geographical Journal*, 119 (2), pp. 121-52; Swinney, G. N. (2007) 'The Scottish National Antarctic Expedition (1902-04) and the founding of Base Orcadas', *Scottish Geographical Journal*, 123 (1), pp. 48-67.

[43] Speak (2004).

[44] The *Challenger* expedition of 1872-6 was the scientific expedition that laid the basis of oceanography. It was commanded by George Nares and supervised by Charles Wyville Thomson. John Murray directed the publication and instructed Bruce's work.

[45] Speak (2003).

[46] Fridtjof Nansen (1861-1930) was an explorer, scientist, diplomat, and humanitarian. He reached a record northern latitude during his *Fram* expedition from 1893-6, after which he retired from exploration. He made innovations to polar techniques, equipment, and clothing. He was also an important Norwegian national figure both as a scientist and a politician.

Antarctic in 1901-4.[47] President Markham replied late and offered only a minor post, by which time Bruce had turned to Scottish patrimony and was planning his Scottish National Antarctic Expedition to take place in 1902-4. Markham viewed him as a rival to Scott, which only served to fuel Bruce's aversion to English institutions. Bruce's Antarctic expedition was a success, but due to a lack of funds, it was never repeated. In 1906 and 1907, Bruce returned to Spitsbergen, a less costly alternative, where he carried out comprehensive surveys.[48] To fund his science, he promoted the Scottish Spitsbergen Syndicate and acted as its consultant until his death in 1921. Despite his accomplishments, he never enjoyed the same recognition as other contemporary polar explorers.

This selected technical history of British polar exploration has highlighted a noteworthy aspect. Despite the best character, skill, and intentions of the explorer, the success of a polar expedition very much depended on his network. Parry and Sabine relied on the backing of institutions such as the Royal Society and the Admiralty. Some institutions represented gentlemanly scientific pursuit, whereas others stood for exclusive military tradition and contemporary governmental policy. Scoresby and Bruce got to feel what it meant not to belong to the club. Scoresby suffered at the hands of those who 'disapproved of the marriage of industrial artisanship and polite natural history.'[49] In addition, the Second Secretary of the Admiralty could be very jealous and possessive of non-naval plans for maritime exploration. Bruce felt the bias of funding bodies towards 'British' enterprises, which usually meant English rather than Scottish ventures. His rift with the Royal Geographical Society was partially mended after Markham's retirement, but prejudices and favouritisms persisted during the early industrialisation of Spitsbergen.

2.5 Perceptions of polar adventures

Regarding said *imaginative history* of polar exploration, it is useful to consider what the expeditions meant to those who went and those who stayed at home. Spufford argues that polar fascination was rooted in the concept of the sublime.[50] The idea originated in the eighteenth century before it was transformed by Romanticism and diffused into the Victorian senses. 'Beauty tended to be thought of as regular,

[47] The president of the Royal Geographical Society was Clements Robert Markham (1830-1916). He was credited with reviving the subject of Antarctic exploration at the RGS but believed that the Royal Navy should be in charge of such exploration.
[48] Bruce, W. S. (1908) 'The Exploration of Prince Charles Foreland, 1906-1907', *The Geographical Journal*, 32 (2), pp. 139-48.
[49] Bravo, M. (2006) 'Geographies of exploration and improvement: William Scoresby and Arctic whaling, 1782-1822', *Journal of Historical Geography*, 32, pp. 512-38.
[50] Spufford (1996).

tender, soothing, polished; and the forceful sensations of the sublime were none of these. The sublime was greater, and when one felt sublime one's attention seemed to be at a greater stretch.'[51] The sublime was also terrifying. A volcanic eruption or a hurricane would constitute something mesmerising yet destructive, and 'snow was certainly represented among the chosen agents of catastrophe.'[52] Although subliminal fear and plain fright could be distinguished by common sense, it is questionable whether every explorer possessed the necessary judgment. In fact,

> It can even be argued that the success and failure of different British expeditions of the period reflects the degree to which they were, and were not, imaginatively captured by a vision of the Arctic as bleak, blank, hostile. Those explorers least able to perceive the Arctic as it was – indifferent rather than harsh, full rather than empty, a problematic dwelling space rather than a moral playground – were also least likely to survive there.[53]

2.7 Ernest Mansfield (1862-1924). (Source: The Northern Exploration Company, Ltd. (1911) The Northern Exploration Company (N.E.C.), Norwegian Polar Institute, Tromsø.)

Victorian literature adopted a specific language based on the sublime and notions of doom. 'With advances in geology, meteorology and climate studies they were able to dress fantasies of apocalyptic destruction in scientific clothes – to invent new delicious ways for their world to end.'[54] Ernest Mansfield's *Ice Maiden*, admittedly Edwardian, is a late example of this type of writing.[55] Mansfield (Fig. 2.7) was the promoter of the Spitzbergen Mining & Exploration Syndicate, Spitzbergen United, and the Northern Exploration Co. He penned the pseudo-documentary when he overwintered in Bell Sound in 1908/9. During this time, he experienced the Arctic in terms of Spufford's subliminal juxtapositions of light and darkness, motion and standstill. He wrote,

[51] Spufford (1996) p. 19.
[52] Spufford (1996) p. 25.
[53] Spufford (1996) p. 58.
[54] Spufford (1996) p. 24.
[55] Mansfield, E. (1910) *Astria: the Ice Maiden*, London: Lonsdale Press.

> What a number of experiences I have had [...]! Some were awful, others terrifying! [...] During the storm everything was black – except the sea. That was white! [...] But I was in a lonely world, with everything silent, except the roaring, raging sea! [...] I was cut off from the world; with no hope of a word from anyone, or news from anywhere. Absolutely alone, in an uninhabited country, frozen harder than adamant![56]

Although the protagonist of *The Ice Maiden* is imaginary, the opening chapters read like Mansfield's autobiography. In fact, they may well be, if he appreciated that 'truth is stranger than fiction.'[57] They recount the arrival of the prospector on Spitsbergen and his preparations to winter. Practicality is often off-set with sentimentality. So when the sun is about to set, the author writes, 'I knew this was the last time Sol would call this year, so I hoisted my Union Jack to greet him.'[58] Through a series of events, the protagonist enters a world of ice. His social contemplations centre around relationships between men and women and his own (un)faithfulness, while his scientific thoughts focus on prolonging life so that an inventor could see the full development of his ideas. In the realm of ice, there exists a bell that, when rung, relates a person's unspoken desires to a servant; a telegraph that not only transmits a message but also materials and people; automobiles made of platinum that fly. Air is rare and valuable, whereas radium is used for fuel and lighting. Gold is medicinal. (Hear, hear!) The list of wonders made possible by ice goes on. 'Ice is the king of nature, and the guardian of the world.'[59] Having previously fallen in love with the queen, the protagonist may marry the princess of the realm of ice but only after a period of reflection back in his own world. At Camp Bell, an actual place described in Chapter 3, accidental unconsciousness overpowers him, preventing him from reflecting or indeed returning. His fate is to share his incredible story with disbelieving listeners.

Mansfield's tale recounts not only part of his personal experiences on Spitsbergen but indicates his self-image as a man and prospector as well as his social and scientific preoccupations. Clearly, there lies a danger in taking alleged fiction too literally, but Mansfield was proud of his background and achievements and saw no reason to downplay or change the basic facts. Thus, *The Ice Maiden* represents the deep-seated sentiments and dreams of an Arctic adventurer, if not Arctic exploration *per sé*. Spitsbergen was a virgin country full of promise to those who were courageous and resourceful enough to accept the challenge.

[56] Mansfield (1910) Dedication.
[57] 'Truth is stranger than fiction, but it is because fiction is obliged to stick to possibilities; truth isn't.' Mark Twain.
[58] Mansfield (1910) p. 31.
[59] Mansfield (1910) p. 67.

PART II

THE ARCHAEOLOGICAL IMAGE

3 The Archaeology of Arctic Mining

3.1 Introduction

Chapters 3 and 4 present some of the archaeological results of the LASHIPA expeditions to Spitsbergen between 2004 and 2010 (Fig. 3.1). This chapter concentrates on the physical remains of two British mining companies, the Spitzbergen Coal & Trading Co. and the Spitzbergen Mining & Exploration Syndicate.[1] The strict division into a largely descriptive account of the expeditions' observations, the *site narratives*, and subsequent *site interpretations* is a choice to enable re-interpretation by others and thereby stimulate scientific dialogue on this topic. The site interpretations address the questions of why the companies were present on the archipelago, how they chose to operate, and what factors contributed to the abandonment of their sites. The findings disclose valuable information about the companies' local networks in particular and create the archaeological image of British mining on the islands.

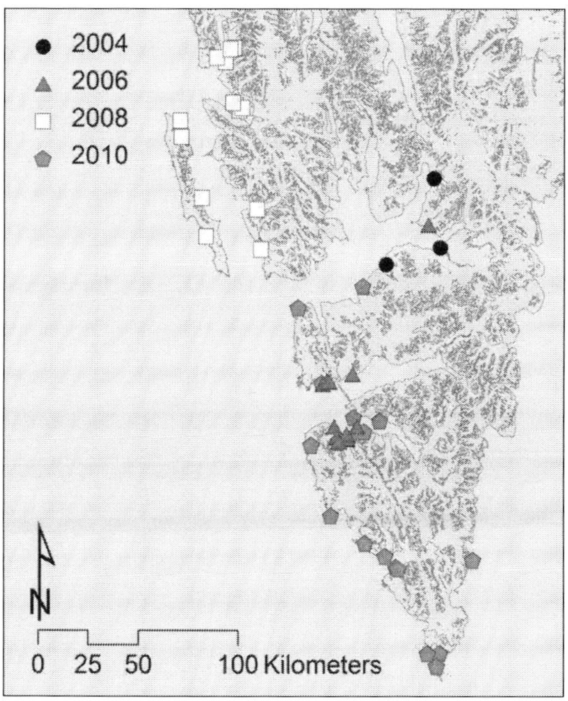

3.1 Location map of archaeological sites visited during the LASHIPA expeditions between 2004 and 2010. The sites depicted have relevance to this study, but not all are treated in the text. (Map: F. Kruse.)

[1] As opposed to the following chapter, which focuses on two *exploration* companies.

3.2 The Spitzbergen Coal & Trading Company

In August 2004, the International Field Course in Arctic Industrial Heritage (hereafter referred to as LASHIPA 1) undertook a one-day survey at **Advent City**, the former settlement and mine of the Spitzbergen Coal & Trading Co. Over 20 participants worked intensively on a range of allocated archaeological tasks. Their field data and subsequent report[2] formed the basis of the following site narrative. Due to its proximity to Longyearbyen, the gateway to Spitsbergen for most visitors, Advent City is one of the archipelago's better-known historical industrial sites. It lies approximately five kilometres to the north of the capital on the north-eastern shore of Advent Bay (Fig. 3.2). The bay opens out into Spitsbergen's central and largest fjord, the Icefiord. Advent City is therefore particularly exposed to any weather conditions arriving across the Icefiord from the west.

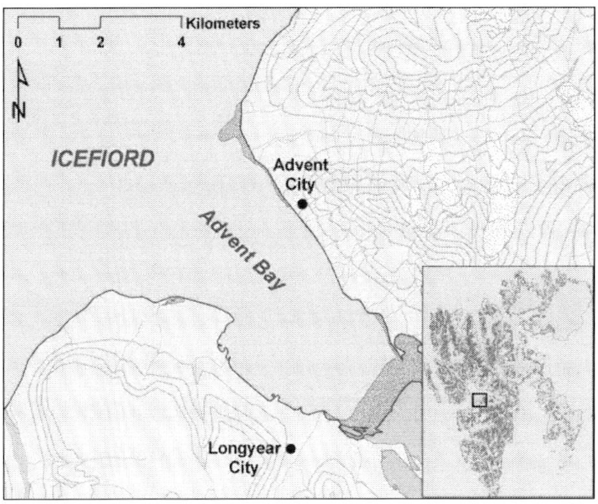

3.2 Location map of Advent City. It lies to the north of Longyearbyen, formerly known as Longyear City. (Map: F. Kruse.)

To date, there has been no indication of archaeological remains predating those of the coal mining operations. Mining began with the trials of a Norwegian predecessor in 1901. The Norwegians erected the first permanent structures on the site in 1903. From its formation in 1904 until the end of summer in 1908, the Spitzbergen Coal & Trading Co. attempted to develop the coal claim. The settlement and the mine were then abandoned, although the company was only dissolved in 1918. Most buildings were reused at another mine nearby.[3]

[2] Martin, P. Avango, D., Hartnell, C., Martin, S. R., Mishkar, L., Oglethorpe, M., Rosness, G., Tennant, E., and West, I. (2006) *[LASHIPA 1] Industrial heritage in the arctic: research and training on Svalbard, August 2004,* Michigan: MTU. The LASHIPA 1 report is available online: http://www.rug.nl/research/arctisch-centrum/pdfs/lashipa1.pdf.

[3] Johannessen, L. J. (1997) *Hiorthhamn. Coal mining under difficult conditions*, Longyearbyen: Governor of Svalbard.

3.2.1 Site narrative

In summer, the site is most easily accessible by sea. Overland travel is hindered by frequent rivers and the absence of roads. In winter, the mode of transport is often determined by snow fall and the formation of ice in the bay. The site overview (Fig. 3.3) provides little indication of the depth of the water or any submerged obstacles.

3.3 Overview of Advent City. (Photo: L. Mishkar, LASHIPA 1, 2004.)

The shore is lined by a narrow cobble beach. From here, an approximately 200-metre wide plateau rises towards the foot of the hills. The steep hillsides display near-horizontal layers of sedimentary rock. A modern building, substantial concrete remains, and other anthropogenic remnants mark the site of the former settlement. Immediately to the south-east of the ruins lies a melt-water gully. Such gullies may witness flash floods when the snow thaws in spring but often dry up as summer progresses.

The archaeological remains encountered during the LASHIPA 1 survey[4] are summarised in the site map (Fig. 3.4). The former mine lay above the 100-metre contour. It was marked by two openings. The larger (1a) had previously been lined with stone, but it was now blocked by fallen rock. The smaller (1b, Fig. 3.5) had been supported by timber, which had since collapsed under natural creep. The ground in front had been levelled to create an operating platform approximately two metres wide. Immediately below this platform was a low retaining wall (1c) made from roughly hewn sandstone blocks. The pithead arrangements were completed by the ruin of a small building (1d) with sandstone walls and a gabled wooden roof. Although the workings could no longer be entered, a large piece of decaying machinery on the slope below arguably provided some information about the conditions underground. The machinery proved to be the body casing of a disc coal cutter (1e, Fig. 3.6).[5]

[4] Martin *et al.* (2006) pp. 33-50.
[5] Carlyle, M. (2012) E-mail to Frigga Kruse, 18 July.

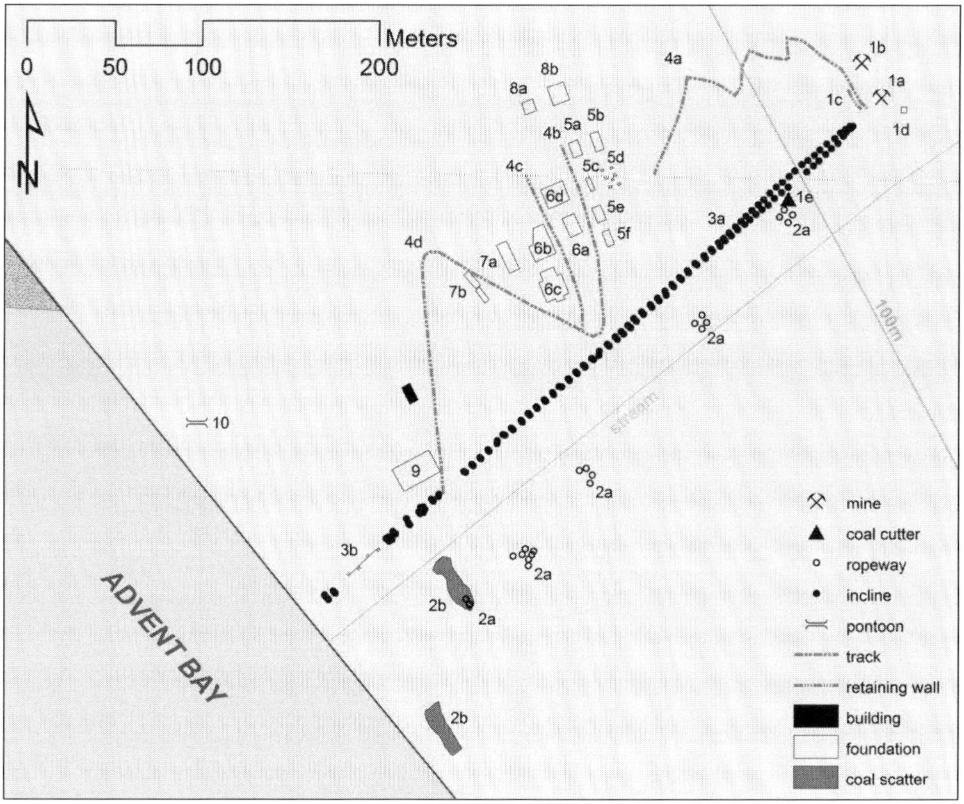

3.4 *Site map of Advent City. (Data: LASHIPA 1, 2004; Map: F. Kruse.)*

3.5 *Timbered opening (1b) collapsed under creep. (Photo: L. Mishkar, LASHIPA 1, 2004).*

3.6 Body casing of a disc coal cutter (1e). (Photo: L. Mishkar, LASHIPA 1, 2004.)

From the former mine, three linear feature systems departed downhill towards the shore. One system comprised five groups of upright but slightly angled sawn-off wooden posts roughly arranged in squares or rectangles (2a, Fig. 3.7). The lowest group was associated with two coal scatters (2b). This system was thought to be the remains of an aerial ropeway.[6]

3.7 This group of sawn-off posts (2a) was the base of a ropeway tower. (Photo: L. Mishkar, LASHIPA 1, 2004.)

The second system comprised two parallel rows of upright, slightly angled sawn-off wooden posts (3a, Fig. 3.8). While the distances between the posts within a row were fairly consistent, the distances between the two rows varied, it being greatest at the foot of the hill, where there was a break in slope. Approaching the shore, there were additional wooden posts next to the concrete structure (9) before these

[6] Martin *et al.* (2006) p. 41.

gave way to a short length of railway embankment with two sets of wooden sleepers still in place (3b, Fig. 3.9). On the embankment lay a thin cover of coal.

3.8 These parallel rows of sawn-off posts (3a) belonged to an incline. Note the positions of the mine (1a), the retaining wall (1c), and the small ruin (1d). (Photo: I. West, LASHIPA 1, 2004.)

3.9 Short railway embankment (3b) with wooden sleepers and a thin spread of coal. (Photo: M. Oglethorpe, LASHIPA 1, 2004.)

At the shore, the melt-water stream, which is oversimplified in the site map above, had since the time of site usage changed direction towards the north-west and presumably obscured any physical remains in that area. This second system was judged to have been a self-acting inclined plane.

The third system comprised several tracks visible as compacted ground cleared of obstacles. One of these tracks (4a) was narrow and zigzagged tightly. It led downhill towards a cluster of building foundations. The zigzagging was most likely the result of both the steepness of the hill and the mode of transport. The feature was concluded to have been the miners' footpath.[7] Three 'streets' originated from within the cluster of foundations. The upper and central ones (4b and 4c, respectively) merged into the lower street (4d) to the south of it, which in turn meandered shoreward. These broader streets were perhaps busier or used by handcarts, horse-drawn carts, or even vehicles.

The streets divided the fourteen foundations of the former settlement into four distinct zones, which were emphasised by differences in size and material of their remains. The first zone was situated above the upper street (4b) and comprised six foundations (5a-f). All but one survived as vague earthworks in the form of levelled platforms and low turf banks. The exception was a pad foundation made of sturdy sandstone pillars (5d, Fig. 3.10).

3.10 Pad foundation of sturdy sandstone pillars (5d) at the top of the settlement. (Photo: L. Mishkar, LASHIPA 1, 2004.)

The central street (4c) granted access to four former buildings in the second zone (6a-d). These were larger and generally better built. Although one comprised little more than earthworks (6a), two had concrete strip foundations (6b and 6c), while the fourth stood out as being the only brick strip foundation on site (6d, Fig. 3.11).

[7] Martin *et al.* (2006) p. 39.

3.11 Brick strip foundation (6d) central to the settlement. (Photo: L. Mishkar, LASHIPA 1, 2004.)

The lower street had serviced two buildings (7a and 7b) in the third and lowest zone. The foundation remains were elongate. While one was again a concrete strip foundation (7a), the other was a simple footing of timber piles (7b, Fig. 3.12).

3.12 The foundation of the lowest lying building in the settlement (7b) was made of timber piles. (Photo: I. West, LASHIPA 1, 2004.)

Another two concrete strip foundations (8a and 8b) in the fourth zone formed the northern perimeter of the former settlement. They could seemingly be reached via both the upper and the central street, but spatially they appeared to be more closely associated with the highest lying houses of the settlement.

Finds among the foundations included a cooking range made in Trondheim, a cauldron, a pot, and a bucket or likely slop pail. Like many artefacts on Spitsbergen, the slop pail had been used for target practice. There is a possibility that any portable items were brought onto the site at a later stage.

Wooden debris was by far the most common, occasionally in the form of wooden cases and boxes.

The most obvious remains were the concrete ruins (9) closest to both the former incline and the shore. The overall size being approximately 20 metres by 14 metres, they appeared to be divided into two sections. The larger section faced the sea and was dominated by massive sandstone and concrete foundations (Fig. 3.13). On the foundations there stood three large concrete plinths (Fig. 3.14). On two of these plinths there were two engines. A third engine remained in its packaging on the shore. The engines were labelled 'The Campbell Gas Engine Co Ltd, Halifax England'. It was therefore clear that they were gas engines in the former engine house. The body of Campbell's 'Little Samson' oil engine lay nearby.

3.13 Sea-facing front of the former engine house (9). (Photo: I. West, LASHIPA 1, 2004.)

3.14 Concrete engine beds and gas engines in the former engine house (9). Note the location of Longyearbyen through the flywheel. (Photo: I. West, LASHIPA 1, 2004.)

The second section had a higher ground level and, unlike the first, no stone walls (Fig. 3.15). It contained a small brick-lined furnace, the fire door still in position. The majority of finds here were machine parts, some having been delivered by Hadfield's of Sheffield. Among the debris were also remnants of coal, coke, and wooden barrels. The white mineral content of the barrels had swelled, bursting the containers, and then hardened. It was most likely cement to make concrete.

3.15 Back of the former engine house (9). Note the exposure to the Icefiord. (Photo: L. Mishkar, LASHIPA 1, 2004.)

Below the engine house, the incline met the waterfront, but no loading facilities were encountered here. Instead, the remains of a floating pier were founded over 100 metres to the north-west. A single barge-like pontoon (10, Fig. 3.16) lay high up on the beach.

3.16 Barge-like pontoon (10) lying on the beach. (Photo: L. Mishkar, LASHIPA 1, 2004.)

3.2.2 *Site interpretation*

The site narrative above has identified several environmental and archaeological components, which can be used to populate the graphic introduced in Fig. 1.1. The results are presented in Fig. 3.17 below and form the basis of the following site interpretation.

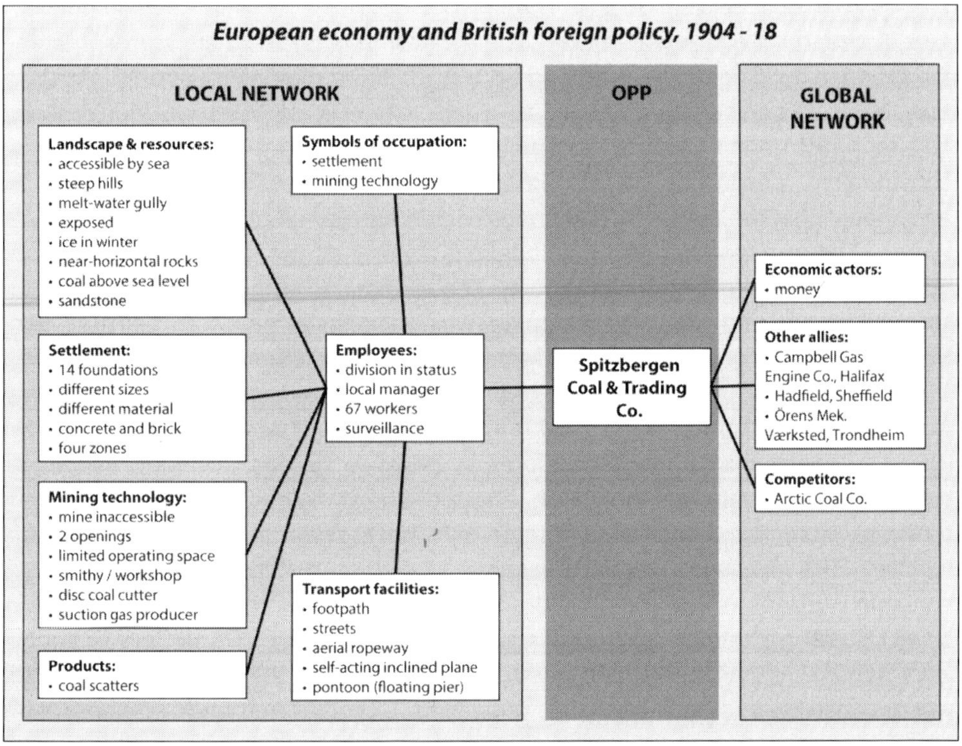

3.17 Environmental and archaeological components are used to populate the actor-network of the Spitzbergen Coal & Trading Co. Notably, no local allies, in-house expertise, or political actors were identified on site. (Chart: F. Kruse.)

Against the background of European market forces and British foreign policy, which find mention in Chapter 5, the Spitzbergen Coal & Trading Co. was active in the Arctic between 1904 and 1908. As far as the motivations behind its presence on Spitsbergen are concerned, its name suggests that it not only intended to exploit coal, but that it also proposed to engage in trading. For this purpose, it continued at the site originally chosen by its Norwegian predecessor. Advent City benefitted from being accessible by sea, freshwater, near-horizontal coal above sea level,

The plan showed an Arctic entrance[19], which opened into an elongated foyer. A door to the right led into the main room. A room to the left of the foyer had substantially thinner walls and could only be entered from the outside. This hinted at a lean-to shed or workshop, which was not insulated. Other archaeological reports were not known prior to the field surveys.

3.3.1 Site Narratives

The site narratives are reconstructed using the results of the LASHIPA 3 expedition.[20] Similar to Advent City, the former sites of the Spitzbergen Mining & Exploration Syndicate were best approached by sea. As seen in Fig. 3.19 above, the route to **Camp Morton** led through a narrow passage to the north of Axel Island onto its leeward side. This natural barrier is the reason why ice often clears only slowly from Lowe Sound. A site overview (Fig. 3.20) shows that the coastline at Camp Morton is dominated by low cliffs. It is nonetheless possible to reach the site because two melt-water gullies had been carved out to meet the shore. By mid-August, the water had subsided, and the gullies afforded safe landing.

3.20 Overview of Camp Morton. Obvious anthropogenic features included two houses and a stockpile of coal. Note the traces of coal up the mountain. (Photo: U. Gustafsson, LASHIPA 3, 2006.)

Above the cliffs lay a gently sloping plateau, which was superseded by steeply rising hills. The hillsides displayed gently dipping sedimentary strata. From a distance, the most obvious anthropogenic features at Camp Morton were two upstanding buildings and a substantial stockpile of coal. A closer look also revealed dark traces on the mountain side, where coal had been worked. A synthesis of the archaeological survey is shown in Fig. 3.21.

[19] An Arctic entrance is an enclosed extension through which a person first enters a building and which traps the cold outside air before the person proceeds inside.
[20] Avango, D., Gustafsson, U., Hacquebord, L., and Hartnell, C. (2008a) *LASHIPA 3, Archaeological expedition on Spitsbergen, August 7-24, 2006*, Groningen: Arctic Centre. The LASHIPA 3 report is available online: http://www.rug.nl/research/arctisch-centrum/pdfs/lashipa3.pdf.

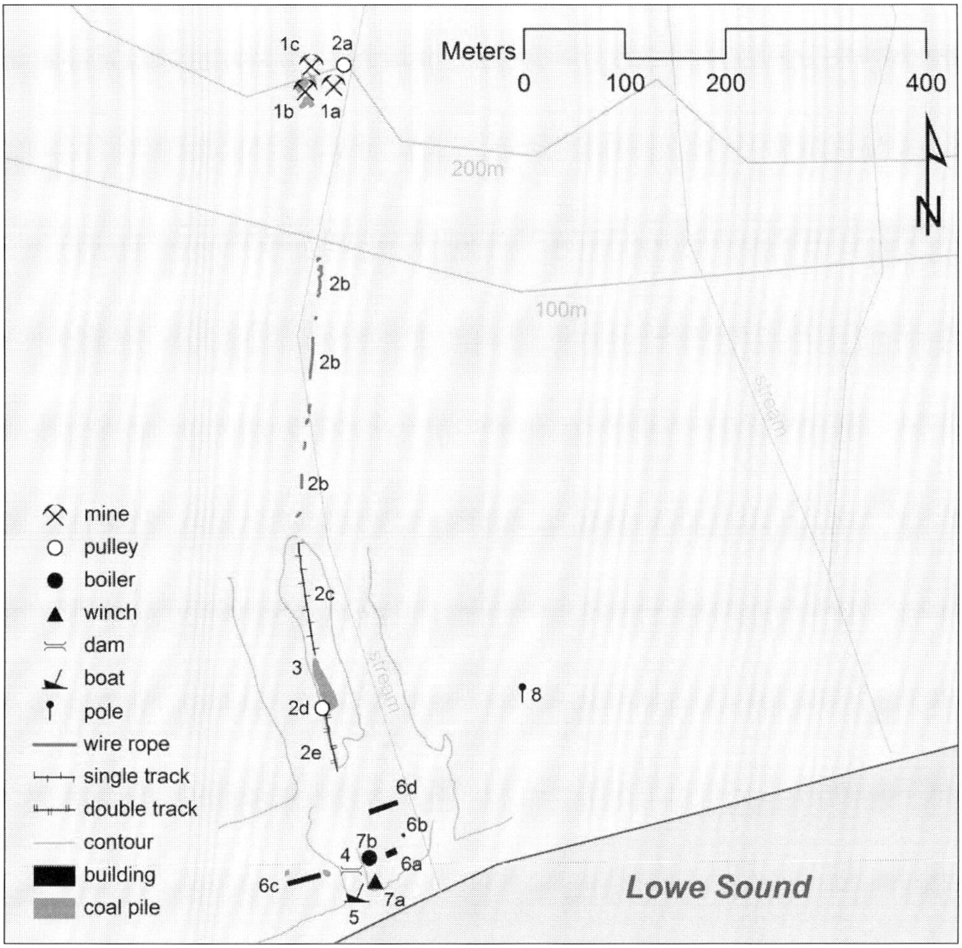

3.21 Site map of Camp Morton. (Data: LASHIPA 3, 2006; Map: F. Kruse.)

Three former mine openings had been recorded (1a-c).[21] They lay on a very steep scree slope. One opening (1a) was situated approximately 150 metres above sea level.[22] It was marked by four horizontal beams, which protruded from the collapsed ground (Fig. 3.22). Some vertical slats had survived, too, as had the bent ends of a narrow-gauge railway. All pointed into the gully. There was no flat surface to stand on. Yet, a thin coal scatter to the left of the portal would have been erased if substantial erosion had occurred. Operating space must have been restricted from the start.

[21] Avango *et al.* (2008a) p. 73-4.

[22] Since the former workings were backed by a rockface, the map creates the impression that they were closer to the 200-metre contour.

3.22 A former mine opening (1a) at Camp Morton. (Photo: U. Gustafsson, LASHIPA 3, 2006.)

A second opening (1b) was recorded a few tens of metres to the west. It had left a linear indentation in the ground (Fig. 3.23). A single rail jutted out of it. The adit was located on a small knoll, where there may also have been a small stockpile. Several rails were placed on top of the remaining coal scatter. Rather than being neatly piled pending utilisation, they gave the impression of having been pulled out of the mine randomly, never to be reused. Not visible in the photograph but hinted at on the map was another coal scatter above this mine, onto which a second stack of railway tracks had in fact been put more deliberately.

3.23 A second opening (1b) on a knoll. Note Camp Morton in the centre of the shoreline below. (Photo: U. Gustafsson, LASHIPA 3, 2006.)

A third adit (1c) lay at a higher altitude than the others. It was also the best preserved with a timbered but collapsed portal and two rails pointing into the air (Fig. 3.24). Although only one track was observed, the portal was wide enough for two parallel tracks. The pointing rails again gave the idea that the ground around them had maybe been worn away, but as previously, a coal scatter survived *in situ* a few metres diagonally below it.

3.24 A third adit (1c) above Camp Morton. Note the workings (1a) in the lower seam diagonally below. (Photo: U. Gustafsson, LASHIPA 3, 2006.)

The workings were associated with a large pulley (2a, Fig. 3.25). By 2006, it was positioned at the erosive edge of the gully.[23] A frame of heavy wooden beams and sturdy metal bars held a large, wooden wheel in place. The base had been sunk into the ground. Enormous roughly hewn sandstone blocks had been dragged in front of it to counteract the weight of the material to be conveyed.

3.25 Wooden pulley (2a) at the edge of the gully. (Photo: U. Gustafsson, LASHIPA 3, 2006.)

[23] Avango *et al.* (2008a) p. 74.

A strand of wire rope (2b) led from the horizontal wheel into the gully. A few metres below the pulley and at about the same height as the lower workings lay a wooden raft with a set of vertical wooden wheels at its front. It had largely been buried under talus, and it was impossible to tell how it had been anchored. The wheels showed signs of wear that suggested that they were rope guides. Although it was partially buried and led through a small ice cave in the gully, the wire rope could be traced over a distance of about 850 metres to the shore. It was associated with sporadic metal rope guides, rail tracks, wooden sleepers, and three types of mine tub. Two-thirds of the way down, a single railway track (2c) survived *in situ* and terminated at the top end of the stockpile of coal (3, Fig. 3.26).

3.26 Single railway track (2c) from the workings to the stockpile (3). (Photo: U. Gustafsson, LASHIPA 3, 2006.)

3.27 Metal pulley (2d) and two parallel tracks (2e) leading from the stockpile (3) towards the shore. (Photo: U. Gustafsson, LASHIPA 3, 2006.)

At the lower end of the stockpile stood another horizontal pulley (2d). This one had a larger, metallic wheel. Two parallel sets of narrow-gauge tracks continued shoreward (2e, Fig. 3.27). Via a cut embankment, the tracks led into the western gully, where they ended abruptly (2e, Fig. 3.28). What the arrangements at the terminus may have been could no longer be discerned. The low rockface below the rails may have been constructed, but it may equally well have been eroded. A little higher up in the gully were the remains of a crude stone dam (4). At the interface of the gully and the beach lay the remains of a wooden boat (5, Fig. 3.29).

3.28 Parallel rails (2e) terminate in the western gully. (Photo: B. Bekooy, LASHIPA 3, 2006.)

3.29 Remains of a wooden boat (5). (Photo: B. Bekooy, LASHIPA 3, 2006.)

The settlement above the cliff consisted of four buildings (6a-d, Fig. 3.30).[24] The buildings were broadly in line with the shore and parallel to each other. The larger of the two upstanding houses was Michelsenhuset (6a), one door facing the sea, another leading out the back. According to the aforementioned calculation, it may have housed 12 people. The restoration work undertaken by *sysselmannen* was clearly visible from the outside, but wooden debris and other refuse, some with historical significance, rendered the interior inhospitable and almost inaccessible. A sign above the seaward-facing door of the smaller upstanding building read 'Camp Morton' (6b, Fig. 3.31). The hut may have provided room for about five men. It was barricaded and locked, suggesting that it was serviceable. It had an outside toilet, which had apparently been added recently.[25]

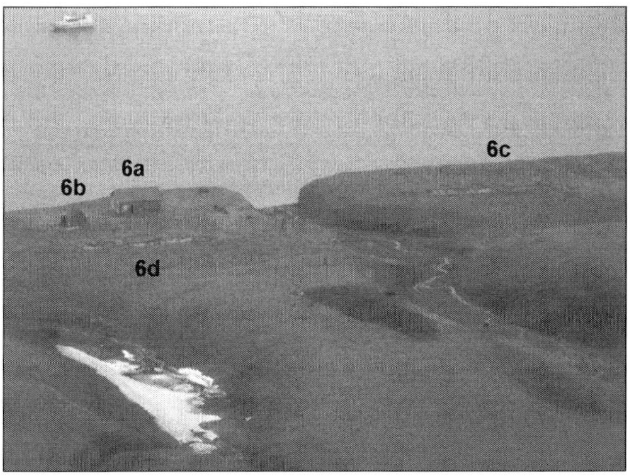

3.30 Camp Morton comprised four buildings: Michelsenhuset (6a), 'Camp Morton' (6b) and two barracks (6c and 6d). (Photo: U. Gustafsson, LASHIPA 3, 2006.)

In addition to the upstanding huts, there were two collapsed wooden barracks. One (6c) suggested 33 workers and was positioned roughly 50 metres north of Michelsenhuset (6a). The ground beneath it had been levelled, and six wooden pegs surrounded it, some with metal wire attached to them. The other (6d) hinted at 35 men and lay across the gully some 100 metres west of the house. It was associated with the remains of two stoves, one of them made by the Norwegian firm O. Mustad & Son, two small coal piles, and four wooden pegs surrounding it. Unfortunately, the LASHIPA 3 expedition did not indicate the barracks' doors and interior partitions on their plans. However, the same type was also found at other sites of the Northern Exploration Co., which offered more detail.

[24] Avango *et al.* (2008a) pp. 68-71.
[25] In mining, the difference between indoor and outdoor toilets could be indicative of the difference between the upper and lower class of mine worker (Porteous (1970) pp. 135-6).

3.31 Seaward-facing facade of 'Camp Morton' (6b). (Photo: B. Bekooy, LASHIPA 3, 2006.)

Indicated on the map above are a winch (7a) and a vertical boiler (7b), which had been components of a portable steam winch. The actual winch was manufactured by Southgate Pile Driver Co. of London (7a, Fig. 3.32). It was fairly wide and had apparently been made according to broad-gauge standards in use on the British mainland. The fact that a narrow-gauge track jutted out from underneath it implied that it had been adapted to fit this and that it had purposefully been moved to its last resting place.

3.32 Winch (7a) and vertical boiler (7b) were components of a probable steam winch. (Photo: B. Bekooy, LASHIPA 3, 2006.)

Some tens of metres away lay the remnants of a vertical boiler (7b, Fig. 3.32), which produced steam as the motive power. A probable third component was a carriage made by William Jones, U. T. S. of London, which had since ended up in the gully among various mine tubs.

Although the beginnings of Camp Morton lie firstly with Michelsen and the site was subsequently used by the Spitzbergen Mining & Exploration Syndicate, there was no evidence besides the symbolic function of the camp and workings that either had claimed the area and the coal seams. The metal sign that was discovered on the ground along the railway tracks read 'N.E.Co. Owned by the Northern Exploration Co. Ltd. London. Claimed in 1905' (Fig. 3.33). Beneath it were the beginnings of a German translation. Approximately 200 metres to the east of the stockpile, a metal pole stuck out of the ground (8, Fig. 3.34).

3.33 Broken claim sign of the Northern Exploration Co. anachronously dated to 1905 (Photo: U. Gustafsson, LASHIPA 3, 2006.)

3.34 Metal pole of a probable claim sign (8) on a hillock above Camp Morton. (Photo: D. Avango, LASHIPA 3, 2006.)

As seen on the map (Fig. 3.19), **Camp Bell** lay on the northern shore of the wide Bell Sound and well away from Axel Island. Cliffs like those at Camp Morton were absent here, and the wide gravel beach appeared to be easily accessible. Yet, the erosive power of the sea had changed the coastline over the last century, which is why the upstanding building had been moved onto higher ground (Fig. 3.35).

3.35 The reconstruction of Camp Bell. Note its former location at the shore on the far right. (Photo: D. Avango, LASHIPA 3, 2006.)

The former location was evident as a depression in the ground surrounded by low gravel banks (Fig. 3.36). According to the measurements of the LASHIPA 3 team[26], it may once have slept nine men. Artefacts included some structural elements such as a sheet of corrugated galvanised steel and parts of a stove made by Ulefos Værk as well as hand tools such as a shovel, a pickaxe head, a drill, and some chisels. These traces will disappear with high water or floods in the near future.

3.36 Former location of Camp Bell. (Photo: B. Bekooy, LASHIPA 3, 2006.)

[26] Avango *et al.* (2008a) p. 50.

The restored Camp Bell was an attractive building (Fig. 3.37) with the floor plan suggested by *sysselmannen*'s report.[27] There was indeed an Arctic entrance, the sea-facing door of which was locked, but which supposedly opened up into a foyer and a main room. To the left was a shed with a separate entrance. The front door, the shed, and the back wall were constructed of newer wood. Long, bleached whale bones were leaning against the porch. The steep mountains in the back of the photograph show sedimentary strata, yet the LASHIPA 3 team noted no obvious signs of prospecting or mining during their visit to the site.[28]

3.37 Reconstruction of Camp Bell. (Photo: D. Avango, LASHIPA 3, 2006.)

3.3.2 Site interpretations

As with Advent City above, the surveyed environmental and archaeological features are particularly well suited to reconstruct the local network (Fig. 3.38). A word of caution: in this instance, the archaeological record is the accumulated result of Michelsen's earliest activities combined with those of the Spitzbergen Mining & Exploration Syndicate and the Northern Exploration Co. More specifically, the physical remains at Camp Morton represent a snapshot of the site when it was abandoned by the latter.[29] Where possible, the site interpretation emphasises the relics of the Spitzbergen Mining & Exploration Syndicate in an attempt to again establish the reasons behind its interest in Bell Sound, its operative choices, and the probable factors that may have caused it to discontinue.

[27] Dahle (1998).
[28] Avango *et al.* (2008a).
[29] As with all archaeological sites, although development ceased several decades ago, site formation processes are, of course, ongoing.

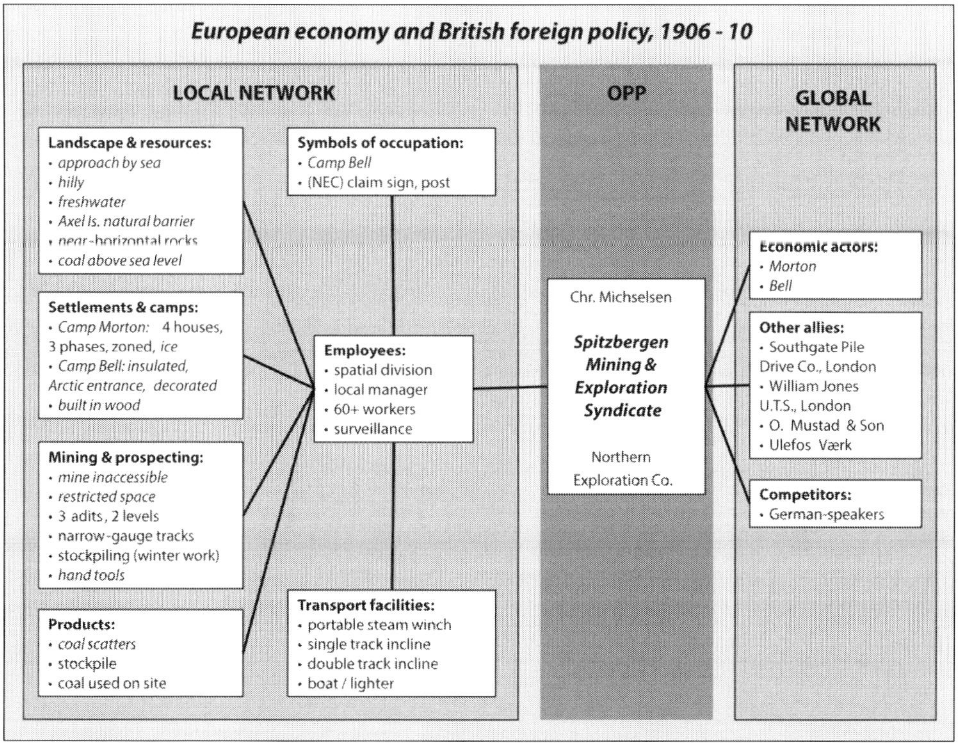

3.38 Environmental and archaeological factors derived from Camp Morton and Camp Bell. Likely aspects in the actor-network of the Spitzbergen Mining & Exploration Syndicate are highlighted in italics. (Chart: F. Kruse.)

Against a similar background as the Spitzbergen Coal & Trading Co. but a couple of years closer to the looming war, the Spitzbergen Mining & Exploration Syndicate conducted its business on the archipelago from 1906 presumably until the Northern Exploration Co. took over the property. The firm's name implied that the group was broadly interested in the extraction of economic minerals and the prospecting for more. Upon the discovery of new deposits, it could decide whether to exploit them itself or involve subsidiary companies to do so. Michelsen had most probably located coal in Bell Sound prior to the syndicate making the area the focus of their activities. The landscape here did not differ much from the aforementioned Advent Bay. It was best approached by sea, there was seasonal freshwater, and the hills comprised gently dipping rocks that contained coal. While Advent City had been exposed, Camp Morton sheltered behind Axel Island. However, access to Camp Bell, the place where Mansfield wintered in 1908/9, was compromised by strong winds arriving across the wide Bell Sound, causing a swell that made landing nearly impossible.

The Spitzbergen Mining & Exploration Syndicate, too, had seemingly decided that housing was paramount to successful mining and exploration. It therefore made use of Michelsenhuset and erected at least one other building at Camp Morton. Both were symmetrical to natural features, in this case parallel to the shore and perpendicular to the gullies, which appears to be a recurring trend on Spitsbergen. The main doors and windows faced the open water as opposed to the hills. Thus, they allowed for a scenic view and the observation of incoming ships rather than surveillance of the workforce. The barracks are thought to have been erected by the Northern Exploration Co., who may have had larger plans for the site in need of more workers and stores. Notably, they were not built next to each other. Instead, one stood a considerable distance to the west and the other to the north of the earlier features. This hints at a hierarchy or division in purpose. All buildings were made from imported wood; no concrete or brick had been used. That wood is no less durable has been proven by the two older structures, which are still upstanding. The barracks have collapsed, which raises the questions if they had only been light constructions and if they had been fit for wintering. The pegs that surrounded them had probably been anchors in a futile effort to keep them upright or to hold the roofs down. Camp Morton will have been disadvantaged by ice clearing slowing from Lowe Sound. This alone could have been a good reason to build Camp Bell, perhaps as an intermediate station for the transfer of workers and stores. If the reconstruction is to be trusted, Camp Bell was a sturdy hut, which had been adapted to the rigors of the climate with thicker walls and an Arctic entrance. The whale bones most likely had a decorative function.

Besides the presence of a mine and the fact that the hillside was never levelled to create more operating space, disappointingly little can be said about the development of the Spitzbergen Mining & Exploration Syndicate. It may have opened one, two, or all three adits at Camp Morton. Alternatively, the Northern Exploration Co. may have extended the original works and driven new headings. The adits have since collapsed, and there is no evidence as to whether room-and-pillar or longwall methods were used underground. Manual labour seems most likely. The Northern Exploration Co. loaded the coal into tubs and pushed these along the tracks. At the surface, the surviving coal scatters point to the existence of temporary stockpiles. These either suggest that others were in charge of transport downhill or that the work was done in stages. The latter finds support in the fact that there was no shelter at height and that surface work may only have taken place when the weather would allow it.

In light of likely site formation processes, the transport system can be attributed to the Northern Exploration Co. Hauling the coal downhill may have been aided by the portable steam winch. Despite the need to raise steam, it was flexible. When the need arose, it may have been used to pull materials ashore. At other times, it may have been stationed along the single track, on which tubs conveyed coal from the mine to the stockpile. The existence of a stockpile gives the impression that this task may have taken place in the winter when it was possible to work in the mine but when the frozen bay would not allow for shipping. The shipping season coincided with constant day light, and the double tracks from the stockpile to the shore are evident for the rush to get as much coal as possible to the market in the short window available. Each company would have been inconvenienced by apparent shallow water at the shore. Once again, no loading facilities survive, but the remnants of a wooden boat, albeit of uncertain origin, may suggest that coal was transferred to a larger ship out in the bay. The stockpile is also evident that the Northern Exploration Co. at least produced a substantial output of coal. Which market it was eventually intended for could not be discerned. However, some was used by the employees on site as the coal scatters next to the western barrack show.

In terms of products, written sources exist that hint at the purpose of Camp Bell, but these have been withheld for Chapter 6 in order to just analyse the physical remains here. No workings were recorded in the immediate vicinity, and there were no stockpiles of any output. There were, however, several hand tools at the original location of the hut, which suggest that the camp may have played a role in prospecting. Although the hut was fit for winter use, prospecting appears to be an unlikely activity in the dark season. It may have housed hunters and trappers instead. Moreover, the camp acted as an important claim marker. While Advent City was a point in the middle of a claim of unknown size, the distance between Camp Morton and Camp Bell gives the impression that a substantial part of the northern shore of Bell Sound had been occupied by the Spitzbergen Mining & Exploration Syndicate. Like the Northern Exploration Co., it may have secured its claim additionally with the help of claim signs, but none were found in the field.

With regards to employees, some comments can be made about Camp Morton, but only when taking all four buildings into consideration. Thus, the comments mostly apply to the continuation of the site by the Northern Exploration Co. A hierarchy or functional division again arises from the type and positioning of the structures. The smaller, British-built hut that could have slept five men thereby retains a managerial role. Michelsenhuset and the two barracks were invariably of lesser importance. The Spitzbergen Mining & Exploration Syndicate may have housed 12 workers in the former, whereas the Northern Exploration Co. may have accommodated more than 60 men in the barracks. Assuming that large artefacts had remained in their original place, the western barrack was associated with a large cooking range, while none was recorded at the northern one. This creates the notion that workers only lived to the west of the gully, while the barrack to the north may have contained stores or workshops. This is supported by the theory that both

Michelsenhuset and Camp Morton acted as physical barriers not so much between the two barracks but between any intruder from the shore and the probable stores. Regarding Camp Bell, the decorative whale bones are noteworthy, since decoration of any kind bears witness to the occupants' pride of place and achievement. It is likely that such ornaments added to social cohesion.

As far as the archaeological signature of the local network is concerned, Michelsen, the syndicate, and the Northern Exploration Co. must jointly be placed at the obligatory point of passage because each added to the physical remains. Evidence for the syndicate's global network firstly comes from the names of the camps themselves, which commonly pay tribute to financial backers. Other allies are again represented in the form of British firms for the provision of machinery and Norwegian firms for the supply of household items. Further enquiry into the Southgate Pile Driver Co. of London showed that this elusive company had once been incorporated with the Southgate Engineering Co.[30] In 1914, it specialised in piling hammers and plant. That year, its proprietor R. H. Annison was a master lighterman and barge owner, who focused on iron, steel, and oil.[31] In fact, he was the first to convey petroleum from vessels to the first petroleum depot. Annison's expertise in barge transportation or his connection with the emerging oil industry may have had a bearing on Spitsbergen. Regarding William Jones, U.T.S. of London, an online search revealed a William Jones & Sons, who were engineers, ironfounders, and boilermakers in Warrington, Lancashire, in the late nineteenth century.[32] They already produced portable steam winches and hoists then, and it seems feasible that they had moved to London in later years. The Norwegian companies identified were O. Mustad & Son and Ulefos Værk, which have not been looked into any further. A last note on the global network relates to the claim sign of the Northern Exploration Co., which had been translated into German. This could either be seen as a general warning to all German-speaking competitors or as a more potent signal to a German rival.

Archaeologically, it is difficult to isolate particular causes that prompted the Spitzbergen Mining & Exploration Syndicate to discontinue on the archipelago. The neglect of the sites in the first instance implies that mining and exploration had not been profitable. Probable causes include that the principal mining site, Camp Morton, was barely reachable in unfavourable ice conditions. This, however, did not deter the Northern Exploration Co. or the Swedish company that for a time

[30] Grace's Guide (2007) *Southgate Pile Driver Co.* Available at: http://www.gracesguide.co.uk/Southgate_Pile_Driver_Co (Accessed: 20 July 2012).

[31] Grace's Guide (2007) R. H. Annison. Available at: http://www.gracesguide.co.uk/R._H._Annison (Accessed: 20 July 2012).

[32] Grace's Guide (2007) William Jones and Sons. Available at: http://www.gracesguide.co.uk/William_Jones_and_Sons (Accessed: 20 July 2012).

operated a mine in Braganza Bay. The few remains that could be attributed to the syndicate could mean that it adhered to a conservative financial policy. It could also mean that it had not invested enough in the development, which would have impeded it, too. It is possible that the syndicate had spent most of its funds on prospecting in the vicinity of Camp Bell instead. This assumption finds support in subsequent events: the syndicate's promoter was instrumental in the formation of the Northern Exploration Co. and opened a gold mine at Camp Miller nearby (Fig. 3.19). It is likely that the syndicate's members expected an Arctic gold rush, but their means ran out before their claims could deliver. There is no evidence that geopolitical reasons such as disputes with Michelsen or the Northern Exploration Co. were to blame, nor competition from other sources.

3.4 Conclusion

Despite the large coal industry in Britain, the archaeological image of British mining on Spitsbergen is created by only two sites: Advent City and Camp Morton. They populate the local networks of the Spitzbergen Coal & Trading Co. and the Spitzbergen Mining & Exploration Syndicate, respectively. The physical remains prove that the companies were not fraudulent; they were formed to put their plans to capitalise on Spitsbergen into practice. Both were met by a distant landscape, in which exposure and ice were problematic. While this may have drawn on their resources and slowed their progress, it may not have been a direct influence on their demise. The sites in Advent Bay and in Bell Sound will have been central to their spheres of influence, but the precise extent of individual claims could not be discerned in the field. The satellite station of Camp Bell may have been a claim hut, which substantially enlarged the property of the Spitzbergen Mining & Exploration Syndicate and from which further prospecting was conducted. The claim sign of another company hinted at additional means by which a claim in the Arctic no man's land could be defended. Advent City was seemingly intended for a large workforce, but the first mining settlement on the islands nonetheless showed signs of overspending and was probably constructed with social as well as geopolitical determinants in mind. The beginnings of Camp Morton were more economical and lay in only two buildings. Although the mines were inaccessible, indirect evidence suggested that these had been operated according to prevailing British models. The mine at Advent City had employed fairly advanced technology, but having been left behind, it raised the question if it had been appropriate. Both firms relied on British manufacture for their machinery, while they drew on Norwegian production for perishable items. Both had put a greater effort into building inclined planes than long-lasting loading facilities. The method of using lighters to supply larger ships in deeper water must have been time-consuming in a short shipping season. Coal scatters and stockpiles suggested that coal had been extracted, but the output during the few years of operation could not be estimated. There were no physical remains of alternative products. The companies must have been aware of the high risks involved. So the small scale and short existence may be evident of a very cautious policy, by which both pulled out at the first signs of trouble. More likely, however, the companies had been overly confident and had underestimated operational difficulties while overestimating the returns. Physical evidence for the head offices, let alone actors in the global network are rare. Support has only been found in the form of toponyms and embossed names on machinery. The Arctic Coal Co. and anonymous German-speakers were possible competitors. Archaeologically, specific reasons as to why the companies should have abandoned their claims could not be discerned. At Advent City, overspending invariably carried the signature of incompetent local management, but the order may have come from higher up. Coal may remain in the stockpile at Camp Morton today because the quality was poor, because it had not been possible to ship it, or because a shift in the market had made it impossible to sell. Whatever the actual cause, the conclusion is that the developments were by and large not conducive to making a financial gain.

4 The Archaeology of Arctic Exploration

4.1 Introduction

This chapter focuses on the material remains of two British exploration companies: the Northern Exploration Co. and the Scottish Spitsbergen Syndicate. Prior to the fieldwork, it was already known that their activities on Spitsbergen had been too widespread to record every site in the time available. The LASHIPA expeditions nonetheless succeeded in collecting more data than can be expressed here. The site narratives are a selection of the most instructive examples. As before, they provide the environmental and archaeological components with which to recreate the companies' actor-networks and consider the research questions in the site interpretations. They give rise to the particular archaeological image of British mineral exploration on Spitsbergen.

4.2 The Northern Exploration Co.

The Northern Exploration Co. was formed in 1910 and engaged in prospecting on Spitsbergen. It covered large tracts of land in search of marketable natural resources and unwittingly created a rich archaeological record. The LASHIPA project visited 34 sites between 2006 and 2010. These can broadly be characterised by resource (Fig. 4.1). The firm was inactive for most of the First World War but renewed its expeditions shortly before the conflict ended in 1918. Its last season occurred in 1927. It sold its properties to Norway in 1932.

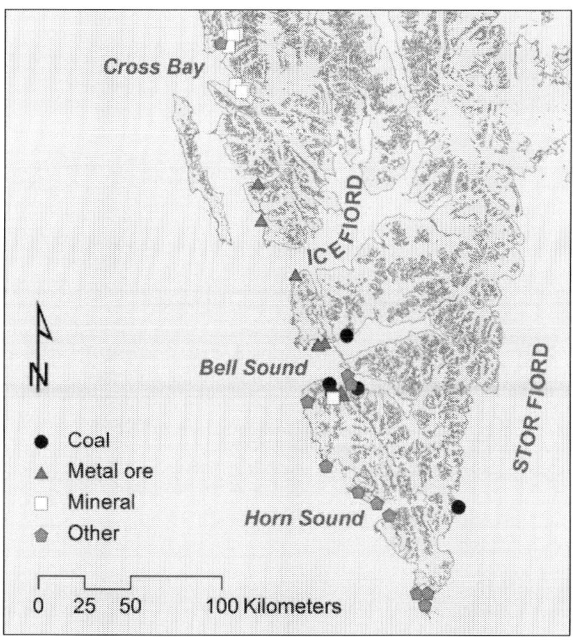

4.1 Location map of 34 sites of the Northern Exploration Co. The sites can be characterised by resource. Obscure or symbolic functions are grouped together as 'other'. (Map: F. Kruse.)

4.2.1 Site Narratives

The map above shows that the archaeological sites lie mainly on the west coast of the island of Spitsbergen. This reflects the present situation, whereby the settlements and the majority of human activities are confined to the area kept relatively ice-free by the northbound Gulf Stream passing to the west of the archipelago. Ice conditions to the north and to the east generally range from unpredictable to impassable. Furthermore, the sites cluster around the accessible waterways, particularly Cross Bay and Bell Sound and to a lesser extent Icefiord and Horn Sound. The coastal landscape comprises raised beaches and marine terraces, while the hinterland is typically mountainous. For the purpose of this chapter, the sites are divided by resource or function. It transpires that mineral prospecting sites were mainly located around Cross Bay; a belt of metal ores was investigated between St. Johns Bay and Recherche Bay; coal mining was restricted to Bell Sound, with one exception on the east coast of Spitsbergen; and sites with other functions lay between Bell Sound and the South Cape.

4.2.1.1 Coal

As shown in Chapter 3, the Spitzbergen Mining & Exploration Syndicate had opened coal at **Camp Morton** but failed to reach the producing stage. Its claim was taken over by the Northern Exploration Co. Although the site was in use before as well as after the First World War, the material remains give the impression that the company did not work the mine to its full potential. Typical of an exploration company, it had probably limited the development to proving the coal seams. The example shows that the boundary between early mining and advanced prospecting is, in fact, blurred. It serves as a reminder to judge exploration sites not only in terms of output but also in terms of attractiveness to subsidiary companies and buyers. Was the site easily accessible? Was the coal seam well opened up? Could the workability of the coalfield be demonstrated? And did some basic infrastructure exist to accommodate visiting representatives and to convince them that the development was ready to commence at any moment? Provided that the representatives were not taken to Camp Morton when ice still blocked Lowe Sound, it may well have been a showpiece of British exploration.

Other coal prospects included Camp Smith, Camp Violet, Calypso Beach, and Davis Harbour (Fig. 4.2).[1] Rossnes reports that **Camp Smith** was built by the

[1] Camp Smith and Calypso Beach appear in the LASHIPA 3 report, the others in Avango, D., Haas, H. R. de, and Kruse, F. (2010) *LASHIPA 9, Archaeological expedition to Spitsbergen, 31 July – 15 August 2010,* Groningen: Arctic Centre. The LASHIPA 9 report is available online: http://www.rug.nl/

Norwegian consul Johannes Giæver in 1904. It was later sold to the Northern Exploration Co. and renamed after one of its directors.[2] When the LASHIPA 3 expedition surveyed the site in 2006, the house with a commanding view over Recherche Bay did not seem likely to survive without intervention (Fig. 4.3).

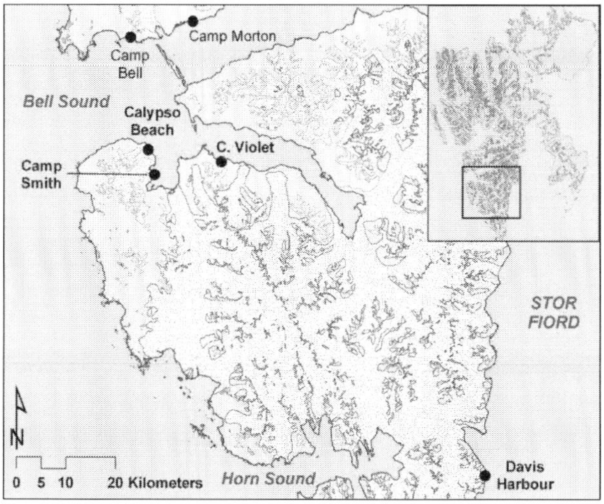

4.2 Location map of the coal properties of the Northern Exploration Co. (Map: F. Kruse.)

4.3 The summer house in Recherche Bay was built in 1904 and later named Camp Smith. (Photo: D. Avango, LASHIPA 3, 2006.)

The expedition recorded over 40 mine tubs at the shore (Fig. 4.4).[3] Their metal wheels were missing. Perhaps they had been salvaged. A construction of sturdy

---

research/arctisch-centrum/pdfs/lashipa9.pdf.

[2] Rossnes, G. (1993) *Norsk overvintringsfangst på Svalbard 1895-1940*, Oslo: Norwegian Polar Institute, p. 46.

[3] Avango *et al.* (2010) pp. 88-9.

wooden beams and boards appears to have been an improvised pier, on which the tubs were landed. They could have been awaiting transfer to another site, or they were purposefully placed in clear view of anyone on the water as a sign of the site being occupied.

4.4 Improvised pier and mine tubs at Camp Smith. (Photo: D. Avango, LASHIPA 3, 2006.)

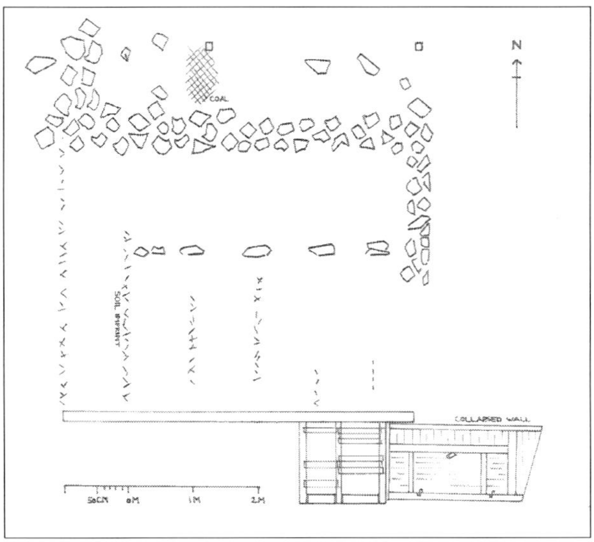

4.5 Building foundation at Camp Violet. Wooden beams had rested partly on the ground and partly on stones that had made up the difference in high. An Arctic entrance faced the water. A small outhouse stood at the back.(Drawing: F. Kruse, LASHIPA 9, 2010.)

According to Rossnes, **Camp Violet** was named by Mansfield after a London acquaintance.[4] A house still stood on the southern shore of Van Keulen Bay in 1936, but when LASHIPA 9 visited the site in 2010, they only recorded two ruins by

[4] Rossnes (1993) p. 52.

a stream.[5] One had been a hunting station made from drift wood. The other had belonged to a well-constructed house (Fig. 4.5). Besides the single-board wall of a small outhouse, no wooden debris remained, suggesting that the house had been removed in its entirety. Two tubs were found half-buried in the colluvium near the beach. A coal scatter along the southern house wall may have been brought from anywhere and at any time. There were no other indications of mining in the vicinity, and the team did not investigate the hinterland. While the location of Camp Violet may have been chosen according to the resource, the water front, the freshwater, and shelter by Cape Ahlstrand, it was not particularly imposing and may only have had a limited symbolic role.

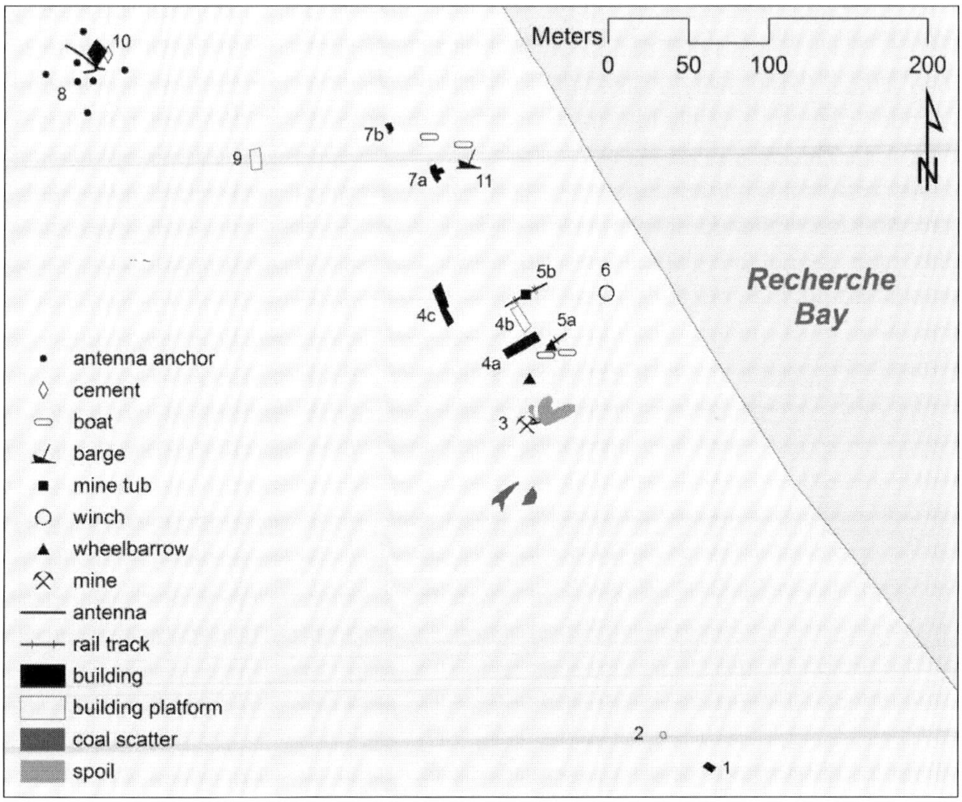

4.6 Site map of Calypso Beach. (Data: LASHIPA 3, 2006; Map: F. Kruse.)

[5] Avango *et al.* (2010).

3, 2010, the view of Mt Hedgehog was obscured by low clouds, which made locating the site difficult.[9] In the end, the team never found the camp or a clear indication of mining. The wide beach was littered with debris that could have had archaeological significance. Wood that could have been building material was found high up on the slopes. Yet, on a quiet day it was easy to forget the natural processes at work. Owing to the retreat of the glaciers over the past decades, the landscape is young, dynamic, and destructive. The erosion of steep mountains and moraines is enhanced by the yearly snow melt, which causes a series of outwash fans (Fig. 4.12).

4.12 Young, dynamic, and destructive: the coastal landscape on the east coast. (Photo: C. Botman, LASHIPA 9, 2010.)

4.13 Approximate location of Davis Harbour, now silted up. (Photo: P. Leminen, LASHIPA 9, 2010.)

[9] Avango *et al.* (2010).

If the archaeological remains were not covered from above, they could have been destroyed by great gales that most likely also blew the wood onto the slopes or by marine floods (Fig. 4.13). The processes that erased the archaeology will already have been at work in the early 1920s. It is imaginable that they contributed to the site's abandonment and have prevented any mining attempt since.

4.2.1.2 Metal ores

The Northern Exploration Co. also prospected widely for metal ores. (Fig. 4.14).

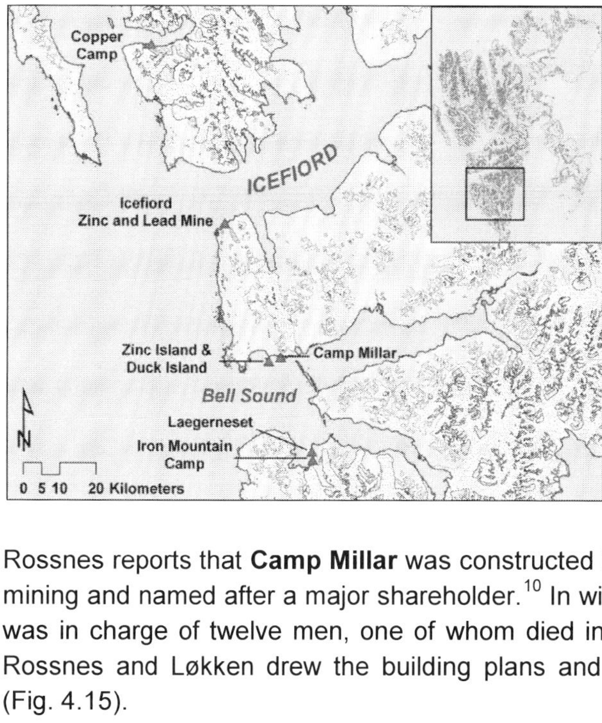

4.14 Location map of the metal ore properties of the Northern Exploration Co. (Map: F. Kruse.)

Rossnes reports that **Camp Millar** was constructed in 1910 for the purpose of gold mining and named after a major shareholder.[10] In winter 1910/11, Arthur Mangham was in charge of twelve men, one of whom died in the following spring. In 1988, Rossnes and Løkken drew the building plans and their relation to the workings (Fig. 4.15).

In August 2006, the LASHIPA 3 expedition recorded the camp and the mine.[11] The smaller hut (Fig. 4.16) had a sea-facing Arctic entrance, which was locked. There were windows on three sides but none the west side. The hut had undergone recent refurbishment. A Norwegian sign above the door read 'Vårsolhytta', subtitled 'Camp Millar'. A whale vertebra was another decorative feature. The porch and the hut were steeply gabled, presumably to inhibit the build-up of heavy snow, and small windows in both gables indicated the presence of a

[10] Rossnes (1993) p. 59-61.
[11] Avango *et al.* (2008a) pp. 54-61.

loft. At the rear, a long stove pipe jutted out of the roof. The dashed rectangle in the drawing below was found to be the wooden floor of a former extension or shed. The team also recorded a toilet at the back. Although there was no shortage of building space, the hut was nestled into a rock outcrop. Cobbles supported the wooden beam foundation. The outcrop was probably better drained and less susceptible to frost heave, seasonal thaw, and bogginess. There were no signs of prospecting. This building was thought to be staff accommodation. Since huts just like it were also found on Marble Island and Storholmen, the type was concluded to be pre-fabricated.

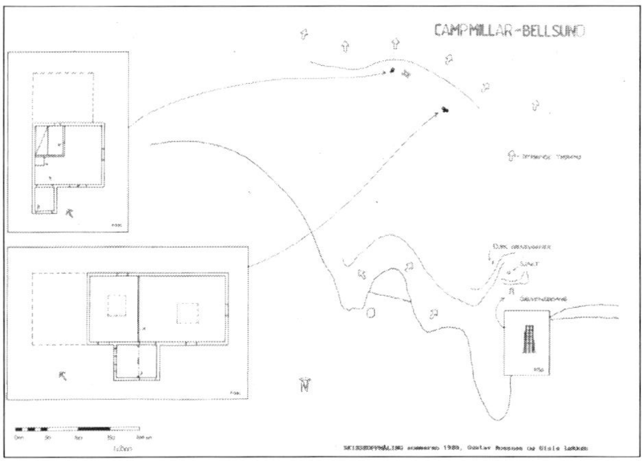

4.15 Plan drawing of Camp Millar in Bell Sound. (Source: Rossnes, G. and Løkken, G. (1988) Camp Millar – Bellsund Skisseoppmåling 1:2,500, Kulturhistoriske Plantegninger, Sysselmannen, Longyearbyen.)

4.16 Pre-fabricated staff house at Camp Millar. (Photo: U. Gustafsson, LASHIPA 3, 2006.)

The larger hut (Fig. 4.17) was positioned 100 metres away and closer to the mine. Its seaward Arctic entrance was also locked. The porch had two large windows as did the front. The east wall and the rear had one opening each. There were no windows in the west wall. The drawing indicated a subdivision in the porch, which may have doubled as a toilet or washroom. It further showed two rooms, which were divided by a wall half as thick as the outer walls, suggesting that the latter were insulated. The site visit proved that the huts were of similar height with comparable gables and lofts. This hut had also been renovated. The two stove pipes in the roof were recent. There was evidence for the existence of a sizable lean-to shed. This was also a pre-fabricated building and in all likelihood housed the workforce at Camp Millar. The type was encountered at Marble Island, too.

4.17 Pre-fabricated workers' barrack at Camp Millar. (Photo: D. Avango, LASHIPA 3, 2006.)

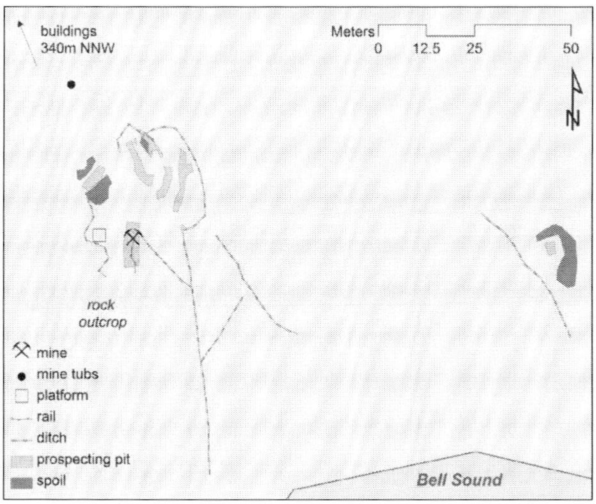

4.18 Site map of the workings at Camp Millar. (Data: LASHIPA 3, 2006; Map: F. Kruse.)

The workings were situated near outcrops of sedimentary rock on the shore of Bell Sound. Selected survey results are plotted in Fig. 4.18. The map reveals two areas where prospecting had taken place and left behind a series of pits, spoil heaps, and ditches. The eastern workings presumably gave negative results and were soon discontinued. Additional features at the western workings included the foundation of a machine next to a stone-lined and timber-supported adit with two sets of narrow-gauge tracks (Fig. 4.19). The mine tubs nearby were most likely used to shift spoil. Since gold is not a bulk commodity, the construction of extensive tracks or a permanent dock was not urgent.

4.19 Collapsed mine entrance at Camp Miller. Note the mine tubs in the top left corner. (Photo: U. Gustafsson, LASHIPA 3, 2006.)

Iron ore was the obvious resource to search for on the rust-coloured flanks of Martin Range on the east side of Recherche Bay. The earliest remains at **Lægerneset** date to the seventeenth century, and the sheltered site with a supply of freshwater has seen frequent re-use. In 1967, Christian Keller recorded it for the Norwegian Polar Institute.[12] He mapped a set of stairs and a building at the cliff edge, which were absent from the detailed total station map made by the Arctic Centre of Groningen University in 1998.[13] They had been eroded. The Dutch map attributed four tent rings and the outlines of two rectangular buildings to the Northern Exploration Co.

In August 2006, the LASHIPA 3 expedition visited the site to find evidence for mineral exploration.[14] The team photographed the dry-stone walls huddled up

[12] Keller, C. (1967) *Ekspedisjonsdagbok for someren 1967*, Norwegian Polar Institute, Svalbard.
[13] Hacquebord, L., Steenhuisen, F., and Waterbolk, H. (2003) 'English and Dutch whaling trade and whaling stations in Spitsbergen (Svalbard) before 1660', *International Journal of Maritime History*, 15 (2), pp. 117-34.
[14] Avango *et al.* (2008a) p. 109.

against the rocks of the headland. On the photos, the tent rings were clearly recognisable. Each was roughly four metre in diameter and comprised traits like round gravel patches lacking vegetation, rings of large cobbles, shallow drainage ditches, and wooden and metal tent pegs (Fig. 4.20). Evidence for one of the buildings survived as a large rectangular platform with shallow ditches and stone settings (Fig. 4.21). A barrack with an extension like the one encountered at Calypso Beach had stood here. The lack of wooden debris suggested that it had been removed whole. Artefacts that pointed towards prospecting at Lægerneset included two wooden wheelbarrows and a number of three-metre drilling rods.

4.20 A tent ring nestled between the ruins of two houses huddled against the rocks at Lægerneset. (Photo: D. Avango, LASHIPA 3, 2006.)

4.21 These earthworks and stone settings belonged to the foundation of a pre-fabricated barrack at Lægerneset. (Photo: D. Avango, LASHIPA 3, 2006.)

From Lægerneset, the hub of prospecting on Martin Range moved to **Iron Mountain Camp**. This location, too, was visited by Keller in 1967. His site plan shows two buildings located next to a small stream (Fig. 4.22). The houses are the same size, but House 2 also has a stone-lined 'patio'. The plan shows a footpath, which leads from the shore straight through House 2 and has an unclear relationship with House 1 before continuing further inland. When compared to other barracks and to Keller's plan drawing of House 1 (Fig. 4.23), House 2 had no doors where it was met by the path. The assumption is that House 1 and the path were concurrent and that House 2 was added later. It was probably the barrack that initially stood at Lægerneset and was transferred to this location when efforts here intensified. Keller does not state where the path terminated.

4.22 Detail of Keller's plan of Iron Mountain Camp (hus = house). (Source: Keller, C. (1967) Ekspedisjonsdagbok for someren 1967, Norwegian Polar Institute, Svalbard, p. 5.)

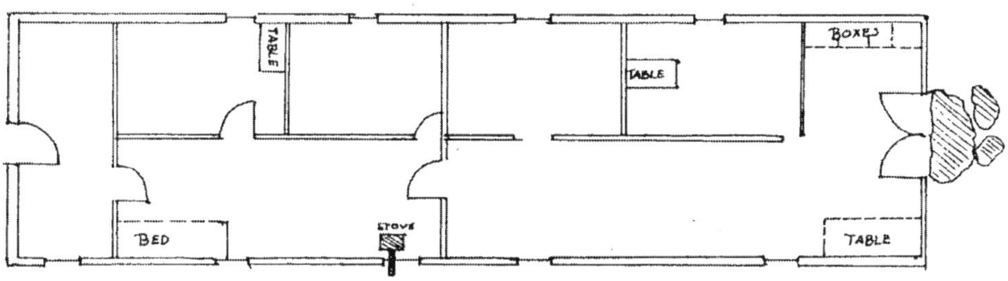

4.23 Keller's plan drawing of House 1 at Iron Mountain Camp. (Source: Keller, C. (1967) Ekspedisjons-dagbok for someren 1967, Norwegian Polar Institute, Svalbard, p. 4.)

The observations from Calypso Beach agreed with Keller's plan drawing (Fig. 4.23). The pre-fabricated barracks of the Northern Exploration Co. had a front door in the gable wall, which led into a small foyer to keep the weather out and any warmth in. One side had four windows, which lit up a room each. The other side had five windows: one in the foyer, two in a possible kitchen or living room, which contained a stove, and two in a large L-shaped workshop. This workshop could be entered through double doors at the rear, where no porch hindered the movement of equipment or machinery. Keller had drawn the outside walls to be thicker than the interior divisions, which probably indicates some insulation.

According to Rossnes, there were four buildings at this location, and he provides a photograph from 1936.[15] Keller's two houses were then upstanding. The extension that initially stood on the stone 'patio' appeared to have been destroyed by wave action, its roof slipping down the gravel shore. Not recorded by Keller was a small hut positioned at right angles to the southern end of House 1. The LASHIPA 3 expedition examined the settlement in August 2006.[16] It stood on a hummocky moraine directly beneath the rust-coloured band at height (Fig. 4.24). This strongly signalled British ownership of the resource. The team verified the presence of the former buildings. House 2 was a pre-fabricated barrack with an extension. The barrack had collapsed, and the extension had vanished. House 1 was a barrack without an extension and was precariously leaning. A foundation and some wooden debris clearly marked Keller's unrecorded hut.

4.24 Iron Mountain Camp on a moraine beneath an outcrop of iron ore. (Photo: D. Avango, LASHIPA 3, 2006.)

The interior of House 1 also matched Keller's plan. Although some of the walls were missing, the division into four rooms and either communal or work spaces was apparent. The large barrack only appeared to have had one stove, which may have been used both for warmth and cooking. Most beams were painted with a large red cross (Fig. 4.25). Such marks usually helped to distinguish and assemble pre-fabricated parts. In this case, the cross is also reminiscent of St. George, the patron saint of England, and raises the question if its maker preferred being

[15] Rossnes (1993) p. 52-3.
[16] Avango *et al.* (2008a) pp. 82-4.

English to being British. The answer could be as simple as the Union Jack being too hard to paint.

4.25 Red St. George's *Crosses on beams at Iron Mountain Camp. (Photo: D. Avango, LASHIPA 3, 2006.)*

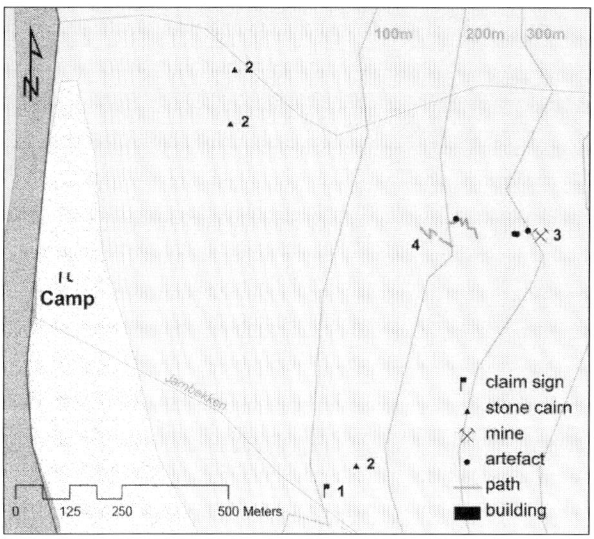

4.26 Map of Iron Mountain Camp and the relationship with its hinterland. (Data: LASHIPA 9, 2010; Map: F. Kruse.)

In August 2010, the LASHIPA 9 expedition re-visited Iron Mountain Camp to investigate its relationship with the hinterland (Fig. 4.26).[17] On a small shoulder on the 100-metre contour, the team found a claim sign of the Northern Exploration Co. (1), damaged but still *in situ*. Stone cairns (2) that once supported additional claim markers were found in other prominent locations. The metal panels had

[17] Avango *et al.* (2010).

disappeared, but colleagues mapping a Dutch site in the vicinity recorded three of them among the remains. The wooden posts had metal tips for beating into the frozen ground. These claim signs formed an unmissable chain between landing sites and the resource.

4.27 Mine opening (3) above Iron Mountain Camp. (Photo: C. Botman, LASHIPA 9, 2010.)

4.28 Small furnace, mining tools, and stockpile of iron ore below the mine on Iron Mountain. (Photo: C. Botman, LASHIPA 9, 2010.)

Almost 1.2 kilometres to the east of the camp at a height of 300 metres, the team discovered a small mine (3, Fig. 4.27). The portal was about a metre high, and banded iron ore was clearly visible. The depth of the mine could not be discerned due to the presence of ice. The miners had placed heavy wooden beams in front of it to create a better foothold. Below the mine, the slope was less steep, and the team recorded a small furnace (Fig. 4.28). It had either been used to fix broken or blunt tools or to smelt small samples of the ore to obtain an assay. Next to the furnace lay some hand plug drills and the remains of a stockpile. The map indicates a very tightly zigzagging footpath (4), via which the miners reached the mine. The ascend was long and difficult, but it appears to have been made on every shift. There was barely any space to build a hut at height.

Malachite, a conspicuous green copper ore, was the focus of the company's activities at **Copper Camp.** The location of the camp and the abandoned works are marked on current topographic maps of St. John's Bay and guided the LASHIPA 5 survey in August 2008.[18] A small wooden hut was plainly visible on the shore (Fig. 4.29). It was made of single boards covered with roofing felt. The hut was barely big enough to sleep a single person. It was more likely used as a communal space while the workers slept in tents. Yet, no tent rings were recorded. The hut displayed a disproportionally large claim sign above the sea-facing door. The metal plaque read, 'The south side of St. Johns Bay from Cape Müller to Osborne Glacier claimed by N E Co London'. Someone had taken the extra care to lay paving stones in front of the door.

4.29 Claim hut at Copper Camp. (Photo: D. Avango, LASHIPA 5, 2008.)

Copper Camp lay close to a glacial stream, which was a source of freshwater and had probably aided the prospectors. In fact, the team found samples of malachite in the outwash. The former workings were encountered at the foot of a moraine approximately half a kilometre from the hut. Five trenches and small pits were recorded. The shape of some pits and the spread of the waste rock around them were thought to hint at the use of explosives (Fig. 4.30). Next to one of the

[18] Avango, D., Aalders, Y., Gustafsson, U., Haas, H. R. de, Hacquebord, L., Hartnell, C., Kruse, F., and DePasqual, S. (2009) *LASHIPA 5, Archaeological expedition on Spitsbergen, 27 July – 17 August 2008*, Groningen: Arctic Centre. The LASHIPA 5 report is available online: http://www.rug.nl/research/arctisch-centrum/pdfs/lashipa5.pdf.

trenches, the team observed the remnants of a malachite stockpile (Fig. 4.31). Its diameter implied its former height, but it probably only represented a fraction of the total yield from the site. Besides a broken pickaxe handle, there were no other features of interest at this location.

4.30 The prospecting pit at Copper Camp was possibly the result of using explosives. (Photo: D. Avango, LASHIPA 5, 2008.)

4.31 A prospecting trench and the base of a stockpile at Copper Camp. (Photo: D. Avango, LASHIPA 5, 2008.)

Hoel claims to have discovered a carbonate-hosted lead-zinc deposit on **Zinc Island** on the north side of Bell Sound in 1913.[19] However, his short description omits the prospecting done by the Northern Exploration Co. in the 1920s. In August 2006, the LASHIPA 3 expedition targeted Zinc Island to assess the extent of this work.[20] The team found it to be a group of small, inhospitable rocks, difficult to

[19] Hoel (1942) p. 386.
[20] Avango *et al.* (2008a) pp. 52-3.

access and challenging to map (Fig. 4.32). Nonetheless, the evidence for trial trenching was extensive. There were also the beginnings of a shaft (Fig. 4.33). This was partially filled with rubble, so the original depth could not be determined. It was unclear how the work had been carried out in the limited space and how much ore, if any, had been obtained.

4.32 Zinc Island as seen from neighbouring Duck Island. Note the four metal pegs in the foreground. (Photo: P. Leminen, LASHIPA 9, 2010.)

4.33 Partially backfilled mine shaft on Zinc Island. (Photo: D. Avango, LASHIPA 3, 2006.)

In August 2010, the LASHIPA 9 expedition took the chance to land on **Duck Island**, which was larger than Zinc Island and afforded more building and operating space (Fig. 4.34).[21] It must once have been possible to walk from the remains of a hut to the south-eastern end, but coastal erosion had since split the island and cut off the camp from another mine shaft (Fig. 4.35). This shaft was lined with wood

[21] Avango *et al.* (2010).

and largely filled with rubble. It is perceivable that the company had sunk it in order to intersect the ore body outcropping on Zinc Island. The metal pins noted in the photograph above, of which a counterpart was recorded on Zinc Island, suggest that the problems of transport to and from the zinc-bearing rocks in heavy seas had been overcome by means of a ropeway or bridge.

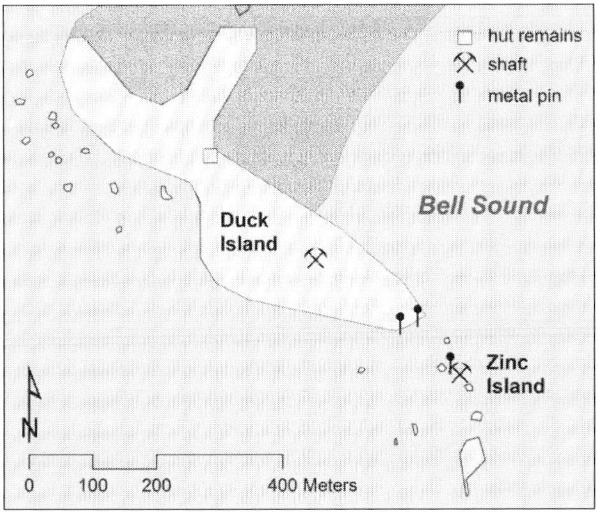

4.34 Site map of Zinc Island and Duck Island in Bell Sound. (Data: LASHIPA 3, 2006 & LASHIPA 9, 2010; Map: F. Kruse.)

4.35 Backfilled mine shaft on Duck Island. (Photo: P. Leminen, LASHIPA 9, 2010.)

Approximately 40 kilometres north of Zinc Island on the southern shore of Icefiord lay the **Icefiord Zinc and Lead Mine**. References to the location are sparse, and the LASHIPA 9 team based itself on primary archival research prior to the expedition in August 2010.[22] Landing was made difficult by the swell and steep marine terraces. On site, the team recorded the remains of two buildings (Fig. 4.36). The larger building (1) had been used as accommodation. Its L-shaped floor plan suggested a porch and a division into two rooms. The smaller building (2) was associated with pieces of machinery and equipment and may have been an engine house.

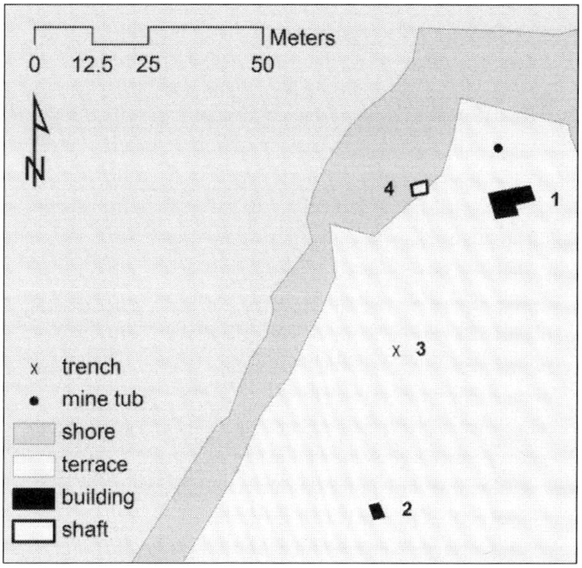

4.36 Site map of Icefiord Zinc and Lead Mine in Icefiord. (Data: LASHIPA 9, 2010; Map: F. Kruse.)

4.37 Concrete shaft (3) with collapsed shafthead frame at the Icefiord Zinc and Lead Mine. (Photo: P. Leminen, LASHIPA 9, 2010.)

[22] Avango *et al.* (2010).

Between the buildings lay an ellipsoidal trial trench (3). It was possibly earlier than the mine shaft (4), which had concrete walls and a ladder fixed to the inside (Fig. 4.37). The shaft had since been exposed by the sea and filled with water. The wooden shafthead frame had collapsed. Other artefacts included some lengths of railway track and the base of a mine tub, but it was difficult to know if these stray finds signalled the beginning or the end of work at the mine.

The historical sources also hinted at trial works further inland. Although the approximate location was known, the most direct route was obstructed by swampy ground. At a distance of about one and a half kilometres, the team found a prospecting pit surrounded by spoil (Fig. 4.38) with a broken hoe and a short ladder nearby. The scale of these works was very limited.

4.38 Minor trial works some distance inland from the Icefiord Zinc and Lead Mine. (Photo: P. Leminen, LASHIPA 9, 2010.)

4.2.1.3 Minerals

Minerals such as marble and asbestos formed a third group of natural resources which the Northern Exploration Co. sought to exploit. The company's marble-related activities were confined to the north-west of the archipelago (Fig. 4.39). Today, Cross Bay and Kings Bay are known for their striking geology, lush vegetation, and vibrant bird cliffs. En-route to one of the glaciers that reach the sea, cruise ships commonly approach **Camp Zoe** at the mouth of Tinayre Bay. The hut was erected in 1911 by the Norwegian Henry Rudi on behalf of Mansfield and named after Mansfield's daughter Zoe.[23] The hunter and trapper overwintered here and guarded the company's claims in the area, for which he received 50

[23] Johansen, B. F. (ed.) (2011) *Cruise handbook for Svalbard*, Tromsø: Norwegian Polar Institute.

Norwegian kroner per month and half his catch. The hut is still serviceable. Besides being a tourist attraction, it is used by the community of Ny Ålesund in Kings Bay.

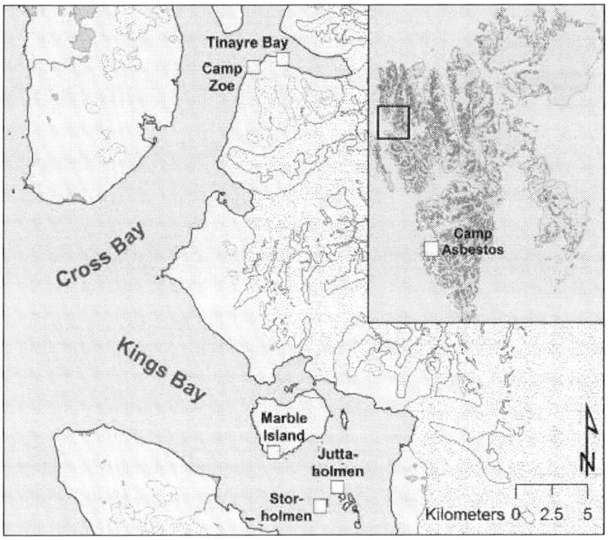

4.39 Location map of the mineral properties of the Northern Exploration Co. (Map: F. Kruse.)

4.40 Camp Zoe had a commanding view over Cross Bay. Note the flagpole near the centre. (Photo: D. Avango, LASHIPA 5, 2008.)

The LASHIPA 5 team surveyed Camp Zoe (Fig. 4.40) in July 2008.[24] The small hut was locked. Nearby were a flagpole, a bench and table, and a short footpath to the shore. The flagpole may once have flown the Union Jack. The vicinity lacked any evidence for prospecting, but primary archival research suggested there was marble in Tinayre Bay. No track or footpath assisted the two-kilometre hike, which

[24] Avango *et al.* (2009) pp. 16-7.

incorporated two sizable melt-water streams, the remains of another hunting hut, and a grave. It was likely that Tinayre Bay had formerly been reached by boat.

Tinayre Bay was a small and scenic bay with a glacier at its head. The outcrop of white marble on its southern shore was widely visible (Fig. 4.41).

4.41 Outcrop of white marble in Tinayre Bay. (Photo: D. Avango, LASHIPA 5, 2008.)

The stone had been quarried by opening naturally occurring cracks with flat wedges.[25] This method created fairly irregular blocks of surface rock, which would have been removed by boat. Fig 4.42 represents a prospecting site, which is highly unlikely to have been mentioned in any historical sources and which only a trained eye could pick out as such.

4.42 A subtle prospecting site in Tinayre Bay. (Photo: D. Avango, LASHIPA 5, 2008.)

[25] Avango *et al.* (2009) pp. 18-9.

Someone had cleared this natural plane. The loose rock still lay at its foot. This detail serves as a reminder how easily prospecting sites may be overlooked. Such oversight substantially reduces the archaeological record and diminishes the extent and historical importance of exploration and prospecting. Other features included a second faintly quarried edge, a borehole, some pieces of wood, and a thin spread of coke. The borehole was less than seven centimetres wide and reached 2.65 metres, where it may have been blocked. Both the wood and the coke were related to working as opposed to housing. The top of the outcrop would have been a poor choice for a hut or a tent. The team did not search further afield.

It is a regrettably persistent misconception that the Northern Exploration Co. referred to the settlement and marble quarries on **Marble Island** by the name of Britain's capital city. The name is a later Norwegian construction.[26] Mansfield discovered the island in 1906, which led to the establishment of the quarries in 1911.[27] Rossnes adds that in 1912/13, eleven British men wintered.[28] Quarrying was revived after the First World War but quickly ceased again. The company employed Norwegian hunters and trappers to guard the property in 1922-6 and in 1927/8 before it was sold to Norway in 1932. Prior to 1940, some Norwegians attempted fish processing here. In the 1950s, several houses were transferred to Ny Ålesund across the bay. The material remains on Marble Island have been the subject of previous archaeological research.[29] In August 2008, the LASHIPA 5 expedition re-assessed Marble Island with the research questions of this study in mind.[30] The results are summarised in Fig. 4.43.

The settlement lay at the head of a natural harbour called Port Peirson, which allowed for discharging ships both at the beach and at the cliff edge. It was positioned next to a stream and sheltered by the surrounding outcrops. It comprised eight pre-fabricated buildings of the types encountered at Camp Millar. There were three smaller houses (1a-c). Two were upstanding but inaccessible. They had retained their annexes at the rear. 'Camp Mansfield' was written above

---

[26] Hoel (1942) p. 268-9.
[27] Johansen (2011) pp. 154-6.
[28] Rossnes (1993) p. 111.
[29] Jørgensen, R., Jensen L. V., Rossnes, G., and Løkken G. (1988) *London Kongsfjorden Oppmåling 1:500*. Kulturhistoriske Plantegninger, Sysselmannen, Longyearbyen; Rossnes, G. and Løkken, G. (1988) *London Kongsfjorden, fasader 1:50*. Kulturhistoriske Plantegninger, Sysselmannen, Longyearbyen; Rossnes, G. and Løkken, G. (1988) *London Kongsfjorden, plan og snitt 1:50*. Kulturhistoriske Plantegninger, Sysselmannen, Longyearbyen; Løkken, G. (1988) *London Kongsfjorden Detaljer 1:5*. Kulturhistoriske Plantegninger, Sysselmannen, Longyearbyen; Nash, K. and Løkken, G. (1990) *Oppmåling 1:1,000*. Kulturhistoriske Plantegninger, Sysselmannen, Longyearbyen; Nash, K. and Løkken, G. (1990) *Ny London Kongsfjorden, havnekran oppris av mast 1:50*. Kulturhistoriske Plantegninger, Sysselmannen, Longyearbyen; Løkken, G. and Nash, K. (1990) *Ny London Kongsfjorden, havnekran oppris 1:50*. Kulturhistoriske Plantegninger, Sysselmannen, Longyearbyen; Løkken, G. and Nash, K. (1990) *Ny London Kongsfjorden, havnekran plan rekonstruksjon 1:50*. Kulturhistoriske Plantegninger, Sysselmannen, Longyearbyen.
[30] Avango *et al.* (2010).

the door of one (1a); above the other (1b) was a flagpole. A third (1c) survived as a wooden floor only. There were also the timber foundations of five larger houses (2a-e). Internal divisions in these barracks suggested two double-walled rooms as well as a single-board workshop or a shed that could only be entered externally. The architecture of the smaller houses and their separate, higher location within the settlement again suggest managerial accommodation or distinct functions, while the barracks will have housed the workforce.

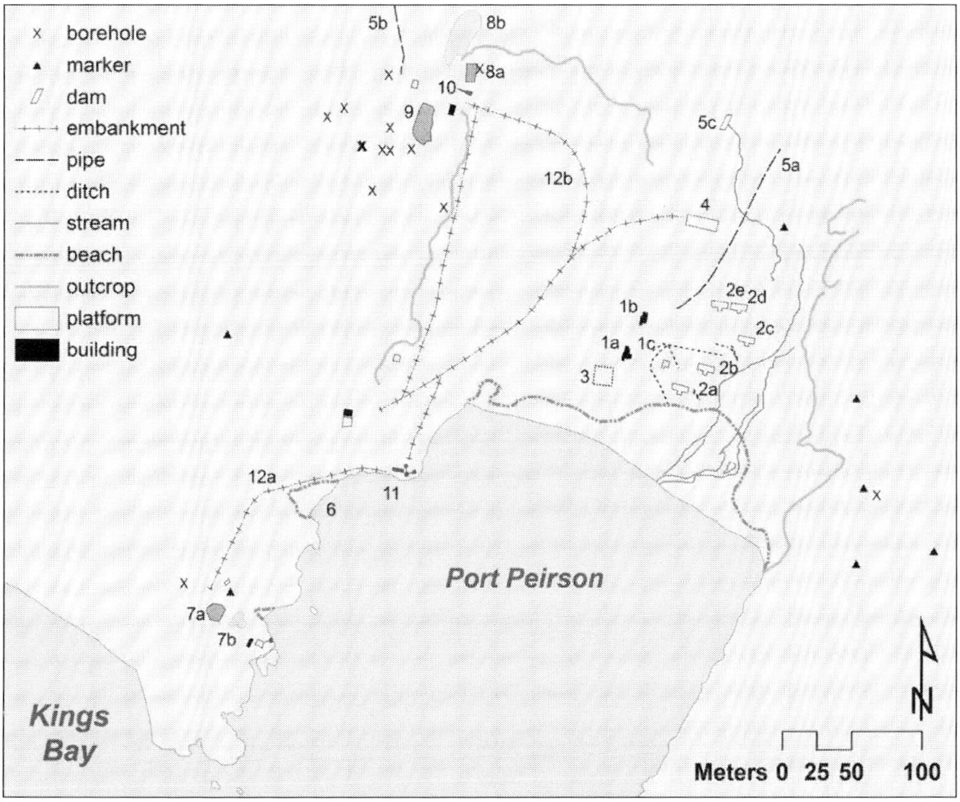

4.43 *Site map of Marble Island in Kings Bay. (Data: LASHIPA 5, 2008; Map: F. Kruse.)*

The settlement encompassed several indicative features. During a recent clean-up, all metal artefacts, mostly stove parts, were gathered on one barrack floor (2a). A large, brick-clad cooking range had been left on another (2b), presumably because it was too heavy, and implied a canteen for the workforce. In front of Camp Mansfield (1a) was a round wooden patio, which resembled the semi-permanent tent base found at Calypso Beach. It may be the only material remains from the time between Mansfield's arrival in 1906 and the first recorded quarrying in 1911.

Also in front of Camp Mansfield were four antenna anchors (3) comparable to those at the wireless station at Calypso Beach. At the back of the settlement, 50 metres from the nearest barracks, lay the remains of a large workshop (4). It was surprisingly far from the shore, where materials will have been delivered, and equally distant from the workings, where machines will have been used. There was no shortage of space, and while smell and noise may have been an issue, it may be that the houses presented a barrier, preventing anyone from stealing. As with the other structures, the lack of wooden debris suggested that the workshop had possibly been relocated. Most remaining machinery had been brought here.

To harness the freshwater, the company had laid a couple of water pipes. One (5a) arrived from an unknown northern source, skirted the workshop, and seemingly led to the staff houses. Another (5b) came from a different location and supplied two quarries. There were thus separate domestic and industrial supplies. Both would have frozen in winter, and if the company wintered, alternatives were needed. It is unclear if a small dam (5c) at the bottom of the waterfall had also been built by the company.

Initial prospecting may have taken place along the marble cliff. Blasting will have been used, but proof was difficult to find. There were several possible coastal extraction sites (6), but the evidence for early methods was vague. In fact, it was impossible to date any quarry using archaeological techniques alone. One quarry (7a, Fig. 4.44) had been placed back from the cliff and was fairly small. The two-metre-high face showed the marks of flat wedges. The aim of the works may have been to obtain blocks to test the deposit. The beginning of a spoil heap (7b) blocked the direct access to the sea, so transport must have been over land to a different location.

4.44 This quarry (7a) on Marble Island was worked using flat wedges. (Photo: D. Avango, LASHIPA, 2008.)

A third quarry (8a, Fig. 4.45) was remarkably regular. Approximately 50 square metres had been opened by a channeler, a special marble-cutting machine. This method was verified by the presence of two unfinished grooves about 0.9 metres apart, which also implied that plans for the quarry were never realised, and a length of broad-gauge rail on which the machine would have run.

4.45 This quarry (8a) on Marble Island was cut by channeler, but the blocks were never extracted. (Photo: D. Avango, LASHIPA 5, 2008.)

4.46 This quarry (9a) on Marble Island was probably blasted. (Photo: D. Avango, LASHIPA 5, 2008.)

The majority of boreholes on site had been logged to the west of the fourth quarry (9a). Whether the boreholes were positioned in a set grid was not immediately apparent. They were usually vertical with one angled exception found on the side of an outcrop. This quarry (9a, Fig. 4.46) was roughly ellipsoidal and with circa four metres the deepest on site. On the flat, sloping plane was the mark of a drill hole, made to place a shot. The coarse surface exposed the presence of a fault, a weakness in the marble. The spoil from the quarries (8b) was dumped nearby.

A wealth of machinery and equipment to split, hoist, and transport the marble had been left on site after the workings were abandoned. Most had been collected in the former workshop (4). The steam-powered channelers and associated machines are unique to the archaeological record of Spitsbergen (Fig. 4.47). While a detailed description of each is outside the scope of this chapter, several displayed the names of manufacturers, for example Sullivan, Ingersoll-Rand, R. Garrett & Sons at Leiston Works, and Schram, Harker & Co. in London.

4.47 Machinery and equipment to extract marble in the former workshop (4) on Marble Island. (Photo: D. Avango, LASHIPA 5, 2008.)

At least two cranes had assisted with the hoisting of stone and other loads. One (10) was a steam-powered mobile crane made by Taylor & Hubbard of Leicester. It had been stripped of its fittings and was parked on a length of broad-gauge rail. The other (11) was a stationary crane at the cliff edge. The size of this crane suggested that large ships were moored, lighted, and loaded on this natural quay. The stationary crane comprised one terminus of an elaborate transport system based on a board-gauge railway (Fig. 4.48). The railway only survived as earthworks and embankments. Most rails had been removed with the exception of those already mentioned and a small stack in front of the warehouse. One line (12a) led towards the southern quarry, but it had never been completed. Another circuit (12b) linked the crane, the northern quarries, and the warehouse. Different branches appeared to have been laid according to the shortest route and the slightest slopes, but it was impossible to ascertain the sequence and why there was not a direct connection between the northern quarries and the warehouse. The choice of gauge must have been determined by the machinery. There were, however, no broad-gauge tubs. A small array of narrow-gauge tubs was piled at the cliff, one possibly being a stone car, but there was no evidence of narrow-gauge rail on site.

4.48 Stationary crane (11) at the cliff and railway embankment (12a) leading towards the southern quarry (7a). (Photo: D. Avango, LASHIPA 5, 2008.)

The Northern Exploration Co. probably used claim signs to consolidate their occupation of Marble Island. There were several possible markers in the form of small stone cairns. The more prominent ones, however, belonged to other nationalities. The Italians had erected a monument to the Nobile expedition of 1928.[31] The inscription of a small wooden board in the vicinity of the workshop had weathered, but it may have read 'SNSK A/S 1936 NK2'. It had therefore been placed by Store Norske after Norway had taken over the claim.

Imperfect records of the aforementioned quarries hitherto comprised the corpus of marble quarrying on Spitsbergen. The development has been decried as a failure, but research on these workings is still incomplete. In addition to Tinayre Bay, there are other sites at which marble has been explored (Fig. 4.49). Primary documents revealed a location approximately 1.6 kilometres north-west of Port Peirson. In summer 2011, LASHIPA asked colleagues at the Netherlands Arctic Station in Ny Ålesund to investigate this and thereby re-discovered a forgotten quarry (Fig. 4.50). Part of the face displayed a series of closely-spaced drill holes characteristic of the plug and feather method. This method was used with the intension to obtain regular blocks, but the remainder of the quarry was quite uneven. The spoil formed a bank between the face and the cliff. Absent evidence for transport over land implied that any marble blocks had perhaps been hoisted out of the quarry and directly into a ship moored at the cliff.

[31] The Italian engineer Umberto Nobile had planned a series of airship flights from Ny Ålesund in Kings Bay around the North Pole. His crash on the ice on May 25, 1928, led to international rescue efforts, during which Roald Amundsen and others found their deaths.

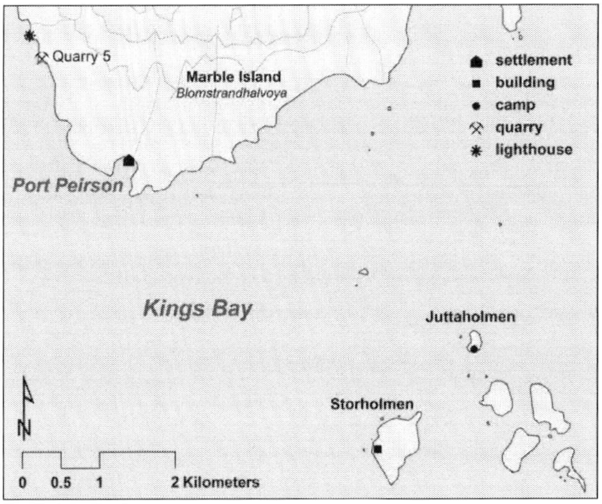

4.49 Location map of other marble sites in Kings Bay. A quarry was re-discovered on the west side of Marble Island; a staff house stood on Storholmen; and Juttaholmen comprised breccia workings and a camp. (Data: LASHIPA 5, 2008; Map: F. Kruse.)

4.50 This re-discovered quarry on Marble Island had been worked by plug and feather method. (Photo: N. Miedema, 2011.)

On **Storholmen,** a raised path made of turf led from the beach directly to a prefabricated staff house.[32] The house, the rear extension, and the outdoor toilet were intact. Because the building was no longer in use, the sea-facing door was unlocked and stood open. Beyond the porch, the house comprised a single room downstairs, where a stove had formerly occupied one corner, and a roomy loft, which contained some leftover eider down. The extension appears to have been the kitchen and housed a small cooking range and a large sideboard. Clear evidence for prospecting could not be seen, although there were hand tools, one manufactured by the Hardy Patent Pick Co. of Sheffield. Occasional stone cairns

[32] Avango *et al.* (2010) pp. 29-30.

may once have been claim markers, but the house was clearly visible from Kings Bay and would have been an effective claim hut. The west-facing position of the house and the distance to the other islands do not suggest a direct link between this building and the breccia exploitation on the latter.

There were at least two quarries of red carbonate breccia in the cliffs on the south side of **Juttaholmen.** In August 2008, the LASHIPA 5 team used one of them as a landing site and only scaled the cliff with difficulty. Both of them appeared to have been worked at sea level (Fig. 4.51). The team noted that there was little room to manoeuvre at high tide.[33]

4.51 Breccia quarry on Juttaholmen. Note the limited space above the high water mark. (Photo: D. Avango, LASHIPA 5, 2008.)

A former camp was located some distance away from the quarries in the shelter of high rock outcrops on the island. It had probably encompassed three tents. One tent ring survived in the form of a vague soil imprint approximately 3.4 metres in diameter and six wooden tent pegs. Another comprised a low bank of turf roughly 3.1 metres across and two wooden pegs (Fig. 4.52). The turf will have been used to seal the bottom of the tent, and a gap in the ring suggested an entrance to the north-west. The third tent ring was not round but roughly square, measuring circa 3 metres by 2.7 metres. On three sides, this tent had also been sealed with sods; the fourth side was open. The turf-less tent may not have stood there as long as the others. The round tents may have been for sleeping, while the square one had a different function such as storage or preparing food. Nearby was also a small rectangular pit, which may have been the outhouse or garbage pit. There were no other signs of the men who had worked and lived on the island.

[33] Avango *et al.* (2010) pp. 31-2.

4.52 *Tent ring made of turfs. Note the wooden pegs of a second tent immediately behind. (Photo: D. Avango, LASHIPA 5, 2008.)*

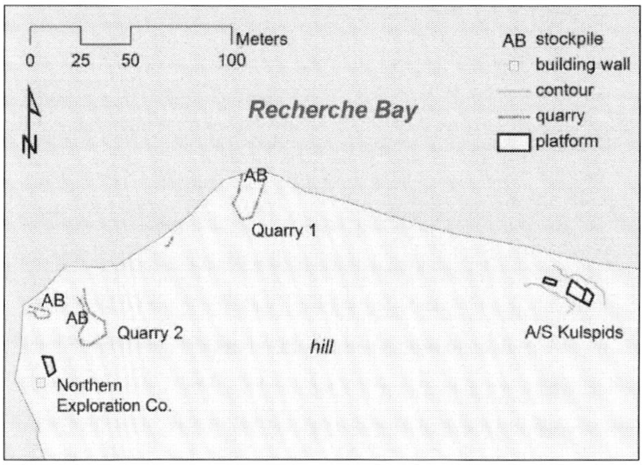

4.53 *Site map of formerly contested asbestos claim in Recherche Bay. (Data: LASHIPA 3, 2006 & LASHIPA 9, 2010; Map: F. Kruse.)*

In contrast to the widespread marble workings, the asbestos quarries were restricted to **Camp Asbestos** in Recherche Bay (Fig. 4.53). Rossnes relates that the deposit was investigated after the First World War by both the A/S Kulspids and the Northern Exploration Co.[34] The Norwegians built a house in 1918. The British followed suit in 1921. Rossnes states that this was a small, single-board claim hut fit for summer use only. Between 1918 and 1922, the claim was heavily disputed. In 1927, it was granted to A/S Kulspids. By 1993, the British hut was gone, the Norwegian one badly decayed. In August 2010, the LASHIPA 9 expedition disembarked at Camp Asbestos to establish if the former claim dispute was evident in the archaeological record.[35]

[34] Rossnes (1993) p. 46.
[35] Avango *et al.* (2010).

The team landed on the gravel beach below the A/S Kulspids hut. There were in fact two huts on a narrow rocky ledge above the water, which benefitted from the protection of the outcrop behind. The ledge may once have been broader, but neither had the foundations been eroded nor had it been subjected to the natural creep of the hill. The structures had collapsed completely and indicated just how much wood can be expected if a house had neither been removed nor used as firewood. Freshwater was not noted. Less the 200 metres to the west, the team recorded a first quarry (Fig. 4.54). The asbestos veins appear to have been worked manually, a single pickaxe head perhaps confirming this, until a pit resulted. A stockpile was collected at the seafront to be loaded into boats. At the time of the survey, this stockpile was slipping into the sea.

4.54 Quarry 1 at Camp Asbestos. Note the stockpile of white asbestos at the shore. (Photo: P. Leminen, LASHIPA 9, 2010.)

4.55 Quarry 2 at Camp Asbestos. Note the stockpile centre right. (Photo: F. Kruse, LASHIPA 9, 2010.)

The second quarry was mapped further west (Fig. 4.55). Its walls showed traces of asbestos veins. At the bottom were water and wooden debris. The wood might have been used for shelter, a boardwalk, or another installation. Hoisting seems unlikely as there was no large spoil heap and the stockpile comprised hand specimens only. A ditch connected the pit and the shore. It was strewn with planks, but it was too rugged to be a path and at the wrong angle to be a drain. A vein discovered near the shore had possibly been followed inland before being quarried. There were other trial works and stockpiles along the shore, which held no additional information.

The LASHIPA 9 expedition located the building platform and a wooden wall of the former British hut (Fig. 4.56). The platform had been levelled and surrounded by large stones, which would have supported a wooden foundation. The wall was ca. 4 metres long and 2.5 metres high. Single boards had been nailed onto posts, but the wall could easily have been insulated by adding a second layer of boards and stuffing the space between them. A likely explanation for the hut's demise was that creep had weakened the structure and the parts had been dispersed by storms. Alternatively, either company destroyed it after A/S Kulspids being granted possession. The beach below the hut was the poorer landing site.

4.56 Remains of hut of the Northern Exploration Co. at Camp Asbestos. (Photo: P. Leminen, LASHIPA 9, 2010.)

4.2.1.4 Other

Besides the company's sites which were established to explore coal, metal ore, or mineral deposits, there were those sites that had multiple users or ambiguous functions. With the exception of Ebeltoft Haven, these were found towards the South Cape of Spitsbergen (Fig. 4.57).

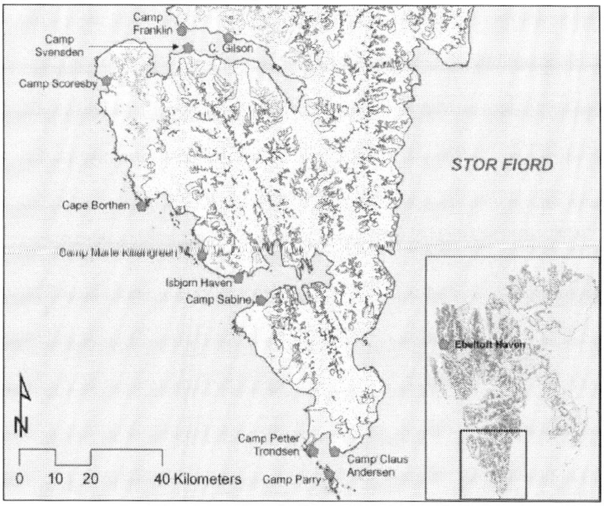

4.57 Location map of other sites connected with the Northern Exploration Co. (Map: F. Kruse.)

Ebeltoft Haven in Cross Bay has seen human usage during all historical periods on Spitsbergen, including English whaling, Russian hunting, and Norwegian trapping.[36] Of particular interest to this research is the construction of the German meteorological station in 1912. It motivated a geopolitical act of the Northern Exploration Co. in 1918. Dege stated that the station comprised a two-story house, a small balloon hangar, a long storage shed, and a rotatable hut for the kite winch.[37] Contemporary photographs suggest the approximate location and arrangement of buildings as well as the presence of an antenna.[38] Norwegian trappers contracted by the Northern Exploration Co. wintered at Ebeltoft Haven between 1911 and 1914.[39] After the outbreak of war, the Germans departed, and Dege thought the Royal Navy destroyed the station in 1915.[40, 41] Near the end of the conflict, the Northern Exploration Co. targeted the site and reported that,

---

[36] Johansen (2011) p. 149.

[37] Dege, W. (1962) 'Deutsches Observatorium Ebeltofthafen – Spitzbergen. Zur 50. Wiederkehr der 1. Überwinterung 1912/13', *Polarforschung,* 32 (1/2), pp. 136-40.

[38] Lüdecke, C. (2001) 'Das deutsche Observatorium in Ebeltofthafen, Crossbai, Spitzbergen (1912-14) – ein Besuch im Sommer 2000', *Mitteilungen der DMG*, 2/2001, pp. 25-7; Steinhagen, H. (2007) ,Verlauf und Ergebnisse der Spitzbergenexpedition von K. Wegener und M. Robitzsch, 1912 – 1913', *DACH Meteorologen-Tagung*, Hamburg, 10-14 September, 8 p.

[39] Rossnes (1993) pp. 111-2.

[40] Dege (1962) p. 137.

[41] The *Columbella* patrolled Spitsbergen and Bear Island between November 1914 and August 1917. It received support from the *Acacia* between May 1915 and July 1916. Both logbooks recorded Lat. 79.2 as their furthest north, Ebeltoft Haven lying at Lat. 79.9. There is no mention of the station being visited, let alone destroyed. Although independent action by British sailors cannot be discounted, it is unlikely. Available at www.naval-history.net (Accessed: 2 August 2012.)

> The German shore looked desolate, and it was littered with the debris left in seeming haste, but it offered a safe landing. Once ashore we were confronted with the German occupation posts, which we immediately pulled up – the first by Mr. Salisbury-Jones – and replaced with British substitutes, bearing the following inscription: Northern Exploration Company Limited. September 3rd 1918. [...] Having formally dispossessed the Germans of the most northern territory in which they have ever had a footing we completed the act of annexation by hoisting the Union Jack and then re-embarked for another destination.[42]

Norwegians acted as company watchmen in the winters of 1924/5 and 1926/7.[43] It is not known if the site served a purpose after Norway obtained it in 1932.

Lüdecke inspected the site in 2000.[44] She noted ample relics and listed several graves, three small huts, a camp site, a roof that seemed to have blown away, and several refuse piles comprising rusted tins and broken glass. She did not, however, find any evidence for the dwelling, the hangar, the shed, and the kite winch. Lüdecke put this down to abandoned buildings commonly being relocated. She interpreted some bricks to be the hearth of the former house and discovered a spool of piano wire, which would have been used to fly the kite. Subsequently, the LASHIPA 5 team arrived in August 2008 with a particular interest in the British actions to expel the Germans.[45] The survey largely verified Lüdecke's observations, but the team noted not one but numerous camp sites mostly characterised by tent rings of large cobbles. Although these showed the popularity of the location with tourists, there were one or two examples, where tents had caused circular patches of reduced vegetation, suggesting semi-permanence. Yet, it was not possible to link these to prospecting activities or territorial claims.

Although the German meteorological station had comprised four buildings, their locations could no longer be established with certainty. Lüdecke had overlooked some very discreet depressions. The Germans had seemingly placed a raft foundation on the ground, which led to deformations of the surface over time. The raft was removed, leaving barely a trace. Similarly, Lüdecke missed the remains of the antenna. These comprised four short wooden pegs with some metal wire fastened to their tops, which had been arranged in a square. Next to these anchors lay metal braces, which were used to tighten the wires. The bricks Lüdecke mentioned were found immediately east of the antenna. Set in concrete, they gave the impression of being the base of a generator or other machinery.

Still the German remains were more easily confirmed than any subsequent British activities. The substitution of national claim markers may have been evident in three wooden boards. A small one lay above the beach near the spit. An

---

[42] Hoel (1966) p. 450.
[43] Rossnes (1993) p. 112.
[44] Lüdecke (2001).
[45] Avango *et al.* (2009) pp. 12-5.

inscription was missing, and it may only have been part of a crate. A large one lay in the proximity of a metal tripod (Fig. 4.58). Despite the lack of a text, it appeared purpose-made and had once stood on two legs. A third one was for all intents and purposes a grave marker, but the survival of an original was highly unlikely. A well-meaning tourist had possibly interfered with the grave, but it was not unthinkable that a claim marker had once been secured among the cobbles. The metal tripod was supported by two cement barrels and a boulder. The barrels in particular were thought to indicate the presence of the British. The tripod may have held the principal claim board or a flag pole flying the Union Jack. In all, the evidence for the British annexation of Ebeltoft Haven was circumstantial and inconclusive.

4.58 Remains of two possible claim markers. (Photo: D. Avango, LASHIPA 5, 2008.)

With the exception of **Camp Svendsen** on Cape Ahlstrand, which was visited by the LASHIPA 3 expedition in August 2006, the remainder of sites on the map above was surveyed by LASHIPA 9 in August 2010.[46] According to Rossnes, several Norwegian hunting stations were connected to the Northern Exploration Co. The station at **Isbjorn Haven**, for example, had displayed a claim sign of the company, who had also acquired a hut at **Cape Borthen** in 1920.[47] The team intended to ascertain if the link between the trappers and the company found expression in the archaeological record. The hut at Isbjorn Haven no longer existed. The one at Cape Borthen had been rebuilt and subsequently abandoned. Its surroundings lacked any indication of a British influence.

An archival document unexpectedly disclosed a list of so-called discovery points, which the company desired to register with the mining commissioner or

[46] Avango *et al.* (2010).
[47] Rossnes (1993) p. 41 & p. 47.

bergmesteren in 1926.[48] Most of these points were supposedly marked by a house in the company's possession and covered a claim of ten square kilometres. The team wanted to prove the validity of this list and chose several camps to investigate. Some carried the names of Norwegian hunters. **Camp Gilson** honoured a company employee. Of particular interest were **Camp Parry**, **Sabine**, **Scoresby**, and **Franklin** (Fig. 4.59 - Fig. 4.62). As seen in Chapter 2, these men were famous British explorers and Arctic heroes. The LASHIPA survey revealed how their status compared to the reality of the sites named after them.

4.59 Camp Parry on South Cape Island. (Photo: P. Leminen, LASHIPA 9, 2010.)

4.60 Camp Sabine in Goose Harbour in Horn Sound. (Photo: P. Leminen, LASHIPA 9, 2010.)

[48] Light, C. (1926) Letter to the Mining Commissioner for Svalbard, 24 February, Bergmesteren for Svalbard, 1926 Northern Exploration Company, Regional Archive, Tromsø.

4.61 Camp Scoresby in Dunder Bay. (Photo: P. Leminen, LASHIPA 9, 2010.)

4.62 Camp Franklin on Eider Island in Van Keulen Bay. (Photo: P. Leminen, LASHIPA 9, 2010.)

The camps were for all intents and purposes Norwegian stations in locations suitable for hunting and trapping during the winter months. Economic minerals were not immediately obvious to the LASHIPA team, who recorded no signs of prospecting, let alone mining. The derelict camps did not live up to the grand expectations aroused by their names. Both Camp Parry and Camp Franklin were positioned on older building foundations and had been rebuilt at least once. How they would have looked in 1926 could no longer be discerned. Although the Northern Exploration Co. evidently had a longstanding connection with Norwegian trappers and had acquired numerous hunting stations by 1926, the archaeological record is ultimately unclear about their individual functions and significance. Collectively, however, the huts served as a chain of claim markers over the great stretches of land claimed by the company.

4.2.2 Site interpretations

The site narratives give rise to a whole host of environmental and archaeological components in the actor-network of the Northern Exploration Co. A summary is presented in Fig. 4.63. The context is the global market and British foreign policy between 1910 and 1927. Against this background, the company sought to explore the economic minerals of Spitsbergen. Exploration differed from mining in that companies did not engage in active mining but tried to secure an area and prove a resource in order to involve subsidiary mining companies or buyers. The pre-war activities of the Northern Exploration Co. were distinctly economic. The focus was on gold and coal in Bell Sound and on marble in Kings Bay. Only at the end of the war did it carry out the geopolitical act of expelling the Germans from Ebeltoft Haven. The focus then shifted to iron ore in Recherche Bay. While all camps and huts acted as claim markers in the no man's land, their importance even grew after the ratification of the Spitsbergen Treaty. Being in possession of remote hunting stations justified further prospecting in those areas.

The Northern Exploration Co. chose to operate across a wide geographical spread, possibly as far as its capital would allow. The ground explored must have been in great excess of the sites chosen for further tests. If the results were promising, the company needed to demonstrate effective occupation on a regular basis. The symbolism of any workings was strengthened by the presence of a hut or more. The company took over Norwegian huts and erected new ones. These included pre-fabricated buildings. In addition, the firm protected its claims by chains of claim markers in strategic locations along the claim boundaries. The most promising sites such as Camp Morton and Marble Island became local centres of bigger size and greater efforts. After the war, the centres were Marble Island and Calypso Beach as shown by the cutting-edge wireless technology. Although the latter had a small mine, its importance may have been as a relatively ice-free location from where to administer all activities and distribute materials. A general manager for Spitsbergen must have been in frequent communication with site supervisors to coordinate the efforts of a large workforce necessary to inspect and maintain the claims and to continue the explorative efforts.

The few previous archaeological studies of the Northern Exploration Co. have largely concentrated on its huts rather than their relationship with the environment and the workings. Yet, important sites like Ebeltoft Haven, Tinayre Bay, Juttaholmen, and Iron Mountain do not have huts. Instead, the company frequently placed buildings at considerable distances away at the shore in a visible spot more suitable for protecting the claim. This was probably the case at Camp Smith and Camp Violet, where workings may yet be found much further afield, and

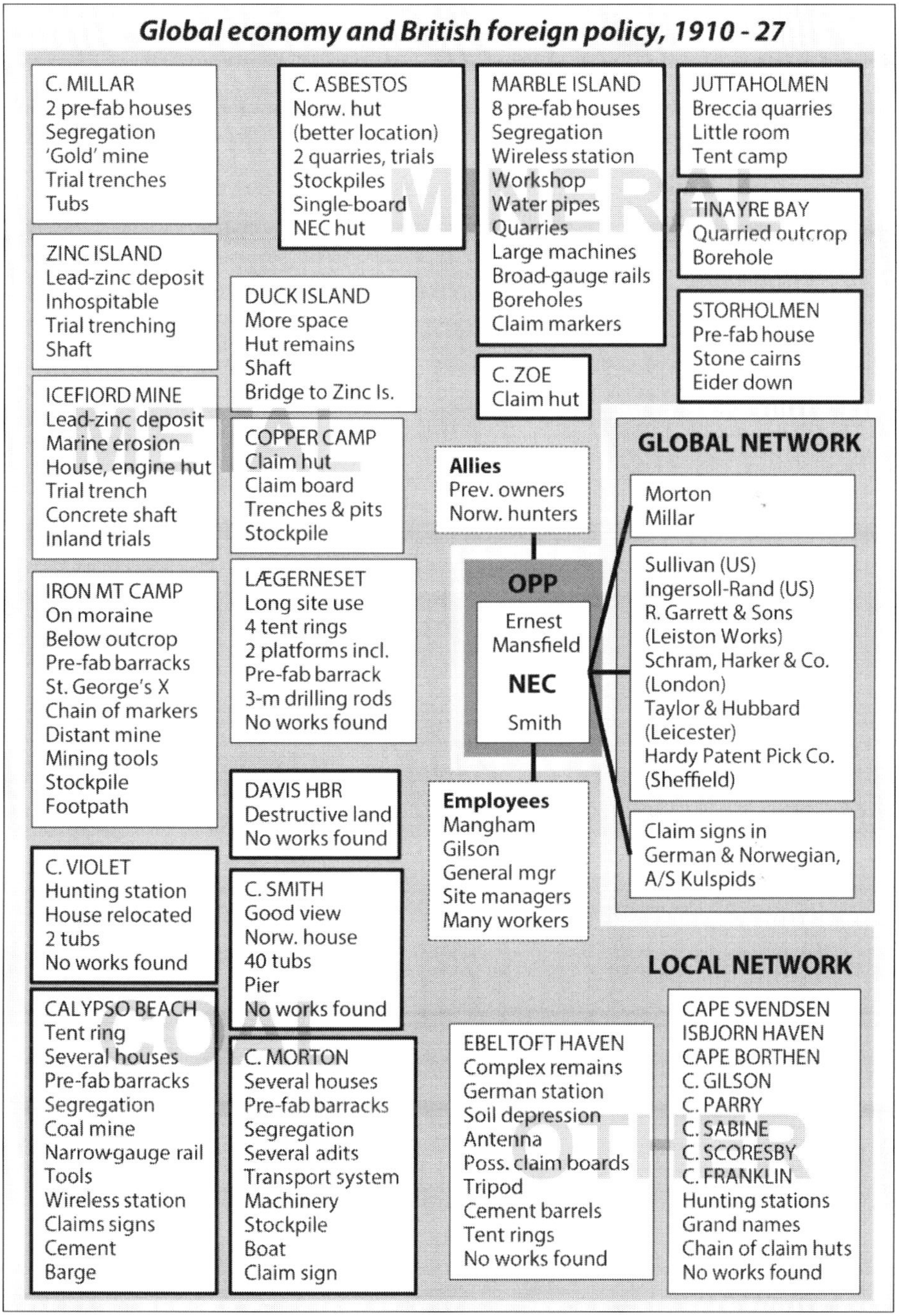

4.63 Selected environmental and archaeological components of the actor-network of the Northern Exploration Co. (NEC). (Chart: F. Kruse.)

definitely at the Norwegians hunting stations taken over in the 1920s, where the existence of the hut permitted additional prospecting.

The pre-fabricated buildings were insulated and fit for habitation in winter, but why would an exploration company winter? Any snow cover coupled with the lasting darkness and bad weather would hinder prospecting. Camp Morton and Calypso Beach arguably had underground workings, which could be developed in winter, but pre-fabricated buildings also existed on Marble Island, at Lægerneset, and at Iron Mountain Camp. Wintering could have had the additional purposes of proving to third parties that it was possible and to make the most of the connection with Norwegian trappers, providing them with a base and sharing the profits. Year-round activity would have been beneficial to demonstrate effective occupation and to keep claim-jumping at bay.

Archaeologically, the foremost reason as to why the Northern Exploration Co. sold its properties to the Norwegian Government in 1932 was that it was unsuccessful. No resource was proven to the extent as to involve a subsidiary mining company. In some instances such as Copper Camp or Zinc Island, there may not have been enough ore. At Davis Harbour and the Icefiord Mine, development may have failed due to environmental reasons. Camp Asbestos was successfully contested by a Norwegian firm. Yet, Camp Morton, Marble Island, and Iron Mountain seemingly had potential. It could not be discerned if local determinants like poor quality or global factors like a shift in the market were to blame. As with Advent City, the question arises if the shallow quarries on Marble Island stood in economical proportion to the settlement and machinery. Mismanagement and overspending are much less obvious if the aim was to investigate different types of marble and create an attractive development. In fact, Marble Island and Tinayre Bay were the only sites, where boreholes were sunk to prove the deposits at depth. With hindsight, the output from Marble Island did not cover the outlay. In fact, the Northern Exploration Co. invested much more into its widespread activities than it got in return.

4.3 The Scottish Spitsbergen Syndicate

At about the same time as the Northern Exploration Co., the Scottish Spitsbergen Syndicate began to explore the natural resources on the islands. Its promoter was William Speirs Bruce, who took part in four expeditions to Spitsbergen before being instrumental in founding the syndicate in 1909.[49] He prospected on its behalf another five times up to and including 1920. Following his death, others took care of the company until its properties were sold a Scottish development company in 1950. During the LASHIPA project, six sites associated with either Bruce, the syndicate, or both were visited and surveyed (Fig. 4.64).

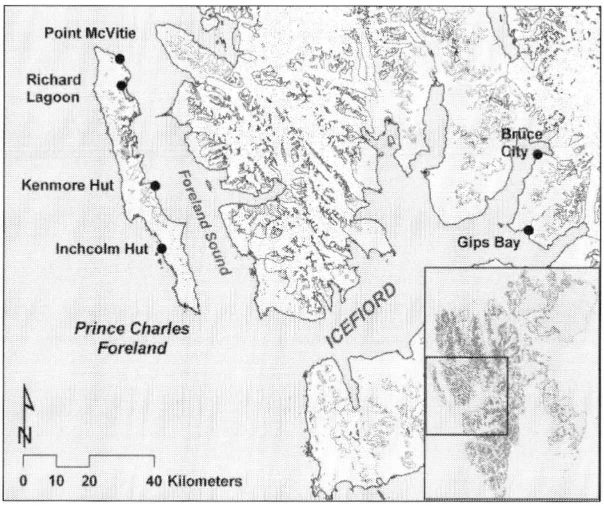

4.64 Location map of six sites of the Scottish Spitsbergen Syndicate visited during three LASHIPA expeditions between 2004 and 2008. (Map: F. Kruse.)

Using readily available sources, these sites could preliminarily be dated.[50] It transpired that Point McVitie and the Inchcolm hut on Prince Charles Foreland had been Bruce's base camps in 1906 and 1907, respectively, while the location of the Kenmore hut played a role from 1909 onwards. After the British being absent from the islands throughout the First World War, the Scottish Spitsbergen Syndicate renewed its expeditions in summer 1919. Bruce City appears to have been the centre of operations then, while Richard Lagoon was one of the areas visited by a prospecting party. Prospecting may also already have taken place at Gips Bay, but the hut here is said to date from 1921.

[49] Speak, P. (2003) *William Speirs Bruce, polar explorer and Scottish nationalist,* Edinburgh: National Museums of Scotland, p. 137.
[50] Bruce, W. S. (1908) 'The exploration on Prince Charles Foreland, 1906-1907', *Geographical Journal,* 3 (2), pp. 139-48; Hoel (1942); Rossnes (1993); Johansen (2011).

4.3.1 Site narratives

4.3.1.1 Prince Charles Foreland

Bruce's account of the survey work on Prince Charles Foreland, which he carried out under the auspices of the Prince of Monaco, was accompanied by a preliminary topographic map.[51] It showed the location of the 1906 base camp at the north end of the island. The 1907 base camp was situated on the west coast, where the Foreland Laichs meet Antarctic Bay. According to Hoel, Bruce did not use the relevant place-names until 1909.[52]

In mid-July 1906, Bruce and two other Scots were landed approximately five kilometres from the northern tip of Prince Charles Foreland. At Point McVitie, they arranged their stores, set up their instruments, and charted the surroundings of the base camp. The team then worked and camped further afield. Travelling over rough ground was difficult, so they initially took only their instruments, some provisions, and sleeping-sacks. They did not take a tent, although it would have been more comfortable. The men occasionally returned to base for supplies and eventually launched a boat to reach the northerly Vogel Hook and their fourth camp site, at which they did pitch a tent. By the end of August, the party had surveyed most of the northern third of the Foreland. Bruce described the landscape here as 'almost absolutely barren'.[53] The only geological observation referred to Vogel Hook comprising cliffs of hard quartzite.

In 1907, the Prince of Monaco supported a second Scottish expedition to Prince Charles Foreland. In the second week of June, Bruce and four others, of whom one was a Norwegian sailor, disembarked about 20 kilometres from the southern point and set up their base camp on the west coast. This base camp comprised the wooden Inchcolm hut. Their survey work was hampered by bad weather, and overland journeys were again arduous. The men completed the survey of nearly the whole of the west coast and collected scientific specimens, which included rock samples and fossils. Bruce did not mention any economic minerals in his paper.

In August 2008, the LASHIPA 5 expedition visited the locations of the base camps in search for evidence of territorial occupation and mineral prospecting. At **Point McVitie**, the landscape was almost featureless, and the site may have been chosen for relative shelter, safe landing, and freshwater (Fig. 4.65). There were no

[51] Bruce (1908).
[52] Hoel (1942) p. 201 & p. 285. Richard McVitie had been a supporter of Bruce's Antarctic work. Inchcolm is a tiny island and early Christian centre in the Forth of Firth in Scotland.
[53] Bruce (1908) p. 141.

rock outcrops nearby. The team recorded the ruin of a wooden hut, a refuse pile with parts of at least two stoves, the remnants of a sunken barrel, and two stone settings too flat to be cairns.[54] Two picks were also logged. Rossnes identifies the hut to be the hunting station built by Henri Rudi in 1921, later known as *Lewinhuset*.[55] This and other huts on the Foreland have since been assessed by *sysselmannen's* cultural heritage department in 2007.[56] Subsequently, specialists from the Norwegian Folk Museum produced detailed architectural drawings and descriptions in June and July 2008.[57] Their work was immediately available to the LASHIPA 5 team. Nonetheless, explicit evidence for a Scottish connection with the site could not be discovered.

4.65 Lewinhuset in almost featureless surroundings at Point McVitie. (Photo: D. Avango, LASHIPA 5, 2008.)

Exposure makes landing on the west coast of Prince Charles Foreland hazardous. As suggested by Captain S. Henningsen of the *MS Farm*, the LASHIPA 5 team therefore debarked on the east coast and crossed the island on foot along the same six-kilometre route Bruce was forced to take over a century ago. The featureless landscape was remarkable. The location of the former **Inchcolm hut** benefitted from high, dry ground and freshwater lakes. A building platform, a wooden hut, an outhouse with a flagpole, two modern scientific stations, and a

---

54 Avango *et al.* (2009) pp. 33-4.
55 Rossnes (1993) pp. 102-3.
56 Hoem, S. (2007a) *Befaringsrapport 2007 Prins Karls Forland – Fredete hytter mm.*, Longyearbyen: Sysselmannen på Svalbard.
57 Planke, T. and Wammen, M. (2008) *Dokumentasjon av fangsthytter på Prins Karls Forland.* [Online]. Available at: http://www.sysselmannen.no/hoved.aspx?m=44267&amid=2380979 (Accessed: 8 November 2011).

couple of stone cairns were signs of frequent human reuse of the site (Fig. 4.66).[58] This has been elaborated on by Rossnes.[59]

4.66 Signs of frequent reuse at the former site of the Inchcolm hut on Prince Charles Foreland. (Photo: D. Avango, LASHIPA 5, 2008.)

The earliest building was the Inchcolm hut erected in summer 1907. In autumn, Norwegians arrived and observed the hut and two tents in the place where they intended to put a hunting station. They began to construct their hut on September 9 and completed it within a week. It was insulated with a turf bank, and because Bruce's hut was seemingly a light construction unfit for winter use, the hunters surrounded it with turf, too. Two men frequented the station during their wintering on the Foreland in 1907/8. In summer 1908, Captain C. Eriksen arrived to take them home. He had probably been given the Scottish hut in exchange for his services and now pulled it down to remove it to Kings Bay. The Inchcolm hut therefore stood on the Foreland for about a year. The Norwegian hut changed hands in the mid-1930s. It then lay in ruins and a new hut was built in the late 1940s. The LASHIPA 5 survey in 2008 did not record any obvious Scottish remains. The building platform of turf, wooden beams, and several finds were for all intents and purposes the relics of the first hunting station. The stone cairns may have been survey points, flagpoles, or claim markers. Evidence against any territorial occupation by the Scots is the fact that their hut was quickly made over to a Norwegian, who subsequently relocated it.

Bruce supposedly erected a wooden hut, which Rossnes refers to as *Brucehytta*, at Dawespynten in summer 1909.[60] The east-facing façade was three metres wide, while the roof was two metres high in the front and 1.8 metres in the

[58] Avango *et al.* (2009) pp. 39-41.
[59] Rossnes (1993) pp. 101-2.
[60] Rossnes (1993) p. 101.

back. The Scots called the closest landmark Point Napier.[61] The hut was known as **Kenmore** after a village in the Scottish Perthshire.[62] For many years, this place-name appeared on official maps and has only recently become obsolete, destined to disappear from living memory. Rossnes further reports that in 1919, the syndicate built a second hut, and for a time the two huts may have co-existed. Norwegian hunters wintered in 1921/2 and in 1924/5, but they were apparently never contracted as watchmen.

4.67 The widely visible Kenmore hut from Foreland Sound. (Photo: D. Avango, LASHIPA 5, 2008.)

A hut was visibly located on a marine terrace above a gravel beach (Fig. 4.67). Planke and Wammen stated that the place was very windy and generally swampy, the raised ground around the hut being the driest spot.[63] The hut measured nearly eight metres by five metres and comprised two sections. The southern one had completely collapsed and although some building material was missing, the wooden floor was largely intact. The northern one was in reasonably good shape (Fig. 4.68). The Norwegians deduced that the hut had once been five prefabricated segments long by three prefabricated segments wide with a pitched roof. The segments had been painted with a universal 'SsS HUT 4' and an individual mark such as '1C' or '4A'. Based on the observation that only one side of the dividing wall had been insulated, they concluded that the southern section must have been an unheated vestibule with work and storage space, while the northern one included the kitchen and the sleeping quarters. Additional features were the remains of a porch, suggesting the entrance to have been from the south into the

[61] Hoel (1942) p. 108 & p. 224. After Captain Frank Bryce Napier, who not only captained the expedition vessel in 1909 but also in 1919 and 1920.
[62] Hoel (1942) p. 232.
[63] Planke and Wammen (2008) pp. 71-9.

vestibule; one surviving window in the former kitchen, which allowed a good view of Foreland Sound; one chimney but the remnants of two stoves; and evidence for a loft. The specialists attribute the partial collapse to a fault in the design. The jointing was not strong enough to withstand the fierce winds.

The LASHIPA 5 team noted firstly the relative proximity of hills and the possibility of natural resources there.[64] Secondly, Kenmore was situated near the midpoint of the east coast of the Foreland in full view of passing vessels in Foreland Sound. Hoem previously found the floor to have been painted red.[65] If the outside of the hut had sported the same colour, it would have greatly enhanced its visibility. The economic potential and strategic advantages of the position may have outweighed its environmental drawbacks. The team recorded a drainage ditch, which confirmed seasonal bogginess. They contemplated if an earlier hut had been incorporated into the present hut. However, the experts' opinion and the fact that similar barracks survive at Bruce City led to the conclusion that the 1909 structure had gone, if it ever existed, and that the current building must have been one of at least four pre-fabricated huts that the syndicate imported in 1919.

4.68 The southern section of the pre-fabricated Kenmore hut had collapsed, while northern one remained upstanding. (Photo: D. Avango, LASHIPA 5, 2008.)

This being 'SsS HUT 4', a small table labelled 'SsS HUT 3' had not been intended for this site. Were the plans adjusted upon arrival on Prince Charles Foreland or was it an oversight? Where was Hut 3? Had the table been superfluous or had that hut fallen out of use for it to be brought here? Other finds included a leather boot, a knife, and a broken whetstone. They pointed towards the household but were largely inconclusive. It is interesting that a stove called 'Modern Mistress' should

[64] Avango *et al.* (2009) p. 38.
[65] Hoem (2007a) p. 49.

have been imported from Britain instead of nearby Norway.[66] Clearly, the name contained a sexual bias, and it can only be imagined what the Arctic explorer who used it thought of it. A couple of picture frames were a rare glimpse at interior decoration, and they had probably contained pictures of loved ones back home. While broken glass bottles are common on many sites, small glass vials and the cork stoppers of large jars were perhaps related to scientific studies. There were also a broken tool handle and the head of a pickaxe. Even if the latter was used to dislodge geological specimens, it is not known if these were of scientific interest or economic value.

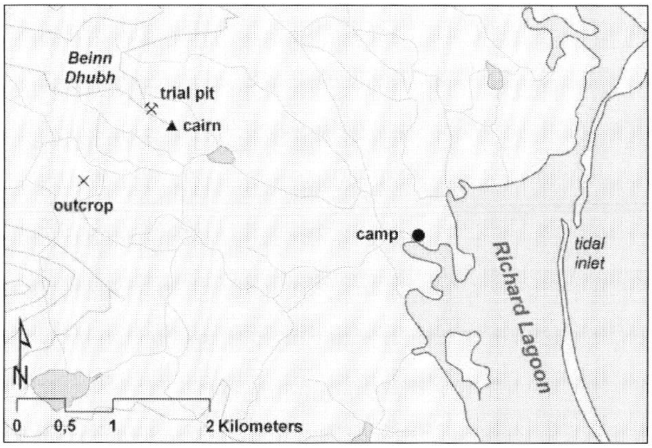

4.69 Location map of a prospecting camp in Richard Lagoon and nearby indications of mineral exploration. (Data: LASHIPA 5, 2008; Map: F. Kruse.)

An account of prospecting camps around **Richard Lagoon** and the exploration of magnetite in 1919 was found during archival research.[67] The lagoon was surveyed by Bruce in 1906 and 1907 and named after Dr. Jules Richard.[68] The LASHIPA 5 team targeted this fourth location on the Foreland in August 2008 (Fig. 4.69). They

[66] Smith and Wellstood Ltd manufactured it at their Columbian Stove Works at Bonnybridge in Scotland. A catalogue from 1911 outlines the features of both the 'Mistress' and the 'Modern Mistress' ranges. (Smith and Wellstood No. 6/11 C. S. W. Catalogue (1911) Smith and Wellstood Ltd, A1419.070, Falkirk Archives, Falkirk.) The latter had a larger cast iron oven, a larger hot plate with five openings, a deeper fire chamber better suited for ashy coal or wood fuel, and a hinged boiler top to facilitate filling. Some models had a copper water boiler within the body of the range, and although this feature might have been useful in the Arctic, 'No. 28', which cost £4 12s without utensils and £6 11s with, did not possess it. The introduction of the 'Modern Mistress' had a profound effect on Scottish island communities ('Modern Mistress' (2005) *Fianuis Shúil (Perspective)*, Series 1, episode 5. mneTV.). People remember the arrival by boat, the assembly of the feet, easy maintenance and cleaning, and the fact that more tasks and comforts were possible at the same time while the stove also effectively heated the cottage. Perhaps the Scottish Spitsbergen Syndicate had chosen the range for similar reasons.

[67] Scott, J. (1919) *Report to the Scottish Spitsbergen Syndicate on the expedition to Prince Charles Foreland, 19/8/1919*, Scottish Spitsbergen Syndicate papers MS 311, Scott Polar Research Institute, Cambridge.

[68] Hoel (1942) p. 360. Richard was associated with the Prince of Monaco's scientific work and accompanied his journeys to Spitsbergen, on which Bruce will have met him.

entered the lagoon through the narrow tidal inlet, which was tricky for a small boat, let alone larger vessels. The account referred to a camp to the north of the lagoon and another one on its western shore. Upon landing, the team perceived an open, watery plane to the north and therefore turned towards the western hills, thereby chancing on a former tent camp near a stream.[69] Among the wooden debris, it comprised a tent ring (Fig. 4.70). It was made of a central cobble that once supported the tent pole, the hint of a circular bank accentuated by a number of stones, and several wooden tent pegs. The tent ring was associated with two tool handles (T), one sharpened to a point and the other a broken pickaxe handle. There was also the lid of a wooden barrel (B) that had most likely held provisions. Not shown in the drawing are two armfuls of driftwood, one next to the central cobble and the other just outside the bank. The unused firewood implied that the tent had a small fire or a stove. The team concluded a British bell tent had been used by prospectors.

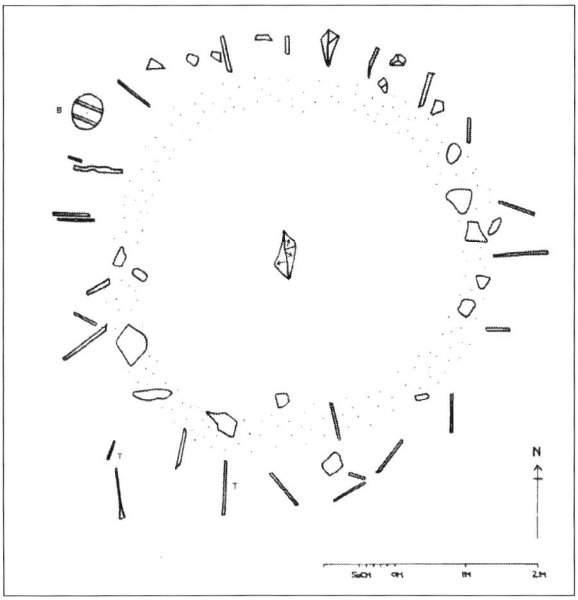

4.70 This tent ring at a prospecting camp in Richard Lagoon constitutes the author's favourite site on the archipelago. Its beauty as well as its archaeological value lie in its informative simplicity and the fact that it captures a moment and a purpose in time, having been undisturbed for almost a century since. (Drawing: F. Kruse, LASHIPA 5, 2008.)

Other finds at Richard Lagoon included a wooden oar, which suggested transportation by rowing boat. Since the camp lay some distance from the nearest landing site, it raises some questions. It is unlikely that the prospectors would have dragged their boat here. Instead, it is plausible that they left it at the shore but took the oars, which would have rendered the craft unusable to anyone else and

[69] Avango *et al.* (2009) pp. 35-7.

therefore safe. This oar had subsequently been broken and discarded. In addition to the tool handles, there were a couple of light-weight picks, which also pointed at prospecting. These picks have so far been found without handles; dismantled they may have fitted into rucksacks. Among the scattered wood of broken boxes, the team found a refuse pile of small bones, rusted tins, and glass bottles and jars. An item that stood out was a small leather pouch, which had perhaps housed a survey instrument. The maker's name could unfortunately not be deciphered.

The camp gave no clear indication of an association with the hinterland. The LASHIPA 5 team therefore aimed for the nearest conspicuous landmark, which was a black hill approximately three kilometres away. Bruce had christened this with the Gaelic name Beinn Dhubh, which indeed translates to Black Hill or Svarthaugen.[70] The colour hinted at coal; the topography lent itself to placing a claim marker or would at least provide a very good view of the surroundings. Thus, the team made a second chance discovery: on the promontory were both a prospecting pit (Fig. 4.71) and a stone cairn.[71] Instead of coal, the pit had been sunk through a dark, slaty rock and into a rust-coloured deposit. Extracted cobbles and boulders were streaked with reddish iron ore. The pit was not very deep, presumably because not much ore had been extracted. Several tools had been left in and around it: a long pry bar, a broken spade handle, a possible plug drill, another metal bar, and a sledgehammer.

4.71 Prospecting pit with discarded tools. (Photo: D. Avango, LASHIPA 5, 2008.)

The well-built stone cairn stood on the promontory with an excellent view into the heart of the Foreland. It may have been a claim sign, but it was more likely to have been a survey marker. In fact, a trig point still appears on modern maps and

[70] Hoel (1942) p. 419.
[71] Avango *et al.* (2009) pp. 35-7.

signals a height of 106 metres above sea level. The view revealed a sizable rust-coloured outcrop (Fig. 4.72) approximately one kilometre to the south-west, which was equal to about four kilometres west from the camp. The team did not visit the outcrop. They assumed that Bruce had seen this hill during his surveys and may have applied the Gaelic version of 'red hill'. Yet, he made no mention of it in his paper, and an official English name does not seem to exist. The colour hints at the presence of iron ore and will have made the outcrop an obvious feature to explore. From a distance, the team discerned several rocky ledges, which could either have been natural targets for investigation or had even resulted from such trial works. It was unlikely that the recorded camp had been the base for daily commuting. Instead, another camp site might exist on the red hill in addition to other claim markers, perhaps flagpoles or claim cairns and boards.

4.72 The rust-coloured outcrop of Okerhaugen. (Photo: D. Avango, LASHIPA 5, 2008.)

In the light of Okerhaugen's potential, the question arises why a small pit had also been dug on Svarthaugen. Geologically speaking, there may have been good reason to infer iron ore at this height, but the promontory was small and profitable workings at this location were hard to imagine. Therefore, the pit may have guarded the prospect. As the LASHIPA 5 team had experienced, Okerhaugen was not immediately visible from the lagoon, but Svarthaugen was. Anyone wielding a pair of binoculars from a passing ship or a landing craft would have seen the trial works, the iron ore, and the tools, which gave the message that the place had been investigated, that a natural resource had been found, and that the owners were at the time working the deposit or at least intended to return to work it. This constituted effective occupation on Spitsbergen, although it could not be enforced in the no man's land and relied heavily on honesty and decency. Equipment from Okerhaugen may have been brought to Svarthaugen and placed deliberately around the pit at the end of the season. With hindsight, it is regrettable that the team did not also check whether the iron ore was in fact *in situ* – or whether it had also been brought across and spread around to make a statement.

4.3.1.2 Bruce City

Bruce City, or Brucebyen, is a cluster of buildings nestled between two lakes at the head of Klaas Billen Bay to the west of Gerrit River and Nordenskiöld Glacier, which terminate in Adolf Bay (Fig. 4.73). The first houses were built by the Scottish Spitsbergen Syndicate in 1919 in the hope to discover a coalfield of commercial value. Bruce City and Bruce's hut were names given by the Scots John Mathiesen and Robert Neal Rudmose Brown in 1920.[72] The syndicate influenced the archaeological landscape in Klaas Billen Bay twofold: it left wide-spread physical remains as well as extensive place-name evidence. The LASHIPA 1 expedition of 2004 was able to survey the camp and record two boreholes in the vicinity. Their data was consulted to produce the following narrative.[73]

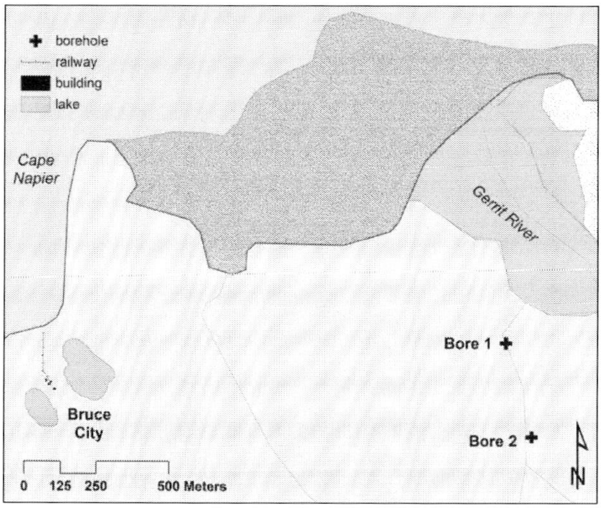

4.73 Location map of Bruce City and two boreholes in Klaas Billen Bay. (Data: LASHIPA 1, 2004; Map: F. Kruse.)

The syndicate may have chosen the location based on earlier investigations and according to the merits of accessibility, resources, and economic potential. It made sense to stay west of Cape Napier, seeing that the water beyond it was much shallower. Nonetheless, larger ships needed to anchor out in the bay while only smaller boats could approach the shore. There were no signs of a dock, but the

[72] Hoel (1942) p. 83.
[73] Martin *et al.* (2006) pp. 64-72.

projected terminus of a narrow-gauge railway hinted at the landing site. A narrow gravel beach lay at the bottom of a marine terrace. The Scots needed to scale the steep slope to get onto the wide plateau above. Bruce City was positioned on a crest between two sizable freshwater lakes.

The railway left the beach at a right angle, but it turned at the foot of the slope to climb it at a lesser gradient. For about a hundred metres, the track led southward before it turned to the south-east. The turn was marked by two switches (Fig. 4.74). If the switches had been carefully planned, they hinted at phased building at Bruce City, whereby the track had in fact originally led southward, where materials were needed first. The rails were then adjusted for a second and last phase. Two mine tubs parked on the track were used for the transport of material and provisions. They were probably pushed manually.

4.74 Railway with two switches leading towards four buildings at Bruce City. (Photo: D. Avango, LASHIPA 1, 2004.)

Following the LASHIPA 1 expedition, *sysselmannen* carried out renovations at Bruce City in 2005 and 2007.[74] The four buildings at the camp were upstanding and serviceable. In Fig. 4.74 above, the building on the far right (1) was known as the Russian hut because it belonged to the Trust Arktikugol. *Sysselmannen* owned the blue-green cabin (2). The third barrack (3) functioned as a scientific station and is partially hidden by the shed (4) to the left of the track. The barracks were of the Kenmore type. They had initially been five pre-fabricated segments long by three segments wide with a pitched roof. However, each had been altered over time and displayed an array of original and modified features.

[74] Hoem, S. (2005) *Rapport istandsetting 2005. Brucebyen, Billefjorden – diverse istandsetting av flere hytter,* Longyearbyen: Sysselmannen på Svalbard; Hoem, S. (2007b) *Rapport istandsetting 2007. Brucebyen, Billefjorden – diverse istandsetting,* Longyearbyen: Sysselmannen på Svalbard.

The Russian hut stood on a gravel foundation, and its pre-fabricated segments were clearly visible (Fig. 4.75).[75] The seaward gable wall had an Arctic entrance. The south-west wall comprised three windows, while the windows in the north-east wall had been blocked. At the rear was a shed, which could be entered from inside and outside. Two chimneys were positioned fairly centrally on the roof.

4.75 The Russian hut at Bruce City in 2004. Left: Arctic entrance and windowless wall. Right: Shed in the rear and three-windowed wall. (Photos: D. Avango, LASHIPA 1, 2004.)

4.76 Bruce City after the fire in August 2010. (Source: Grini, S. O. (2010) Uforsiktig omgang med varme var årsak til at en av fire hytter i Brucebyen brant ned forrige tirsdag. Available at: http://www.svalbardposten.no/nyheter/b%C3%B8telagt-hyttebrann (Accessed: 4 July 2012)).

[75] Martin *et al.* (2006) p. 67.

In August 2010, a group of tourists spent a night in the approximately 90-year old cabin – and accidentally set it on fire.[76] It is only a small comfort that both the LASHIPA 1 expedition and *sysselmannen* had undertaken their respective surveys before the catastrophe occurred. Valuable material remains were lost (Fig. 4.76). The example highlights the fact that cultural heritage is indeed a non-renewable resource in need of timely and meticulous archaeological recording.

Sysselmannen's hut had been completely renovated.[77] If any pre-fabricated segments survived, they were hidden beneath a new exterior. Although the look had been changed, the dimensions still suggested a common origin with the other barracks.[78] The shore-facing gable wall no longer comprised an Arctic entrance, only a small window. The door had been moved to the north-east side, where one large window remained. The other side and the rear also encompassed a window each. There was still a central chimney. The scientific station was comparably simple. Weathering patterns suggested that the gravel foundation had once been higher. The pre-fabricated segments were obvious. There was a door without a porch in the sea-facing gable wall. The north-east side had seemingly never had windows, while the three windows towards the south-west were boarded up. The rear gable wall was featureless. A small chimney was located towards the back of the hut. The shed was of a similar construction style but only two segments long. The shore-facing gable wall may have had two doors or a former door had been moved into the present position. A second door with a tiny window above it was located in the south-west side. The other side and the rear were featureless. The site had seen such intense re-use that most artefacts were inconclusive. *Sysselmannen's* reports showed the typical stencilled mark of the Scottish Spitsbergen Syndicate in the porch of the Russian hut as well as under the roof of the shed.[79] Yet, the mark was noticeably absent from the majority of photographs. Behind the scientific station was a large pile of metal artefacts, which incorporated several lengths of casing and other drilling equipment.

Bruce City was placed back from the sea possibly to protect it from untimely coastal erosion. All buildings were aligned approximately north-west to south-east and positioned slightly off-set. Their front doors faced the sea, while their side windows would either have looked down Klaas Billen Bay, from where

[76] Pospech, A. (2010) *Fire in Brucebyen.* Available at: http://www.photographypospech.com/apps/videos/ (Accessed: 4 July 2012); Kvile, K. E. (2010) *Brannen oppklart.* Available at: http://www.svalbardposten.no/nyheter/brannen-oppklart (Accessed: 4 July 2012); Amundsen, B. (2010) *Bøtelagt for hyttebrann.* Available at: http://www.svalbardposten.no/nyheter/b%C3%B8telagt-hyttebrann (Accessed: 4 July 2012).
[77] Barr, S. and Rosted, A. M. (1982) *Brucebyen, Billefjorden. Registrering av kulturminner på Svalbard. Field report C8-040,* Longyearbyen: Sysselmannen på Svalbard.
[78] Martin *et al.* (2006) p. 68.
[79] Hoem (2005) p. 6; Hoem (2007b) p. 9.

ships with men, materials, and messages would have approached, or into Adolf Bay, which was scenic. Scenery may have been an important asset for an exploration company. Although the buildings stood close to each other, they could not be looked into from neighbouring windows. If the railway is anything to go by, the Russian hut was built before the others. It is perceivable that the management moved into the completed accommodation prior to the workforce, which perhaps remained in tents until the barracks were finished. However, tent rings have not been found. It is unclear if the barracks had separate functions and if these included workshops or stores. The smaller shed may have been an outhouse or store. Lastly, there were no stone cairns or other features that implied flagpoles or claim signs. Besides Bruce City itself, Scottish declarations of territorial occupation did not survive in the archaeological record.

Leaving Bruce City, the LASHIPA 1 team searched for prospecting sites of the syndicate nearer Gerrit River.[80] The first one comprised a length of drilling rod, which stuck vertically out of the ground (Fig. 4.77). The assumption was that it marked a finished borehole. The borehole may have been sunk when the tributary to Gerrit River followed a different course or when its bed was at least completely dry, but since then, water had obscured the form and function of the associated artefacts. Tools and drilling equipment lay in and partially around two wooden boxes, but it was not clear if the scattered wood and the metal wheels had been part of a machine or a cart. Some of the wood was painted with 'No.1 Plant'. A simple tripod made from three round beams had been discarded.

Fig. 4.77 The site of bore 1 comprised a borehole, some equipment, and a rudimentary tripod. (Photo: M. Oglethorpe, LASHIPA 1, 2004).

Some 320 metres upstream, the LASHIPA 1 team recorded a second site (Fig. 4.78). It was widely visible because someone had re-erected the wooden tripod of

[80] Martin *et al.* (2006) pp.69-70.

a more sophisticated drilling rig. Furthermore, there was a small wooden hut. Its walls had been stencilled with labels such as ‘No. 1’ and ‘R. SIDE’ to allow for speedy assembly. It had probably protected an engine and was filled with equipment including a small furnace to repair any damage to the rig on the spot. The concentration of materials piled up inside hinted at a clean-up at the site. The shack had possibly also provided shelter for the drill crew, but it is unlikely that the men slept here. They may have used tents or returned to Bruce City at the end of each shift. Behind the shack lay a substantial pile of drilling rods, mostly three metres long but also some one-metre lengths, and occasional three-metre casing. A wooden box next to it was intended for core samples. Fully packed, it would have been very heavy. Additional wooden boxes were also stencilled ‘No.1 Plant’.

4.78 Bore 2 consisted of a better rig and a hut to protect the engine. (Photo: M. Oglethorpe, LASHIPA 1, 2004.)

4.3.1.3 Gips Bay and Gips Valley

Gips Bay is a wide bay at the mouth of the large, ice-free Gips Valley between Klaas Billen Bay and Temple Bay. The area is named after the gypsum beds in the surrounding mountains.[81] It embraces an array of industrial relics, not all of which date back to the Scottish Spitsbergen Syndicate but are more recent. Rossnes describes that the area was not visited much by hunters; it was, however, claimed by Bruce in 1906 and later transferred to the Scottish syndicate.[82] The Scots built some huts and sank boreholes into the Kulm coal. The first hut in Gips Bay was built by the Swedish AB Isfjorden-Belsund in 1911 and later moved to Mimer Bay.[83] In 1921, the Scots erected a second hut, which eventually decayed. Drilling equipment and a tractor with a trailer were brought ashore on June 29, 1922.

[81] Hoel (1942) p. 153.
[82] Rossnes (1993) pp. 79-81.
[83] Avango (2003); Avango (2005).

Rossnes reproduces a photograph of the Scottish hut flying the Union Jack while two men busied themselves with the vehicle. That same summer, the syndicate constructed a hut in Gips Valley for the investigation of coal.[84] It appears that the Scottish huts in Bünsow Land were not particularly popular among Norwegian hunters. Rossnes only reports one wintering at Bruce City in 1922/23 and the use of the hut in Gips Bay as a subsidiary station in three consecutive winters between 1933 and 1936.

In 2002, *sysselmannen* carried out an impact assessment, which did not take the Scottish sites into account but which described the later activities. Finn Coal Development carried out a drilling programme in Bünsow Land from 1970 until 1980 to assess the natural resources.[85] The firm concluded that ca. 10 million tons of coal could be extracted. In 1982, the Norwegian Ministry of Industry contacted *bergmesteren* with regards to the claims of Finn Coal Development, which also implied the involvement of the Arctic Exploration Co. AS of Oslo.[86] By the end of the decade, the British firm Northern Resources Ltd. had bought the rights and commissioned an extensive environmental survey to be carried out by the Norwegian Polar Institute.[87] Northern Resources intended to commence mining in the early 1990s, but due to financial problems, they quit before any work could be done.[88]

The LASHIPA 3 expedition surveyed Gips Bay and Gips Valley in August 2006.[89] The following site narratives are based on their records. The team followed a road from the shore approximately 17 kilometres up the valley. Along it, they logged a series of boreholes. This chapter concentrates on four areas of Scottish origin (Fig. 4.79). Later remains have been sidelined.

Area 1 was situated at the coast. Presumably, the hut depicted in Rossnes' historical photograph once stood here and guarded the claim. The modern hut encountered was of similar shape and size with an additonal lean-to shed. It was the focal point for other features (Fig. 4.80). Three large, rusty hand drills lay next to it. An old Fordson tractor was in a state of terrible decay.[90] Two two-wheeled

---

[84] In addition, Rossnes lists Scottish huts on the east side of Bjona Haven dating to 1919 and on the north side of Adolf Bay, but these were not visited during the LASHIPA project.

[85] Governor of Svalbard (2002) *Plan for nye verneområder på Svalbard – Konsekvensutredning*, Longyearbyen: Governor of Svalbard, p. 46.

[86] Ulfstein (1995) p. 337.

[87] Brekke, B. and Hansson, R. (1990) *Environmental atlas, Gipsdalen, Svalbard.* (3 vols.) Oslo: Norwegian Polar Institute.

[88] Governor of Svalbard (2002) *Plan for nye verneområder på Svalbard – Konsekvensutredning,* Longyearbyen: Governor of Svalbard, p. 46.

[89] Avango *et al.* (2008a) pp. 17-38.

[90] The tractor was a Fordson Model F first made by Henry Ford in the United States in 1917. (Ford tractors (2012) Available at: http://www.ssbtractor.com/features/Ford_tractors.html (Accessed: 9 July 2012)). In 1918, mass production began to meet a large order from the British Government. From 1919, the Fordson was also manufactured in Ireland. The Fordson was affordable to the average farmer and

carts made by H. Wadsworth & Son of Halifax in England were parked at the foot of a raised wooden platform.[91] The dirt road led from here up into the valley.

Area 2 was situated approximately 12.5 kilometres up Gips Valley. It comprised material remains from both the Scottish and the Finnish period (Fig. 4.81).[92] The Scottish remains centred around a small, wooden hut (Fig. 4.82). The doorway and two windows may have been original, but there had also been some renovation and efforts to better insulate the hut. The portable finds included a sturdy wooden bench with the stencil mark 'SSS Hut 1', the stove 'Modern Mistress No. 28', and an array of drilling equipment. The hut and an associated borehole had been placed on higher ground than the surroundings, where rockhead may have been closer to the surface and where erosion from the passing stream was less likely. Two outhouses were recorded along the edge of the high ground.

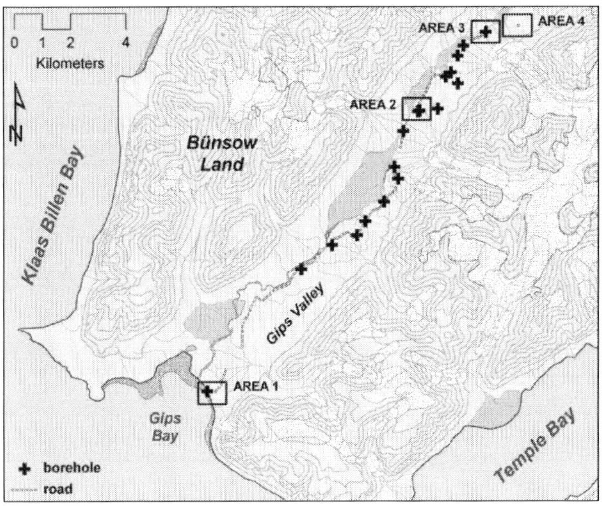

4.79 Location map of four areas of Scottish activities in Gips Valley. (Data: LASHIPA 3, 2006; Map: F. Kruse.)

---

cheaper than horses, thereby changing the industry, but its problems included the lack of weight, which caused wheel slippage in some conditions, and the habit of flipping over backwards if the plough encountered any obstructions. (Wik, R. M. (1972) *Henry Ford and grass-roots America,* Ann Arbor, MI, USA: University of Michigan Press.)

[91] The carts and some machinery at Gips Bay were made by Henry Wadsworth and Son of Halifax in England. Wadsworth was a master wheelwright, who retired already in 1878, after which his eldest son Robert ran the firm until his own death in 1922. (Glover, D. C. (2012) *Cartwright history of Orange Box building.* Available at: http://www.halifaxcourier.co.uk/community/nostalgia/cartwright-history-of-orange-box-building-1-4259949 (Accessed: 9 July 2012)). A connection with the Scottish Spitsbergen Syndicate cannot be proven, but it is notable that this was not a Scottish but an English firm. Perhaps the Scots had exploited the imminent end of the business and purchased this English equipment at a particularly good price.

[92] Avango *et al.* (2008a) pp. 33-7.

4.80 Hut in Area 1 with a tractor, two carts, and a raised platform. (Photo: U. Gustafsson, LASHIPA 3, 2006.)

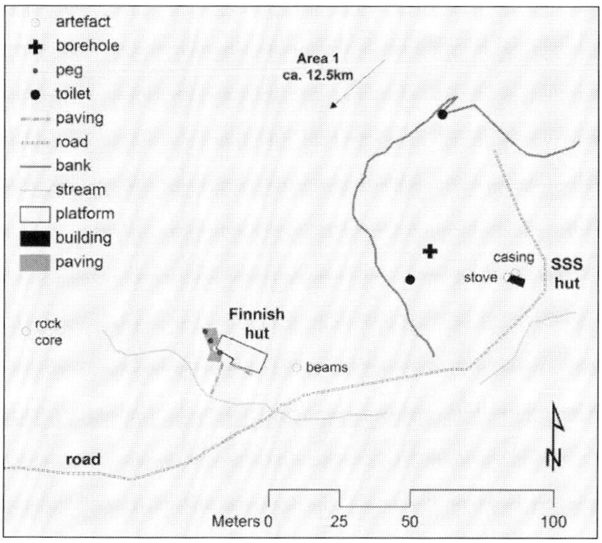

4.81 Site map of Area 2 showing a Scottish hut and a Finnish hut foundation. (Data: LASHIPA 3, 2006; Map: F. Kruse.)

The Finnish remains included a large, rectangular foundation kept dry by a series of shallow ditches. An area in front of the former house had been paved, and a flagstone path led down to the stream. A former employee had communicated the original set-up of the site and the extent of the prospecting activities to archaeologist Dag Avango.[93] A Finnish borehole was not recorded, but a short distance to the west of the hut, rock core was found, which suggested that more than 65 metres of sandstone had been penetrated at this location without coal being encountered. The core was not significant enough to be taken away.

[93] Avango, D. (2011) Conversation with Frigga Kruse, 11 November.

4.82 The Scottish hut in Gips Valley had seen frequent re-use. (Photos: D. Avango, LASHIPA 3, 2006.)

The remains in **Area 3** mirrored that encountered at the second prospecting site near Bruce City (Fig. 4.83).[94] The same three-sided shack, no door, no window, was labelled 'No. 2'. Both sides and the back were again pre-fabricated and individually marked for easy assembly. The whole had once been packaged in 'Box No. 24'. The roof had collapsed, and someone had used a thick wire to keep the hut in place, but it will not be long before it crumbles.

4.83 Erosion threatens this Scottish prospecting site in Gips Valley. (Data: LASHIPA 3, 2006; Map: F. Kruse.)

As before, the limited interior space was cramped with equipment and materials. Among the items was the familiar small furnace, which had even retained its bellows. The metal wheels of mine tubs stood out because indications of a narrow-gauge railway had not been found in Gips Valley. They were believed to have been salvaged scrap metal to be smelted for quick repairs on site. Other finds such as a

[94] Avango *et al.* (2008a) pp. 31-2.

workbench, machinery parts, hand tools, and an oil can verified that heavy work had taken place here and that the shed was not meant to accommodate people.

The hut was open towards the side where drilling had occurred. The borehole was marked by a metal pipe sealed off with a piece of wood. At the time of drilling, the ground would have been flat and stable. Since then, the Gips River had changed its course, and the Scottish remains were threatened by erosion. Between the shed and the borehole lay the beams of the tripod. Two low trestles suggested that heavy rods and casing were kept off the ground during the operations. Behind the hut was a large collection of three-metre drilling rods and casing. Several narrow wooden cases seem to have been core boxes specially made for indicative one-metre lengths of core such as layers of coal. No core was found in the surrounding, nor were tent rings or other forms of dwelling. Finn Coal Development may have re-used the site for their purposes, perhaps for the temporary storage of equipment. Eight red jerry cans, on the other hand, may have been placed at this recognisable landmark by the drivers of snow scooters.

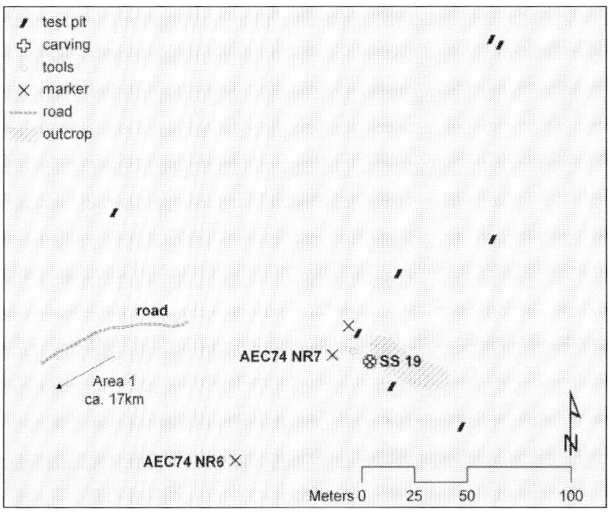

4.84 Area 4 centred on a rock outcrop with a claim marker. (Data: LASHIPA 3, 2006; Map: F. Kruse.)

Roughly 17 kilometres from the coast, **Area 4** was the furthest inland locality surveyed in the LASHIPA project (Fig. 4.84).[95] Although the team was aware of other sites throughout Gips Valley, they could not reach these due to natural obstacles and time constraints. The site did not encompass any noticeable huts or tent rings. Instead, it was centred on a rock outcrop with the eye-catching engraving 'SS 19' (Fig. 4.85). The symbol had been pecked out and most likely

[95] Avango *et al.* (2008a) pp. 28-31.

marked a Scottish claim. The team inferred that the number stood for the year 1919, but it could also be a consecutive number in a system, implying that there were 18 others and that there may have been more. As with the prospecting pit on Prince Charles Foreland, tools such as a sledgehammer, two pickaxes, and a pry bar had been placed around it. Furthermore, there were the discarded metal pole of another marker and a galvanized food container that could be kept warm with hot water.

4.85 The engraved outcrop and the tools functioned as a claim marker. (Photo: D. Avango, LASHIPA 3, 2006.)

4.86 The pit proved that coal existed at a shallow depth 17 kilometres into Gips Valley. (Photo: D. Avango, LASHIPA 3, 2006.)

Around the outcrop of sandstone were several test pits, which had unearthed coal at very shallow depths (Fig. 4.86). The prospectors may have obtained the necessary geological information from the outcrop, which made boreholes superfluous at this location. Whether they considered this to be a favourable position for a coal mine could not be ascertained. That this location was of interest to different waves of prospectors can be seen in the relics of at least two broken claim markers, which read 'AEC74 NR6' and 'AEC74 NR7'. AEC stood for the Arctic Exploration Co., the partner of Finn Coal Development, and the year in which the numbered boreholes were sunk.

4.3.2 Site interpretations

The site narratives once again disclosed environmental and archaeological actants and actors, this time in the network of the Scottish Spitsbergen Syndicate (Fig. 4.87). It is immediately noteworthy that despite the syndicate being active for over four decades in the changeable context of global economy and politics between the years 1909 and 1953, this representation of its network is not overly crowded.[96]

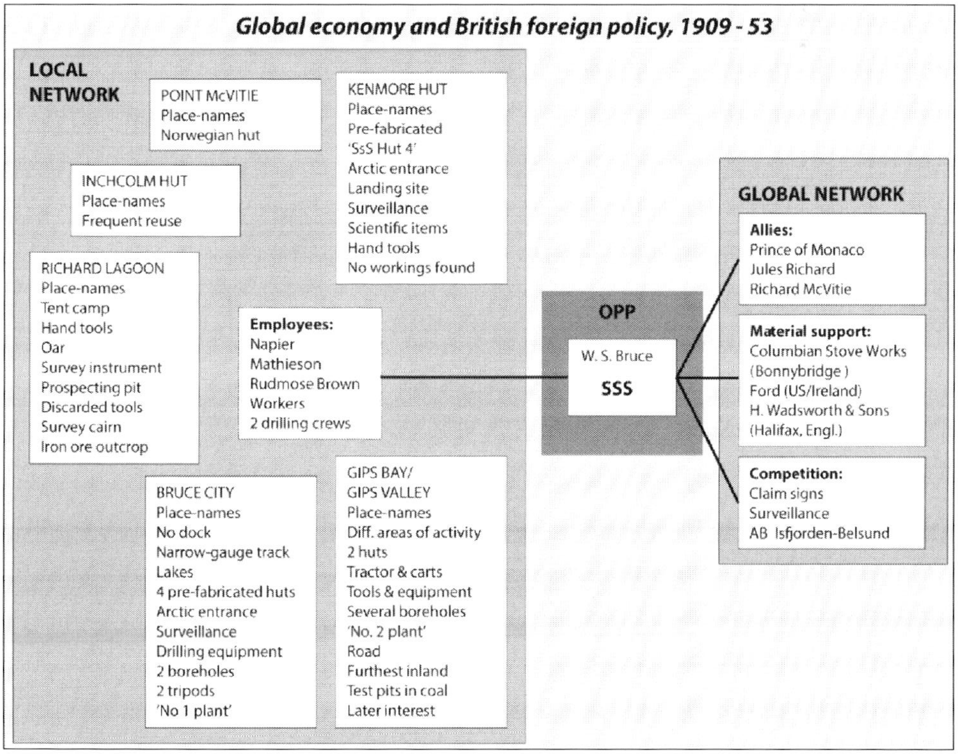

4.87 Selected environmental and archaeological components of the actor-network of the Scottish Spitsbergen Syndicate (SSS). (Chart: F. Kruse.)

[96] Some known sites have not been visited in the LASHIPA project, but these would not have changed the graphic significantly.

The archaeological record reveals little about the composition of the syndicate. The fact that William Speirs Bruce was initially at its centre is known from historical sources. Similarly, the names of employees such as Napier, Mathieson, and R. N. R. Brown appeared in Hoel's *Place-names* during the preparation phase of the fieldwork. The archaeological signature of the workforce is faint, especially in the light of camps and huts not being representative of mobile prospecting parties, who may have stayed in tents. Yet, tent rings were practically non-existent. However, it is clear that the syndicate made use of specialists in the form of at least two drilling crews to operate the three proven drilling rigs. It is notable that no local allies of the syndicate could be identified. There appears to be no connection with earlier Norwegian site users or contemporary trappers.

Bruce's interest in the Arctic no man's land stemmed from his science. His base camps on Prince Charles Foreland in 1906 and 1907 were surveyed during the LASHIPA project, but they were inconclusive. No direct evidence for mineral prospecting or territorial occupation could be found. Since an early hut at Kenmore had not survived, physical remains dating to before the First World War are extremely rare. The remains dating from after the war indicate an economic interest in the respective areas. The natural resources investigated included iron ore on the Foreland and coal in Klaas Billen Bay and in Gips Valley. Although the latter also comprised an abundance of gypsum, the LASHIPA project did not record that the syndicate also targeted this.

Whether Bruce or the syndicate had a political interest in the archipelago could not be discerned archaeologically. The syndicate certainly erected Bruce City and other huts for the practical purpose of shelter and storage. The buildings would also have been an expression of ownership of the surroundings and the resources. However, there were only few huts, and none gave the impression of being only a claim hut. Yet, the syndicate engaged in far-reaching topographic surveys, which resulted in a wealth of Scottish place-names. If needed, the toponyms would have been very effective tools to stress British efforts and rights on the islands.

The topographic surveys, which may have produced useful base maps, will have entailed mineral prospecting. Unlike the Northern Exploration Co., the syndicate did not only open up promising surface deposits. At Bruce City and in Gips Valley, it chose to investigate the coal at depth. It therefore sunk a series of boreholes, which will have produced reliable information about the quality and extent of the coalfields that would not have been revealed by trial trenching only.

The syndicate seemingly did not waste funds on unnecessary surface works, and sinking boreholes was a cost-effective method. Mismanagement and overspending are therefore unlikely reasons as to why operations were discontinued. It is possible that the coalfields and other resources were simply uneconomical. It is also possible that the syndicate had in fact done everything an exploration company could do to prove the minerals on its claim. If it did not convert to active mining itself, it needed a subsidiary company or a buyer.

Information about its global network is sparse. Some backers have been identified from place-name evidence, but these originated from Bruce's scientific network. The manufacturers evident in the archaeological record may just have been suppliers without an active role in the business. Competitors have only been revealed indirectly through the syndicate's use of claim markers and related symbolism. Despite previous works attesting to the claim dispute with the Swedish AB Isfjorden-Belsund, later AB Svenska Spetsbergen, these could not be verified by archaeological methods.

The syndicate lasting for over four decades possibly means that either economic actors continued to provide funding despite the fact that Spitsbergen did not deliver the promised raw materials, or that the firm had no creditors to pay off. Perhaps the syndicate had lost interest when it did finally sell its properties. The sale presumably nowhere near recovered the original outlay.

4.4 Conclusion

The archaeological image of British mineral exploration on Spitsbergen is composed of approximately 40 sites of two exploration companies, the Northern Exploration Co. and the Scottish Spitsbergen Syndicate. With the exception of the extreme north, these sites were situated all along the west coast of the island of Spitsbergen. Although this coast is deemed the most accessible part of the archipelago, the companies' spheres of influence were so great that they will have experienced the full range of environmental peculiarities at the different locations. Adaptations, however, were infrequent.

On Spitsbergen, there was no limit to the size of a claim, provided a company could express effective occupation and defend the property against claim jumpers. The wide spread of especially the Northern Exploration Co. was a strategy to occupy large areas as quickly as possible in order to beat competitors to the potential natural resources. One reasoning behind it would have been that if an area proved barren, there would surely be economic minerals in another. Additionally, the company with the largest holdings could hope to influence the sovereignty of Spitsbergen in its favour.

The challenge lay in unquestionable occupation. The Northern Exploration Co. relied on extensive settlements, camps, and claim huts, some with vague practical purpose and not always associated with workings. In addition, it erected countless claim signs. These arguably lost much of their effectiveness by being dated to 1905, five years prior to the company's formation. Nonetheless, it put great efforts and resources into establishing a chain of symbols along the west coast and presumably on the east coast, though no remains have been found.

Based on archaeological materials, the claims of the Scottish Spitsbergen Syndicate were confined to Prince Charles Foreland, Klaas Billen Bay, and Gips Valley, which were not as large as those of the Northern Exploration Co. but still substantial. Many of the same symbols of occupation were evident but on a much lesser scale. The syndicate was not as showy and refrained from unnecessary huts or surface works. Instead, it carried out far-reaching topographic surveys and bestowed its areas with place-names still in use today. It put much emphasis on mapping to express the rights to its claims.

The firms' claims and facilities not only needed to address effective occupation and thorough exploration, they also needed to be attractive to potential subsidiary companies or buyers. Especially those sites of the Northern Exploration Co. that verged on active mining must be seen in this light. The company went to great length, even involving American machinery, to prove the deposits short of reaching the producing stage itself. The Scottish Spitsbergen Syndicate was less extravagant, seemingly never wintered, and relied instead on data provided by its boreholes.

The archaeological record reveals two exploration companies of very different character and strategy. Being exploration companies, they presumably never aimed for their own permanence on Spitsbergen. Their buildings were sufficient for prospectors and to guard the claims, and their workings proved the resources but never reached production. The products of the exploration companies were the claims themselves as well as the information about the resources, their workability, and their potential profitability. Ultimately, however, these products were inadequate to attract third parties. After the companies abandoned the sites, the claims were never worked. Reasons for this now need to be sought in the historical sources.

PART III

THE HISTORICAL IMAGE

5 The Spitzbergen Coal & Trading Co. Ltd. (1904-18)

5.1 Introduction

Chapters 5 to 8 present the results of extensive archival research. This chapter is based on the contemporary documents and secondary literature concerning the Spitzbergen Coal & Trading Co. Founded in 1904, the company was the first of the British firms to exploit the mineral resources on the archipelago and the only one with a predecessor. The section on the company's formation therefore includes a brief account of the preceding Norwegian activities before Englishmen initiated a new business and provides a chronological overview of subsequent events. These events are characterised firstly in terms of the global network followed by an assessment of its local counterpart. Combined with the archaeological image of Chapter 3, the historical data creates the overall picture of the company's development. The discussion returns to the questions why the Spitzbergen Coal & Trading Co. had been attracted to Spitsbergen, how it chose to proceed, why it was dissolved, and what it ultimately achieved.

5.2 Formation and chronological overview

The Norwegian predecessor was the A/S Bergen-Spitsbergen Kulgrubekompani or simply Bergen Co.[1] In 1901, seven men arrived on the northern shore of Advent Bay and located coal on the hillside approximately 100 metres above sea level. Using metal wire, the party fenced off a claim and erected wooden boards along its boundary. The main board read that the property had been occupied on July 22 and that it was delineated by a boundary between Advent Bay and Sassen Bay. In addition, the men built a shed to store tools and explosives. They then left for Kings Bay, where they found coal on the southern shore and took possession of it in a similar manner. The total output that year was five tons of coal, which was taken to Norway for tests.

In 1903, an expedition of 15 under the leadership of the Norwegian engineer Stener August Fangen arrived at Advent Bay at the beginning of June. The group anchored close to the place that would become Advent City. The water was too shallow for the ship to approach the shore, so equipment and supplies were discharged in rowing boats, which was an unassuming task in calm weather (Fig. 5.1). Although a kitchen and a mess tent were put up (Fig. 5.2), the men mainly lived and slept on board.

[1] Unless otherwise stated, details of the Bergen Co. from: Hoel (1967a) pp. 553-84.

5.1 Unloading stores in Advent Bay in calm weather. (Source: Photo Library np003420, Norwegian Polar Institute, Tromsø.)

5.2 Kitchen and mess tent of the Bergen Co. Note the narrow-gauge railway track in front. (Source: Photo Library np003149, Norwegian Polar Institute, Tromsø.)

During the summer, the men drove an adit of 80 metres into the coal seam. They levelled the gradient at the mine and built simple retaining walls (Fig. 5.3). The pithead arrangements included a railway track into the mine, some winding gear, and a small smithy. The simple ropeway (Fig. 5.4) transported the coal downhill to the coast. By the end of the season, the Bergen Co. had extracted approximately 40 tons of coal. Some filled the stores of a Hurtigruten ship. Some was used on the expedition's own journey from Tromsø to Bergen. An amount of coal also fuelled a locomotive on the Bergen-Voss railway, and the remainder was sent to Bergen's gas plant for further tests.

5.3 Levelled space at pithead creating room for railway tracks, winding gear, and a stone-built smithy. (Source: Hoel (1967a) p. 560.)

5.4 Lower terminus of the ropeway, coal dumping being in progress. (Source: Hoel (1967a) p. 557.)

Armed with favourable reports, Fangen travelled to England in autumn 1903. He had formerly been employed by Emerson Muschamp Bainbridge and now met the coalowner in London to promote the Arctic coal mine. Besides making a financial contribution, Bainbridge put the Norwegian in touch with colleagues and friends, who would be instrumental in the formation of the new company. Of these, Jacob Kruuse Müller Hessler was a ship owner and timber merchant as well as the Norwegian consul in West Hartlepool in North East England. William Black had been one-time secretary and manager at Nunnery Colliery Co. Ultimately, Fangen was able to raise about £4,000 in England.[2] The money was most likely intended for a new expedition in 1904.

Following the meeting with Fangen, Bainbridge felt that the Arctic coal mine was worth pursuing and initiated a mining company. The application to the Board of Trade was signed by Bainbridge, Black, and the colliery agent William Austin Marshall Toyne on May 11, 1904.[3] Bainbridge resided in London, but Black, who would be the managing director, was a Sheffield man. The Spitzbergen Coal & Trading Co. was therefore registered in Sheffield.[4] Besides Bainbridge and Black, the board of directors also comprised the solicitor William Edwin Clegg.[5] The colliery clerk Edwin Vickers Weston carried out his secretarial duties at Kings Chambers on Angel Street in the town centre. The Treasury approved a nominal capital of £2,500, which was divided into 2,494 ordinary shares of £1 each and 120 founders' shares of 1s each.[6]

While the new expedition sailed for Spitsbergen, Black signed an agreement with the Spitzbergen Coal & Trading Co. on June 17, 1904.[7] He was acting on behalf of those who had previously invested in the Bergen Co. The shareholders had transferred their rights to Black, who in turn sold all property, assets, and interests of the adventure to the firm.[8] A condition of the agreement was that the Spitzbergen Coal & Trading Co. would issue fully paid up ordinary shares to the former subscribers, the amount being representative of their previous commitment. Accordingly, 1,969 ordinary shares were issued, of which Norwegian investors received 62 per cent, while the residual 38 per cent went to British subscribers. Although the agreement did not specify, which possessions had been

---

[2] Braastad, J. (nd) *Oplysninger om aktieselskapet Bergen-Spitsbergen og the Spitsbergen Coal and Trading Co. Innsamlet hos ing. S. A. Fangen, Kiberg.* Norwegian Polar Institute Box 85, Regional State Archive, Tromsø.

[3] *Application for incorporation* (1904) Spitzbergen Coal & Trading Co. BT31/17239/81000, The National Archives, Kew.

[4] *Certificate of incorporation* (1904) Spitzbergen Coal & Trading Co. BT31/17239/81000, The National Archives, Kew.

[5] Clegg was noted for football, politics, and temperance in Sheffield society. He and his brother were the first brothers capped for England. In his prime, he was a solicitor in the family firm Clegg & Sons as well as a Liberal politician opposed to socialism and the Labour party. He rose to Lord Mayor of Sheffield in 1898 and was among the names that dominated the history of the British Temperance League (BTL), a North of England teetotal and Christian society. (Addy, S. O. and Pike, W. T. (1900) *Sheffield at the opening of the 20th century: contemporary biographies.* Held at: Central Library, Sheffield; *The Times* (1932) 'Obituary. Sir William Clegg.', 23 August, p. 12; Blocker, J. S., Fahey, D. M., and Tyrrell, I. R. (2003) *Alcohol and temperance in modern history. A global encyclopedia. Google books* [Online]. Available at: http://books.google.com/books (Accessed: 6 September 2011)).

[6] *Statement of nominal capital* (1904) Spitzbergen Coal & Trading Co. BT31/17239/81000, The National Archives, Kew.

[7] *An agreement between William Black and the Spitzbergen Coal & Trading Co.* (1904) Spitzbergen Coal & Trading Co. BT31/17239/81000, The National Archives, Kew.

[8] The formal use of the word 'adventure' was presumably fast going out of fashion; it did not reappear in subsequent documents of this or later companies.

signed over and although ownership was seemingly retained by the Norwegians, the Spitzbergen Coal & Trading Co. was ready to commence business.

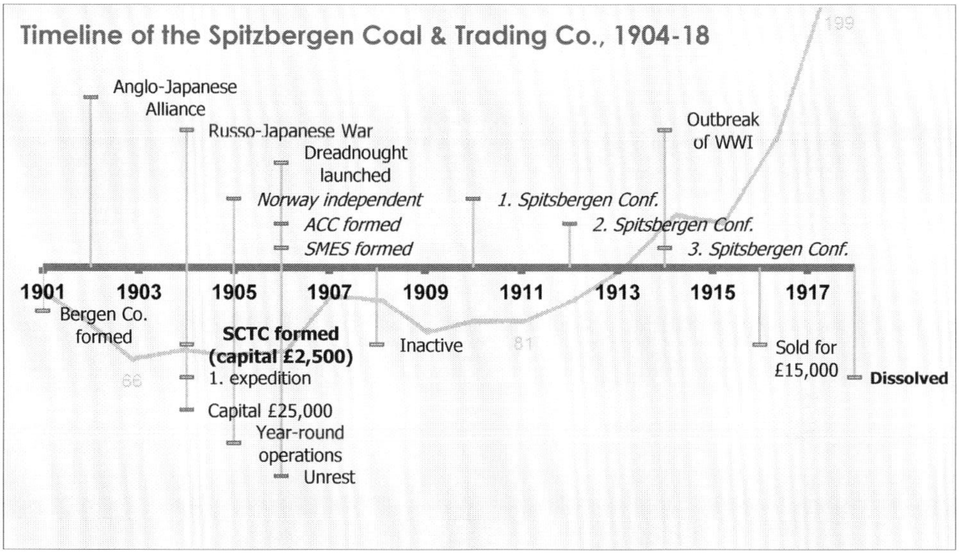

5.5 Timeline of the Spitzbergen Coal & Trading Co., 1904-18. Events below the bar are company-specific. Above the bar, events in italics are relevant to Spitsbergen, while others are thought to have defined the global context. The grey line indicates the indexed British coal price movements, whereby the bar denotes 1913 = 100 and the bottom edge approaches 0. Actual percentages have intermittently been added for clarify. (Sources for the coal price development, see Fig. 2.2 and Fig 7.28; Chart: F. Kruse.)

The chronological overview in Fig. 5.5 summarises the subsequent events. The global context was marked by political upheaval. The Anglo-Japanese Alliance, the Russo-Japanese War, and the launching of the first *Dreadnought* battleship signify Britain's emergence from political isolation, a shift in the European balance of power, and the intensification of the arms race with Germany, respectively. All are thought to have contributed to the outbreak of the First World War. Meanwhile, the Swedish-Norwegian union had ended, which fuelled the aspirations of both countries to be the foremost Arctic nation. Shortly afterwards, the Arctic Coal Co. and the Spitzbergen Mining & Exploration Syndicate were founded, which increased the presence of American and British actors in the no man's land. Three Spitsbergen conferences took place to settle the legal status of the islands, but the third had to be postponed indefinitely at the outbreak of war. Economically, the British coal market had barely stabilised after the peak in 1900, with a temporary high in 1907 and 1908, before the rise into abnormally high prices during the war began. A strong British market with correspondingly steep export rates (export being restricted during the war) meant that importers like northern Norway looked for cheaper alternatives. In those periods, Spitsbergen coal was potentially good business. Against this background, the Spitzbergen Coal & Trading Co. sent an expedition in 1904 and increased its capital to £25,000 later in the same year. Year-round operations commenced with the expedition in 1905, followed by unrest that culminated in a strike in winter 1906/7. From the end of summer 1908, the company was inactive until it sold its properties to a Norwegian entrepreneur during the war. After 14 years in existence, it was eventually dissolved in 1918.

5.3 The global network

5.3.1 Economic actors

During company formation, Black had obtained the 120 founders' shares, which he in his double role as the company's managing director used to build up the corpus of English shareholders. These founders' shares were completely allotted by June 20, 1904.[9] If the number acquired equalled a person's interest, Black was very confident, retaining 22 shares.[10] His lead was followed by the silk manufacturer Fred Davenport (12), the steel manufacturer Robert Colver (10), the iron merchant Robert Smith (10), and Bainbridge (10). Thereafter, new investors had subscribed for less. The Norwegians only obtained a total of six founders' shares between them. They showed very little continued interest in Advent Bay, and their further involvement is in fact invisible in the historical sources. That these financial transactions were essentially designed to transfer certain rights and barely met the company's immediate monetary needs showed the application for a first mortgage over £3,000. This was granted on June 22, 1904.[11]

After the expedition's return, Fangen continued to promote the Arctic development. This task presumably fell to the Norwegian as opposed to the directors because he had first-hand knowledge. With new reports from Bergen's gas plant, the engineer travelled to England in October and once more succeeded to enrol support.[12] At the end of the month, the shareholders decided to increase the nominal capital to £25,000, and the Treasury granted the issue of 22,500

[9] *Return of allotments* (1904) Spitzbergen Coal & Trading Co. BT31/17239/81000, The National Archives, Kew.
[10] Which nevertheless amounted to 22s or £1 2s only.
[11] *Mortgage* (1904) Spitzbergen Coal & Trading Co. BT31/17239/81000, The National Archives, Kew.
[12] Hoel (1967a) p. 558.

ordinary shares at £1 each.[13] Any money raised during their allotment most likely supported the expedition in 1905, which witnessed Fangen's replacement by Percival Muschamp and the first wintering by any mining company in 1905/6. Since returns would not be instantaneous, the company applied for another mortgage over £6,000 from the Union of London and Smiths Bank in November 1905.[14] It thereby accrued a debt of £9,000, but this second mortgage could already be satisfied in August 1906.[15]

A first disruption to the board occurred when Black died in June 1906.[16] At the general meeting held in September, Bainbridge also no longer appeared to be a director. Instead, Clegg had been joined by Ernest Douglas Black, who replaced his late father as managing director, the silk manufacturer Fred Davenport, who had originally invested in the Bergen Co., the estate agent Henry Coverdale and the steel manufacturer Robert Colver, who had both acquired founders' shares in 1904, and the steel manufacturer George Senior.[17] During the meeting, the directors and shareholders agreed to provide the company with the operational capital of an additional £10,000. This was to be issued as 400 preference shares at £25 each. The company did not apply to the Treasury this time, probably because it did not expect its approval. Instead, it created a third mortgage over £10,000.[18]

After the strike in winter 1906/7, Ernest Black took charge at Advent City. The company's resources were nearing exhaustion, and the mine had so far failed to provide a paying output. The company therefore issued a memorandum to reinforce the bond with its debenture holders and shareholders.[19] It estimated that the claim would last 20 years at an output of 50 thousand tons per year. A heading had been driven into the seam without any mining or engineering problems. There were sufficient plant, machinery, equipment, and houses for accommodating and feeding the staff, the efficient working of the colliery, and the delivery of coal to passing steamers and whalers. 70 persons, including a doctor, a manager, and an under-manager, had recently been employed on site, which had been proven safe to winter. At present, the directors forecast an output of 1,000 tons per week, which a second coal cutter could double. Their agent in Norway had connections with the country's commercial and shipping industries, and although recent orders had been

[13] Hoel (1967a) p. 558; *Notice of increase in the nominal capital* (1904) Spitzbergen 1897-1905 FO83/2147, The National Archives, Kew.
[14] *Mortgage* (1905) Spitzbergen Coal & Trading Co. BT31/17239/81000, The National Archives, Kew.
[15] *Summary of share capital and shares* (1906), Spitzbergen Coal & Trading Co. BT31/17239/81000, The National Archives, Kew; *Satisfaction of a mortgage* (1905), Spitzbergen Coal & Trading Co. BT31/17239/81000, The National Archives, Kew.
[16] *London Gazette* (1906) 'William Black, deceased', 9 October, p. 7043.
[17] Hoel (1967a) p. 563.
[18] *Mortgage* (1905) Spitzbergen Coal & Trading Co. BT31/17239/81000, The National Archives, Kew.
[19] *The Spitzbergen Coal & Trading Co. Limited. Memorandum.* (nd) William Olssons Archives 1899-1922 F2:20, National Museum of Science and Technology Archives, Stockholm.

cancelled due to unforeseen circumstances, the directors were confident that they would soon be able to deliver large quantities of coal at good prices. They had already spent £50,000, and another £25,000 was necessary to pay off existing debts and carry on the development. In view of this shortage, the directors now offered the debenture holders and shareholders three options: 1) they could issue a prospectus and obtain the necessary capital from the public; 2) they could waive their rights and give a first charge on the undertaking to anyone willing to advance the capital; or 3) the company could dispose of the lot on terms.

The public was never approached, nor did anyone advance the capital. The company had in fact already begun to look for potential buyers, among them the Swedish Government. Albert Senior, the director's son, had recently been to Sweden on other business and now wrote to the prime minister on behalf of the company.[20] The directors were thinking of selling, 'owing to the difficulties they have had with the workmen out there, during the past winter. Unfortunately, those in charge do not understand Norwegian or Swedish, and they have been unable to directly address the men. Furthermore it is a long way off from Sheffield, and the directors now find [it] very inconvenient in every way.'[21] Yet, there was coal ready for shipment, and if the Swedish Government was interested, the company could make a good offer. Senior thought it advisable for Sweden to have a coalfield near Narvik.[22] Although the islands were not navigable all year around, the Swedish would be considerably closer to operations than England, could communicate with the workforce, and understood prevailing conditions. If the prime minister were not interested and could not recommend a Swedish syndicate, the directors would make the same offer first to Norway and then to any others. A sale was not effected, but the company closed a deal with the Swedish Government to deliver 6,000 tons of coal to the railways at Narvik, 4,000 tons at a price of 16s 9d per ton, another 2,000 at 16s 3d.[23] Although the wintering in 1907/8 only produced 3,500 tons, the quality of the coal was said to have improved.[24]

A balance sheet (Fig. 5.6) reveals the precarious financial situation of the Spitzbergen Coal & Trading Co. at the time the mine was discontinued. Typical for a small business, the company had not dealt entirely in cash but had built up an inventory of properties and goods. Such businesses usually owed money to suppliers and authorities, but even if they wanted to, their assets were not easily turned into cash. The company listed its shareholders' equity in the form of capital

[20] Senior, A. (1907) *Letter to Arvid Lindman, 9 August*. Spetsbergenarkivet, The National Archives, Stockholm.
[21] Senior (1907).
[22] Narvik was the largest harbour in North Norway with a direct railway link to the iron ore mines approximately 180 kilometres away in Kiruna in Sweden.
[23] Dole (1922a) p. 359.
[24] Dole (1922a) p. 357.

issued. Its liabilities comprised the long-term debt. In addition, it had accrued expenses and loans for uncertain reasons such as money owed to U. Aagaard, the British vice-consul in Tromsø, and the large sum borrowed from the Sheffield Hallamshire Bank. The company's assets included fixed assets such as the coal claim, plant, buildings, and stores, a schooner being listed separately. The involvement of individuals such as Aagaard, John Bruvik, the company's agent in Bergen, and the former manager Muschamp as well as the Tromsø Privatbank and the Neptune Steam Navigation Co. may be indicative of investments or loans but are largely intangible. The bottom line was that the company had spent more than £55,550. Its net worth was only the equivalent of the capital that had been issued: £24,577. It therefore owed roughly £31,000, which the directors needed to recover before the company could wind up respectably.

Liabilities	£	s	d		Assets	£	s	d
Capital issued (less calls in arrear)	24577	0	0		Property and assets	4792	7	7
Debentures	10200	0	0		Further expenditure, debenture, interest (less coal sold)	35700	14	9
Sundry loans	2056	4	0		U. Aagaard	2881	0	11
U. Aagaard	304	10	0		Tromsø Privatbank	3935	13	5
Sundry creditors	824	2	10		J. Bruvik	4195	15	10
Sheffield Hallamshire Bank Ltd.	17591	6	9		Neptune Steam Navigation Co.	1722	15	3
					P. Muschamp	26	17	4
					Schooner 'Heloise'	760	11	7
					Sundry debtors	933	12	7
					Cash in hand	3	12	4
	55553	3	7			55553	3	7

5.6 The balance sheet reveals the financial situation of the Spitzbergen Coal & Trading Co. by September 1908. (Source: Balance sheet (1909) Spitzbergen Coal & Trading Co. BT31/17239/81000, The National Archives, Kew.)

Due to a lack of funds, the company's determination stagnated, their actions became few and far between, and the situation got increasingly worse. Already in September 1909, the British vice-consul in Bergen had reported of a scandalous state of affairs.[25] The company had been unable to pay two other British vice-consuls their dues. Kjeldsberg in Trondheim was owed £55 and Aagaard in Tromsø £650. Furthermore, it had been unable to pay the workmen's wages, reducing their families to difficulties. Although frequent applications for payment

[25] *Memorandum by Mr. Hertslet respecting British claims in Spitzbergen* (1910) Foreign Office FO881/9813X, The National Archives, Kew.

had been made to Sheffield, none was forthcoming. The local press in Tromsø stated that Norwegian watchmen, who had guarded Advent City for two winters, had not received their wages either. The men were now taking over the movable possessions of the company.

In the period between 1911 and 1916, the company only just fulfilled its legal obligations. It duly filed summaries of its share capital with the Board of Trade, but these did not register any changes, and the long-term debt persisted.[26] Bainbridge had died on May 12, 1911, and the British Foreign Office enquired whether his heirs intended to uphold his claim on the islands.[27] If so, they needed to make a statement as to the boundaries of the territory supported by maps. A similar request was sent to the Spitzbergen Coal & Trading Co. in March 1912.[28] In the eyes of the Foreign Office, the company's insolvency was fast becoming a liability to British foreign relations. The ministry understood that it was the generally accepted custom on Spitsbergen that a claim would lapse if it had not been worked over a period of two or three years. Thus, the company's claim had seemingly lapsed. Unless the directors planned to start up work again immediately, the Foreign Secretary advised them to come to terms with one of the recent Norwegian contenders. He strongly recommended that the company pay their Norwegian creditors without delay. The existing debt 'only served to strengthen the contention that they were not in a position to work the properties which they originally took up in Spitzbergen, and that therefore any title they might have had to those properties could be held to be no longer valid.'[29]

In October 1912, the company replied that it had not abandoned the claim.[30] In its temporary absence, the Arctic Coal Co. had been acting on its behalf. It did not envisage any problems with the Norwegians Carl Meyer and Frederik Hiorth. The latter was in fact a potential buyer, and the directors were in direct communication with him. While they were trying to come to an agreement, the Arctic Coal Co. had been asked to provide caretakers for Advent City. Although the company promised to satisfy the Norwegian loans in due course, by the end of November 1912, the British minister in Norway informed the Foreign Office that the company would no longer be assisted for as long as it owed money to several consular officers. If the Foreign Office had any further contact with the company, it should press for the payment of these debts. If the company did not oblige, the department should refuse to support it.

---

[26] Spitzbergen Coal & Trading Co. BT31/17239/81000, The National Archives, Kew.

[27] *Memorandum by Mr. Hertslet respecting British claims in Spitzbergen* (1913) Foreign Office FO881/10276, The National Archives, Kew.

[28] *Memorandum by Hertslet* (1913). The company only complied in April 1913.

[29] *Memorandum by Hertslet* (1913).

[30] *Memorandum by Hertslet* (1913).

During the war, the property and interests of the Spitzbergen Coal & Trading Co. were eventually taken over by A/S De Norske Kulfelter Spitsbergen.[31] The Bergen-based firm had been founded by Frederik Hiorth, who in May 1916 travelled to England and agreed to a purchase sum of £15,000.[32] The Norwegian proposed to use the English plant and houses at a more sheltered site further into Advent Bay. He would not work the same coal.[33] Hiorth intended to open up a coal seam above the 600-metre contour instead. Interestingly, the higher seam showed a fault above Advent City, which may have had repercussions for the one worked by the Spitzbergen Coal & Trading Co. Unlike the English, the Norwegians were going to construct a substantial dock. Besides the indication of how the plant at Advent Bay came to be reused elsewhere, the history of this Norwegian company, however, is inconsequential to this book.

In a general meeting on August 22, 1917, the stakeholders finally decided to dissolve the Spitzbergen Coal & Trading Co.[34] A liquidator was appointed in October. After an existence of almost 14 years, the company was crossed off the register on January 9, 1918.[35]

5.3.2 Political actors

Bainbridge initially contacted the Foreign Office after the expedition's return in autumn 1904. It was in the interest of the Spitzbergen Coal & Trading Co. to ascertain the legal status of its claim in Advent Bay. In September, Bainbridge therefore informed the Foreign Secretary Lord Lansdowne[36] that he was aware that the islands belonged to no one and that, in light of investments being at risk there, he was anxious to know whether the Foreign Office could afford the shareholders any protection against other countries.[37]

The Foreign Office launched an internal investigation, which showed that three decades ago the Governments of Sweden and Russia had been inclined to claim certain rights if not actual sovereignty over Spitsbergen.[38] At the time, the British Government had assured Sweden that it had no objections to a Swedish

---

[31] *A/S De Norske Kulfelter Spitsbergen* (1916) De Norske Kulfelter, Norwegian Polar Institute, Tromsø.
[32] Hoel (1967a) p. 572.
[33] *A/S De Norske Kulfelter Spitsbergen* (1916) De Norske Kulfelter, Norwegian Polar Institute, Tromsø.
[34] *Resolution* (1917) Spitzbergen Coal & Trading Co. BT31/17239/81000, The National Archives, Kew.
[35] Hoel (1967a) p. 573.
[36] Henry Petty Fitzmaurice, 5th Marques of Lansdowne, was Foreign Secretary between November 1900 and December 1905. During his term, he brought Britain out of isolation by signing the Anglo-Japanese Alliance in 1902 and negotiated the Anglo-French Entente Cordiale in 1904. He witnessed the Russo-Japanese War (1904-5) and the dissolution of the union between Norway and Sweden (1905).
[37] Bainbridge, E. M. (1904) *Letter to Lord Lansdowne, 27 September.* Spitzbergen 1897-1905 FO83/2147, National Archives, Kew.
[38] *Mr Bainbridge. 27 Sept. 1904* (1904) Spitzbergen 1897-1905 FO83/2147, National Archives, Kew.

annexation, provided that its fishing rights were observed. Although the Russian Government disapproved of a Swedish occupation, it was not averse to Swedish colonies on the islands. In 1904, the Foreign Office therefore reasoned that Russia may not object to any companies working the coal deposits there. There seemingly were no precedents to Britain licensing coal exploitation in a no man's land, so the Foreign Office based itself on the working of guano islands that belonged to no one. The British Government had previously declared sovereignty over such islands and granted licenses under the condition that it would not compensate losses caused by the absence of protection or by the abandonment of sovereignty over the territory. However, this form of license only affected islands in which no other nation had shown interest. In the light of both Sweden and Russia having made earlier claims, Spitsbergen had acquired an unprecedented status. Subsequently, the Foreign Office referred to the case of Kerguelen's Land in Antarctica. Although in the possession of France, the French had not objected to a British firm wanting to extract guano there, but the British Government informed that firm that it would do so at its own risk. The Foreign Office concluded that the policy regarding Kerguelen's Land also applied to Spitsbergen. It therefore provided Bainbridge with the answer that 'this island being under no particular flag, any mining operations there must be carried on at the risk of the promoters.'[39]

Bainbridge undertook two more attempts to sway the Foreign Office. He firstly enlisted the help of an influential acquaintance inside the department, who obtained a note which was hoped to assure that British subjects could develop Spitsbergen undisturbed.[40] The Foreign Office maintained that the contents of that note were in essence already known and did not change the fact that the British Government could not guarantee the protection of mining operations on Spitsbergen.[41] Thereafter, the coalowner instructed the company's solicitor to formalise all that had transpired between the company and the Foreign Office. In a statement to Lansdowne, Clegg reiterated that his clients understood Spitsbergen's legal status, but they had nonetheless undertaken further exploration for coal and intended to ascertain if there were other minerals to be worked.[42] In the first instance, they aimed to trade with whalers and other vessels in need of a coaling

[39] Foreign Office (1904) *Draft letter to E. Bainbridge*. Spitzbergen 1897-1905 FO83/2147, National Archives, Kew.

[40] Bainbridge, E. (1904) *Letter to E. Gorst, 17 October*. Spitzbergen 1897-1905 FO83/2147, National Archives, Kew.

[41] *Mr E. Bainbridge. 17 October 1904* (1904) Spitzbergen 1897-1905 FO83/2147, National Archives, Kew.

[42] Clegg and Sons (1904) *Letter to the Marquis of Lansdowne (Foreign Office), 15 November*. Spitzbergen 1897-1905 FO83/2147, National Archives, Kew.

station.[43] Despite the unsettled legal status, the shareholders had agreed to substantially increase the capital, and Clegg proposed that 'if any hostile Sovereign power endeavours to forcibly eject them from their position or commits any other acts detrimental to the interests of the English shareholders to ask for the protection of the British Government to see that their property is safeguarded and their rights are recognised.'[44] Incidentally, the Norwegian shareholders were not mentioned, but Clegg informed that he would send a similar notification to the Norwegian legation to London.[45]

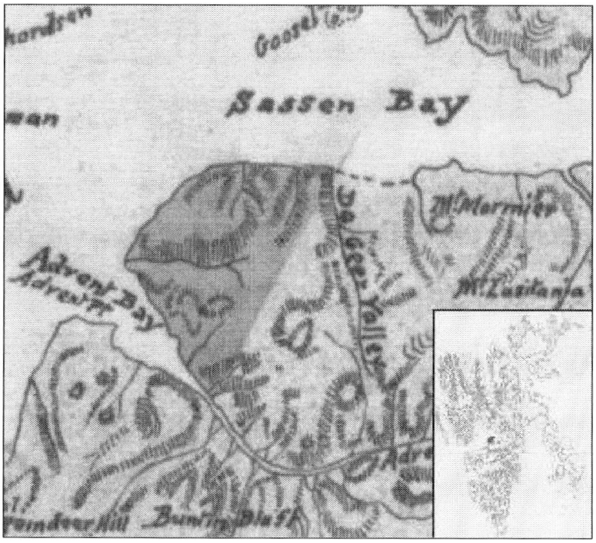

5.7 Detail of plan showing the claim of the Spitzbergen Coal & Trading Co. lodged with the Foreign Office in November 1904. (Source: Clegg and Sons (1904) Letter to the Marquis of Lansdowne, 15 November. Spitzbergen 1897-1905 FO83/ 2147, National Archives, Kew.)

Clegg's statement to the Foreign Office contained a plan outlining the company's claim on Spitsbergen (Fig. 5.7). The territory depicted corresponded to the claim boundaries stated on the Bergen Co.'s claim board of 1901 and must have been the property made over to the English in June 1904.[46] It measured approximately 135 square kilometres. Although it is not clear whether the pink colour was a conscious choice, it was in keeping with the colouring of British imperial maps at

[43] This seemingly innocent allusion to 'coaling station' was arguably aimed at the Achilles' heel of the British Empire, such coaling stations and naval bases being of vital strategic importance to control the trade routes and there being none north of the British Isles, let alone in the European Arctic.
[44] Clegg and Sons (1904) *Letter to the Marquis of Lansdowne (Foreign Office), 15 November.* Spitzbergen 1897-1905 FO83/2147, National Archives, Kew.
[45] Until the Second World War, Norway's bilateral diplomatic missions were called legations and the diplomatic envoy had the title of minister. Since 1945, the missions have been called embassies and the heads of missions have been appointed as ambassadors.
[46] Despite the Bergen Co.'s investigation of Kings Bay, the Spitzbergen Coal & Trading Co. therefore never had any interests there.

the time.[47] Deliberate or not, this small English claim on the Arctic archipelago now gave the impression of being an integral part of Britain's global possessions.

The strike in winter 1906/7 put the company's trust in the British and Norwegian Governments to the test.[48] The board did not hear about it until Davenport, Black, and Bruvik entered Icefiord in spring 1907.[49] They were naturally alarmed, not least of all because Bruvik had already sold several thousand tons of coal in Norway, which would now not be delivered. The company's ship departed for Tromsø, and Davenport sent a telegram to Arthur J. Herbert, the British minister to Kristiania.[50] He described a mutiny of 40 armed men, who had seized rifles and ammunition, subjected the manager to a murderous assault, and threatened the lives of all those loyal to the company. Among the trustworthy staff were seven English men and one woman as well as 20 Norwegian men and two women. The general manager was begging for the immediate dispatch of 20 soldiers, an officer, and someone to collect evidence. The company's steamer was waiting at Tromsø to transfer them to the island, where they were needed to protect lives and property, arrest the ringleaders, and bring them to trial.

The British minister in turn notified the Foreign Office, where Edward Grey was now serving as Foreign Secretary.[51] Although Grey related the strike to the Admiralty, he could not see the purpose of sending a British official to Advent City.[52] Instead, he was happy for the Norwegian authorities to restore order by any means they saw fit; Britain would support them. Instructed by Grey, Herbert deliberated with the Norwegian Foreign Department that it appeared to be best if the company sent its steamer back to Advent City to remove the disaffected workers.[53] He added that the directors should be asked to send the men home before they became a burden in Tromsø. Although he trusted that these measures should be successful, he added that if the Norwegians decided to send a force to

---

[47] National Maritime Museum (2011) *Why is the British Empire coloured pink on maps?* Available at: http://www.nmm.ac.uk/explore/sea-and-ships/facts/faqs/general/why-is-the-british-empire-coloured-pink-on-maps (Accessed: 16 September 2011).

[48] Dole, H. (1922) *America in Spitsbergen. The Romance of an Arctic coal-mine. Part 1.* Boston: Marshall Jones Company, pp. 294-5

[49] Dole (1922a) p. 304.

[50] Davenport, F. (1907) *Telegram to A. Herbert, 5 June.* Norwegian Foreign Department Box 5115, National Archives of Norway, Oslo. The capital of Norway was originally called Oslo. After a fire in 1624, Christian IV decided to rebuild the town and changed its name to Christiania. From 1877, the name was spelled Kristiania. In 1925, it was changed back to the original name, Oslo.

[51] Grey served the longest continuous term of any Foreign Secretary from November 1905 until December 1916. Highlights of his career included the Anglo-Russian Entente in 1907, the Agadir Crisis in 1911 as well as the July Crisis and the outbreak of the First World War in 1914.

[52] Grey, E. (1907) *Memo, 8 June.* Norwegian Foreign Department Box 5115, National Archives of Norway, Oslo.

[53] Herbert, A. (1907) *Letter to Lóvland, 13 June.* Norwegian Foreign Department Box 5115, National Archives of Norway, Oslo.

Advent City, the British Government would appreciate it if a representative could accompany it to ascertain the requirements of its subjects.

On June 26, 1907, the company's ship finally anchored at Advent City.[54] The directors came to an arrangement with the rioters, and the men departed the settlement. On July 5, Black informed Herbert that he was back in Tromsø, this time with 40 strikers on board.[55] The company accused eight of these men of assaulting their manager, theft, and incitement to murder. However, the police in Tromsø did not intervene. Black, therefore, asked Herbert, if the company could expect to get satisfaction through the British Government. He suggested that the case be dealt with severely to prevent the same thing from happening again in the future. The minister transferred his question to both the British Foreign Office and the Norwegian Foreign Department.[56] In view of Spitsbergen's peculiar status, the matter involved such difficult legal questions that he confined himself to passing on the message and waiting for further instructions.

Following the riots, the Spitzbergen Coal & Trading Co. had requested assistance from the British legation in Kristiania, but none was given.[57] Nor could the Norwegian Government send soldiers to keep order among the insurgents because Advent City lay outside Norwegian jurisdiction, and some of the ringleaders were not Norwegian. When the company eventually brought everyone back to Tromsø, the Norwegian courts could only assume jurisdiction related to the men's contracts, which had been drawn up in Norway. Not only had the men not fulfilled these contracts; they actually owed the company for their board during the winter. Naturally, the company had no means of collecting this debt, but they had won an important legal victory. This victory was also relevant to other firms operating on Spitsbergen; it showed that the Norwegian legal system would not support Norwegian citizens in violation of their contractual agreements.

Not until mid-August 1907 was the matter brought up in the House of Commons.[58] The Foreign Secretary was asked, if he knew of a near-fatal attack on a British citizen on the islands, which put British lives and property in danger, and that a request for British assistance had been made. Grey was desired to state, if negotiations had progressed to the point of Spitsbergen being either nationally or internationally controlled or if he had any plans to promote such an arrangement. Grey, in turn, answered that the Government had been aware of the attack, but 'the long interval that elapsed between the outbreak and the date when it became

[54] Dole (1922a) p. 318.
[55] Black, E. D. (1907) *Telegram to A. Herbert, 5 July*. Norwegian Foreign Department Box 5115, National Archives of Norway, Oslo.
[56] Max Müller, W. G. (1907) *Letter to Lóvland, Norwegian foreign minister, 6 July*. Norwegian Foreign Department Box 5115, National Archives of Norway, Oslo.
[57] Dole (1922a) pp. 309-11.
[58] HC Deb 15 August 1907 vol 108 c1577.

known here made it impossible for any British warship to reach the island in time to afford any help, even had the actual situation warranted such a step.'[59] Furthermore, 'His Majesty's Government cannot undertake to initiate negotiations with the object of instituting some kind of control over Spitzbergen. Any British subject going to that island for trade or other purposes must do so at his own risk, and we cannot exercise any jurisdiction or assume any responsibility there.'[60] Grey's words were reproduced in full in *The Times* two days later.[61] They passed practically unnoticed by the British public.

The strike was probably among the causes for Spitsbergen's emergence on the global political stage. The Norwegian Government tentatively enquired if Britain would join a first international conference to discuss the improvement of the conditions on the archipelago. In August 1908, the Foreign Office replied that it would instruct its minister in Kristiania to take part, but he would not be able to commit to anything. Furthermore, 'His Majesty will of course be bound to make it a condition of entering into any arrangement as regards these islands that the existing rights of British subjects are safeguarded.'[62] The British and American ministers to Norway then met to discuss the matter.[63] Besides Arthur J. Herbert and Herbert D. C. Peirce, John M. Longyear, founder of the Arctic Coal Co., was present at this informal meeting. The latter gathered that Sweden, Norway, and Russia were jealous of each other, each intending to control the islands. While Herbert thought it premature to call an international conference, Longyear maintained that concerted action was needed. British and American interests were continuously growing and needed to be protected. Although the British diplomat gave no clue as to what his Government would do, Longyear believed that it would take the same stance as the Americans.

The Foreign Office had developed an independent diplomatic interest in Spitsbergen when the company again contacted Grey in October 1909.[64] Following a reminder that the development at Advent City had cost over £50,000, the new secretary H. Walker, now based at Victoria Chambers on Figtree Lane, brought up the imminent conference. Walker believed that the Norwegians would ask for control over Spitsbergen and suspected that they wanted to acquire all desirable properties on the islands without paying any compensation to the English or the Americans. He requested that their claims, titles, occupancy, development, and

---

59 HC Deb 15 August 1907 vol 108 c1577.
60 HC Deb 15 August 1907 vol 108 c1577.
61 *The Times* (1907) 'Questions not orally answered. British subjects in Spitzbergen', 17 August, p. 8.
62 Langley, W. (1908) *Letter to Jakhelln, 26 August.* Norwegian Foreign Department Box 5036, National Archives of Norway, Oslo.
63 Dole (1922a) p. 426.
64 Walker, H. (1909) *Letter to E. Grey, Foreign Secretary, 27 October.* Norwegian Foreign Department Box 5227, National Archives of Norway, Oslo.

work should be fully recognised. The company would not object to taxes for administration and policing as long as the costs were reasonable. As British citizens, they hoped that Grey would instruct his representative to safeguard British interests. In November 1909 and again in May 1910, the Foreign Office assured the company that their and all British interests would be safeguarded in the conference, the date and programme of which were not yet known.[65]

The first Spitsbergen conference passed without a change in the island's status. By the time Bainbridge died in May 1911, the Spitzbergen Coal & Trading Co. had become an international liability. As shown above, the Foreign Office was predominantly concerned about whether or not the company could pay its existing debts and keep friendly relations with Norway. Consequently, the company kept the contact with the Foreign Office or any other political actor to a minimum before it was dissolved.

5.3.3 The Arctic Coal Co.

When the Bergen Co. arrived in Advent Bay in summer 1901, the Kulkompagniet Trondhjem-Spitsbergen had already occupied a coal claim on the southern coast. The Bergen expedition was shown the extent of this claim and departed to the northern shore.[66] Like the Bergen directors enrolling English investors, the Trondheim Co. mobilised American capital and was soon absorbed into the Arctic Coal Co. Although the two companies developed side by side, subsequent events meant that the Spitzbergen Coal & Trading Co. could never be certain of the Americans' place in its network.

Due to the proximity of the two mines, the American representative William D. Munroe was able to closely scrutinise the Bergen Co. and its successor. In March 1905, he was under the impression that the company had chartered a large steamer to convey a substantial workforce to Advent City in May.[67] In light of the growing scale of the activities, he concluded that he would not be able to get an option on the company's land. Within a month, he discovered that the firm was, in

---

[65] Langley, W. (1909) *Letter to the Secretary, Spitzbergen Coal & Trading Co., 1 November*. Norwegian Foreign Department Box 5227, National Archives of Norway, Oslo; Langley, W. (1910) *Letter to the Secretary, Spitzbergen Coal & Trading Co., 6 May*. Norwegian Foreign Department Box 5227, National Archives of Norway, Oslo.

[66] Bohne, F. (1904) *Letter to O. Jeldness, 21 January*. Arctic Coal Co. Collection, Privatarkiv 101, Box 91, Regional State Archive, Tromsø.

[67] Munroe, W. D. (1905) *Letter to F. Ayer, 22 March*. Arctic Coal Co. Collection, Privatarkiv 101, Box 91, Regional State Archive, Tromsø.

fact, registered in England, and an option on American terms was out of the question for the time being.[68]

Despite the growing activities, Munroe reported that Fangen had been unable to effectively recruit employees. The engineer had used the Norwegian newspapers to advertise for labourers of all classes. Those who applied were made to sign contracts 'the nature of which no man would submit to.'[69] Since the contracts were written in English without a translation, Fangen eventually only enlisted one or two friends of his. The newspapers in Trondheim cautioned their readers, and Munroe speculated that if Fangen employed any men at all, they would possibly be from England.

In August 1905, Munroe reported that, 'the operations of our friends across the Bay are a wonder to behold. I would not care to be a shareholder in it even if I were worth a billion.'[70] He conceded that they were disadvantaged by a poor harbour, but Fangen continued to be their worst drawback. Witnesses had seen him drunk aboard his yacht most of the time while the company's operations were left poorly managed. Although the expedition had arrived in June, only two-thirds of the cargo had been discharged. There had been little prospecting, and the fact that the mine was fairly well opened up was due to a good English foreman. Arthur Mangham[71] had extracted 200 tons of coal shortly after beginning his work. Yet, the mine lay idle in August, the men having been ordered to construct buildings and foundations of machinery instead. This was arguably necessary if 50 men were to winter. Following Fangen's exit, Munroe observed that £20,000 or more had been spent, and nothing had been achieved besides Fangen now suing the Spitzbergen Coal & Trading Co. for five years' worth of salary. Munroe concluded, 'it was the worst case of bad management I ever saw.'[72]

The English directors attempted to salvage the situation shortly after the Arctic Coal Co. was formed in February 1906. In March, Munroe travelled to Sheffield to obtain English miners and equipment. Hearing of this, both Black in

[68] Munroe, W. D. (1905) *Letter to F. Ayer and J. M. Longyear, 20 April.* Arctic Coal Co. Collection, Privatarkiv 101, Box 91, Regional State Archive, Tromsø.

[69] Munroe, W. D. (1905) *Letter to F. Ayer and J. M. Longyear, 20 April.* Arctic Coal Co. Collection, Privatarkiv 101, Box 91, Regional State Archive, Tromsø.

[70] Munroe, W. D. (1905) *Letter to F. Ayer and J. M. Longyear, 10 August.* Arctic Coal Co. Collection, Privatarkiv 101, Box 91, Regional State Archive, Tromsø.

[71] Arthur Mangham was a coal miner and colliery manager from Thorpe Hesley near Rotherham in Yorkshire. Munroe became very well acquainted with Mangham when the latter visited the American mine one day and provided Munroe with a thorough assessment. This gesture proved beneficial. Firstly, Munroe gave Mangham authority, and possible remuneration, to act as his watchman in winter 1905/6 and to eject all intruders from the American site; secondly, Arthur's son Bertrand would become the winter superintendent of the Arctic Coal Co. from 1906 till 1913. Arthur would go on to work for the Spitzbergen Mining & Exploration Syndicate as well as the Northern Exploration Co.

[72] Munroe, W. D. (1905) *Letter to F. Ayer and J. M. Longyear, 16 September.* Arctic Coal Co. Collection, Privatarkiv 101, Box 91, Regional State Archive, Tromsø.

Sheffield and a little later Bainbridge in London made an effort to bring about a merger with the Arctic Coal Co.[73] Such a merger would conceivably pool their resources and share the benefits while they could repair the company's image and prevent American competition and prejudice. Munroe, however, was not interested in an amalgamation. He knew Fangen's substitute, the English mining engineer Percival Muschamp, had not done good work during the first wintering either.

Munroe nonetheless had few grievances with Muschamp. Since the Spitzbergen Coal & Trading Co. had expressed its recognition of the American rights and an amalgamation was entirely out of the question, the two companies exchanged favours amicably. Following the death of Black in June 1906, however, Muschamp was summoned back to England. While he was away, Advent City appeared to have been left without a manager. In July, Munroe reported that things were going badly across the bay.[74] Muschamp's return was seemingly delayed, and the men had been short on food for a month. Word was that the mining engineer would be back soon, and then things were to be done on a big scale. Munroe was confident, however, that the Spitzbergen Coal & Trading Co. could no longer compete with the Arctic Coal Co.

Despite the failed merger, the Spitzbergen Coal & Trading Co. tried to enrol the Americans to advance the legal situation. Secretary Weston informed them that he had contacted both the British Foreign Office and Fridtjof Nansen, the first minister of independent Norway to London, in the hope of getting protection and assistance in governing that portion of Spitsbergen, which his company had taken possession of.[75] The Foreign Office merely repeated that Spitsbergen lay outside its jurisdiction. Weston therefore believed that Nansen, whose country was more closely concerned than Britain, would be more sympathetic if he were approached by both English and American stakeholders.[76] By November 1906, the

---

[73] Munroe, W. D. (1906) *Letter to the Arctic Coal Co., 13 July*. Arctic Coal Co. Collection, Privatarkiv 101, Box 91, Regional State Archive, Tromsø.

[74] Munroe, W. D. (1906) *Letter to the Arctic Coal Co., 28 July*. Arctic Coal Co. Collection, Privatarkiv 101, Box 91, Regional State Archive, Tromsø.

[75] Weston, E. V. (1906) *Letter to W. D. Munroe, 20 August*. Arctic Coal Co. Collection, Privatarkiv 101, Box 91, Regional State Archive, Tromsø.

[76] Weston envisaged, '1. That in our respective contracts with the Norwegian labourers employed by us on the Island we should insert a clause that all offences or conflicts or differences should be disposed of by a Norwegian court in Hammerfest. 2. That our respective Managers should be entitled to exercise the same authority as if appointed by the Norwegian Government. 3. That both companies should either together or separately sign an application to the Norwegian Government asking it to appoint a magistrate or magistrates to preserve law or order on the grounds that the present position of affairs is considered by both companies as unsatisfactory and that it was desirable to have some recognized authority appointed. 4. That in addition to both companies making the application either separately or jointly it is also desirable that other persons should also be obtained by us to join in such applications to the Norwegian Government to attain authority or jurisdiction over that portion of the Island occupied by your company. (Weston, E. V. (1906) *Letter to W. D. Munroe, 20 August*. Arctic Coal Co. Collection, Privatarkiv 101, Box 91, Regional State Archive, Tromsø.)

Spitzbergen Coal & Trading Company had not heard from the Americans. To prompt a response, the directors notified them that the British press had reported on the probability of Norway taking over Spitsbergen. They thought this to be very desirable for both parties. They were curious to learn about the American inquiries in Washington.

Meanwhile, the strike occurred. From across the bay, Bert Mangham, the first winter superintendent at Longyear City, recounted the events.[77] Muschamp, a former officer in the British Army, appeared to have treated the workforce like raw recruits. In autumn, the supply ship had landed more than 14 thousand bottles of beer and other wet goods. A bar had then been established, where the workers spent all they earned. As a result, some were insanely drunk. When disorder broke out early in the season, Muschamp appeared among the men with a gun and threatened them. The men promptly disarmed him, threw him to the ground, and Mangham believed they would have killed him if not more level-headed men had stepped in. Muschamp escaped to his house, barricaded himself, and did not step outdoors in daylight in fear of being shot. During this time, machinery was broken, and the men refused to work. Water tanks were left full and burst when the water in them froze. There were continued rows and fights.

With the disposal of the business in mind, the Spitzbergen Coal & Trading Co. renewed the negotiations with the Arctic Coal Co. in November 1908.[78] The directors stated that the development had cost over £50,000, of which the plant and its transport had amounted to about £17,000. They were prepared to sell the business and the tangible assets for £15,000. If it could not be sold whole, the company was willing to sell all plant, machinery, houses, and other property at Advent City for £8,500. The company had opened talks with other parties, but it felt that the Americans should have the first refusal because the two firms were working in close conjunction. Frederick Burrall, the American manager, put in a bid that appears to have been for certain items only. The company, however, regretted that it could only sell its tangible property as one lot.[79] By spring 1909, the directors had not found any other buyer, so they again addressed the Americans.[80] They asked for only £10,000, but Burrall decided that the property was undesirable. He had heard that the last shipment had been very high in ash and tended to give Spitsbergen coal a bad name.

---

[77] Dole (1922a) p. 294-5.

[78] Clegg, W. E. (1908) *Letter to H. P. Burrall, 17 November*, Arctic Coal Co., Norwegian Polar Institute, Tromsø.

[79] Walker, H. (1909) *Letter to H. P. Burrall, 6 March*, Arctic Coal Co., Norwegian Polar Institute, Tromsø.

[80] Dole (1922a) p. 395.

Advent City had lain idle for about a year when the Americans inspected the workings in summer 1909.[81] Longyear's negative opinion undoubtedly reduced the bargaining power of the Spitzbergen Coal & Trading Co. even further, although the next manager John Gibson may not have mentioned this during his stay in Sheffield in January 1910.[82] So the directors provided him with a confidential document declaring the seam to be a metre thick and the mine to be in a position to turn out 200 tons of good-quality coal per day; an output of 20,000 tons per year or more would already make a profit. A supplement detailed the installations at Advent City.[83] Several items had cost over £1,000 each. These included the gas producer plant, the electric plant, the houses, the colliery stores, general stores, and five lighters. The total expenditure was listed as £11,191.

The Spitzbergen Coal & Trading Co. gave the death of their manager and prime mover as the reason for its readiness to sell the near-perfect plant.[84] To Gibson it seemed that the directors were in fact indifferent towards their claim. They were content to rest on the Foreign Office's promise to safeguard their rights. The American suggested to them that it would be hard to find a purchaser if the title could not be guaranteed and if the laws and regulations adopted in due course would have a material impact on the sale. Yet, he felt that the directors would not act in the matter and 'that they would be very well satisfied if they lost the property in such a manner as to have an excuse for themselves.'[85] The Arctic Coal Co. ultimately considered that it had enough on its hands already and once more refused the English offer.

In May 1912, the Americans dispatched a party to Advent City to dismantle one of the bunk houses there.[86] They rebuilt it at Longyear City to provide sleeping quarters for fifty men. By 1913, the Spitzbergen Coal & Trading Co., unable to receive the asking price for its property, allowed the Arctic Coal Co. to take

---

[81] Longyear observed, 'the vein has as much rock as coal, both being in alternate layers of from four to sixteen inches each. The expense must have been tremendous, and only half the product was salable. There were illustrations everywhere about the place of 'how to do it'. The whole thing was a melancholy looking 'layout' for the money that has been spent. In many places the drifts were partly filled with rock which has fallen from the roof, and in some places we saw crushed timbers showing that the hill above the mine has begun to settle down into the openings. The accumulations of frost-crystals in some places hung down a foot from the roof, and we got many showers of them because of not stooping low enough. The idle electric wires looked like feathery crystal cables, two or three inches in diameter, as if the frost fairies were getting up Christmas-tree decorations.[81] (Dole (1922a) p. 407.)

[82] Dole (1922a) pp. 3-4.

[83] *Actual original cost of plant, machinery, houses, stores, etc.* (nd) Longyear Collection MS-031, Box 4, Folder 20, Michigan Technological University Archives and Copper County Historical Collections, Michigan.)

[84] Dole, H. (1922b) *America in Spitsbergen. The Romance of an Arctic coal-mine. Part 2.* Boston: Marshall Jones Company, p. 4.

[85] Dole (1922b) p. 389.

[86] Dole (1922b) p. 136.

anything it wanted at a reasonable cost.[87] The Americans were interested in a number of lighters, which lay disused on the beach. In winter 1913/4, they took an iron warehouse, an unfinished family barrack, and a stable and put them into usable condition.[88] They also removed all good timbers, tools, iron supplies, cables, props, an iron fan, iron tanks, bricks, water pipes, window sashes, and anything else that might be handy. This had to be done before the Norwegian creditors of the company could seize the property.

5.4 The local network

5.4.1 Claims and natural resources

In spring 1904, the Spitzbergen Coal & Trading Co. purchased the SY *Heloise*.[89] She was 30 metres long, could hold 90 tons, and provided an independent means by which to reach the coal claim on Spitsbergen. Hessler refurbished the yacht in West Hartlepool and loaded her with materials. In Tromsø, men were taken on board, and the party under Fangen left for Spitsbergen at the end of May 1904. Hoel asserts that the claim boundary was revised upon arrival in Advent Bay (Fig. 5.8).[90]

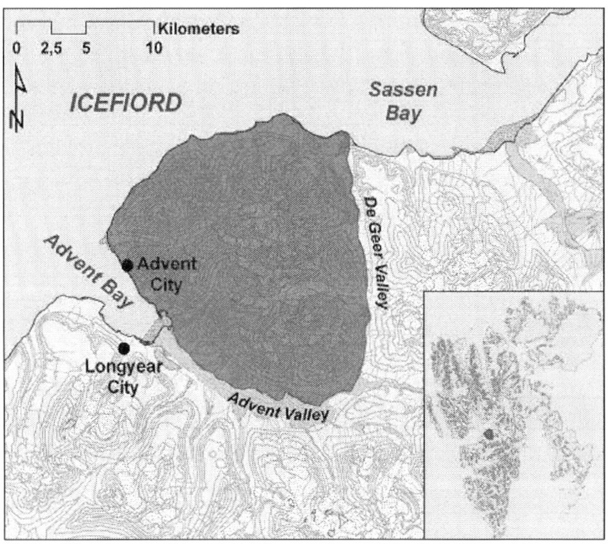

5.8 Map showing the extended claim of the Spitzbergen Coal & Trading Co. in 1904. (Source: Hoel, A. (1967a) Svalbard. Svalbards historie 1596-1965. Part 2. Oslo: Sverre Kildahls Boktrykkeri, p. 560; Map: F. Kruse.)

[87] Dole (1922b) p. 218.
[88] Dole (1922b) p. 234.
[89] Hoel (1967a) pp. 557-9.
[90] Hoel (1967a) pp. 557-8.

The difference between the original boundary and the later one was that of an arbitrary straight line across the mountains that would have been impassable, difficult to secure, and therefore unjustifiable, and the natural passage through the De Geer and Advent valleys. This act essentially doubled the company's territory, which was seemingly never communicated to the directors in Sheffield. Even after the expedition's return in September, Clegg forwarded a plan of the original claim to the Foreign Office.

With a capital of £25,000, the Spitzbergen Coal & Trading Co. fitted out another expedition in spring 1905.[91] The *Hørda*, a wooden steamer less dependent on favourable winds, was chartered in Haugesund and brought to Hull to load machines and material. At the beginning of May, the expedition, geared up for 100 men again led by Fangen, left Hull.[92] The *Hørda* went via Bergen to Trondheim, where Zakariassen was waiting aboard the *Heloise* to be towed northward. The expedition finally departed Trondheim on June 9, 1905.[93] At Easter, it arrived in Tromsø to obtain additional workers and equipment. After a five-day crossing, the party reached Spitsbergen.

5.9 Landing materials using rowing boats and lighters at Advent City in 1905. The difficulty of this task was proportional to the power of the surf. (Source: Hoel, A. (1967a) Svalbard. Svalbards historie 1596-1965. Part 2. Oslo: Sverre Kildahls Boktrykkeri, p. 559.)

According to Hoel, Fangen got attached to the landing site of his earlier expeditions.[94] In 1905, it was christened Advent City and made the base of the company's mining operations. The men began to discharge the cargo, but they

[91] Hoel (1967a) pp. 559-61.
[92] 'Spitzbergen'(1905) *The Colliery Guardian (Indian, colonial and foreign supplement)*, 19 May, p. 111.
[93] Hoel (1967a) p. 559.
[94] Hoel (1967a) p. 559.

soon discovered that they should have positioned the settlement further into the bay, reasons being poor landing conditions (Fig. 5.9) and limited building space. They nonetheless concentrated on tracing the coal seam northwards, but the trials took longer than expected and did not give good results. Although an employee later claimed that they had discovered a second seam higher up (which would in the future be mined by Hiorth), they had run out of time to continue the work.[95] The season had moved on; it became important to unload and erect the buildings before snowfall set in.

Fangen's disruption of the summer work did not alter the plan for the first wintering at Advent City in 1905/6. The mining engineer Percival Muschamp was the company's first winter superintendent.[96] He was assisted by the Norwegian engineer Hovde. Due to bad weather, Muschamp and the winter workforce were unable to disembark before October 23. The ship immediately returned to Norway and barely made it clear of the ice.

The winter marked the beginning of year-round operations at Advent City. The company aimed to mine about 6,000 tons of coal during the dark season. Besides increasing the output from the mine, this was a way to ascertain effective occupation in the no man's land. During his stay, Muschamp additionally staked out a coal claim at Heerodden to the east of the mouth of Green Harbour. The company, however, did not recognise this claim.[97] On June 9, 1906, Muschamp departed the islands on board a whaler. Hoel clearly states that Muschamp then stepped down from his position, but this was not the case. The engineer returned from England to manage the second wintering in 1906/7.

During this second winter, the English foreman Francis W. Matthews sporadically kept a diary.[98] In January 1907, Frank recorded a hunting trip to one of the company's hunting huts approximately seven miles from the settlement. Besides the immediate purpose of providing shelter for hunting parties, the two huts probably also guarded the company's property, especially since they displayed large claim boards (Fig. 5.10). Matthews was after reindeer. He saw three, but it was too dark to shoot. There were plenty of reindeer the next day, yet he and his men only killed one. Thereafter, the company seemed to have made a deal with the Arctic Coal Co., which resulted in a food store of 24 reindeer.

---

[95] Almklow, M. (nd) *Vedrørende The Spitsbergen Coal and Trading Company ltd.s anlæg og drift paa Spitsbergen.* Norwegian Polar Institute Box 85, Regional State Archive, Tromsø.
[96] Hoel (1967a) pp. 560-2.
[97] Arctic Coal Co. (1909) *Letter to the US Secretary of State, 27 December*, Arctic Coal Co., Norwegian Polar Institute, Tromsø.
[98] Diary of Frank Matthews (1906/7) provided by Thompson, A. (2010) Letter to Frigga Kruse, 8 June.

5.10 One of the two hunting hut of the Spitzbergen Coal & Trading Co. Note the company's claim sign on the wall. (Source: Photo Library np032551, Norwegian Polar Institute, Tromsø.)

Under Ernest Black, the workforce extracted coal throughout summer 1907 and the subsequent winter.[99] The few sources that cover this period do not indicate any changes to the company's territory, and it is unclear if further prospecting for coal or other minerals took place. As shown below, animal products provided an alternative income, but it was apparently not enough to sustain the company's operations at Advent City beyond the summer 1908.

5.4.2 Manifestations

Fieldwork has generated a clear archaeological image of Advent City (Chapter 3), to which the documents nevertheless add essential details. The first structures were built by the Bergen Co. in 1903. They included a simple kitchen and mess tent, which were supplied via narrow-gauge rails. At pithead, the Norwegians had levelled the slope, stabilising it with retaining walls, and constructed a small smithy. A testimony of the early coal mine creates a rudimentary impression: the men apparently shot seals and used their blubber for lighting in the heading.[100] A simple ropeway conveyed the coal down the hill.

In 1904, Fangen ordered this simple ropeway to be replaced with a double-acting one in order to improve the transport of coal to the shore.[101] When the managing director William Black arrived on a tourist steamer, staying for eight days, he brought three lighters from southern Norway. Shown in Fig. 5.9 above,

[99] Hoel (1967a) p. 565.
[100] Braastad, J. (nd) *Oplysninger om aktieselskapet Bergen-Spitsbergen og the Spitsbergen Coal and Trading Co. Innsamlet hos ing. S. A. Fangen, Kiberg*. Norwegian Polar Institute Box 85, Regional State Archive, Tromsø.
[101] Hoel (1967a) pp. 557-9.

they replaced the impractical rowing boats to improve the transfer of coal from the shore to the ships in the bay. Although a better canteen had also been put up, the men continued to live on board.

In 1905, the company began the construction of the settlement in earnest.[102] Summer had progressed dangerously, and unloading was made difficult by exposure and a strong surf. Fortunately, the buildings were pre-fabricated, and the company had sent a carpenter to assemble them (Fig. 5.11). At the close of the season, Advent City comprised several workers' barracks, the large manager's house known as *Huset* ('The House'), and a storage building (Figs 5.12 & 5.13).

5.11 Two pre-fabricated workers' barracks and the brick foundation of Huset in 1905. (Source: Photo Library np002341, Norwegian Polar Institute, Tromsø.)

5.12 Five barracks and Huset in mid-September 1905. Note the ropeway towers to the left and in the centre. (Source: Photo Library np002340, Norwegian Polar Institute, Tromsø.)

[102] Hoel (1967a) p. 559.

According to Dole, the pre-fabricated buildings were proof that the company intended to winter.[103] The wooden beams and boards displayed distinguishing marks, which enabled the carpenter to put each piece in its proper place. Despite the simple technique, the single-story houses themselves were not simply made. Because it was necessary to protect the inhabitants against the harsh winter weather, the houses had walls of more than one facing. There were usually two double casings of boards with an air space of several inches between. The houses for the officers even comprised three layers with not only an air space but also a layer made of insulating materials such as felt, cork, or linoleum that conserved heat. Witnesses thought the houses to be comfortably arranged. The workers' barracks were long, hall-like rooms, where several men bunked together. The managerial residences consisted of a row of single rooms.

5.13 Four barracks, Huset, and the store in 'Ida Jackson Street' in mid-September 1905. Due to limited space, the ground had to be levelled first. Note the cart tracks in the street. (Source: Photo Library np002339, Norwegian Polar Institute, Tromsø.)

Between November 1905 and February 1906, Muschamp mainly employed the workers in the open. The men took down the ropeway and replaced it with a single-track incline by the light of paraffin lamps. In the manager's absence in summer 1906, more buildings, machinery, and plant were erected at the expense of the mine.[104] In September, the incoming doctor described a settlement of seven or eight buildings (Fig. 5.14).[105] The workers had also placed a roof over the mine entrance to protect it from storms and snow (Fig. 5.15). Near the landing site, an engine house was in the process of being built, which would provide the houses

[103] Dole (1922b) p. 97.
[104] Hoel (1967a) p. 562.
[105] *Husmoderen* (1906) 'Fra Advent Bay'. Available at: http://polarlitteratur.com/husmod.htm (Accessed: 20 September 2011).

and the workings with electricity. On October 1, 1906, Matthews laid large cables into the mine, but by February 2, the engines were still not working.[106]

5.14 Several photographs of Advent City were turned into postcards from the world's most northerly town in 1906. Note the incline and a water pipe to the left. (Source: Wilse, A. B. (1906) NF.W 05514. Galleri NOR [Online]. Available at: http://www.nb.no/ gallerinor/ index.php (Accessed: 21 September 2011).

5.15 The roofed mine entrance and the incline in 1906. (Source: Wilse, A. B. (1906) NF.W 05515. Galleri NOR [Online]. Available at: http://www.nb.no/ gallerinor/index.php (Accessed: 21 September 2011)).

When the winter's unrest had subsided and the directors eventually landed at Advent City in 1907, they discovered that the settlement had as of yet no harbour.[107] There was only a small pier at the shore, where lighters needed to be loaded to transfer the coal to the ship (Fig. 5.16).

[106] Diary of Frank Matthews (1906/7).
[107] Dole (1922a) p. 304.

5.16 A small pier at the edge of the water. Note the parallel tracks of the self-acting incline. (Source: Wilse, A. B. (1908) NF.W 09358. Galleri NOR [Online]. Available at: http://www.nb.no/gallerinor/index.php (Accessed: 21 September 2011).

5.17 The self-acting incline, the coal tip, and the engine house. Note the smoke coming from the plant. (Source: Wilse, A. B. (1908) NF.W 09356. Galleri NOR [Online]. Available at: http://www.nb.no/gallerinor/index.php (Accessed: 21 September 2011).

Writing from Hertfordshire in August 1907, Muschamp nonetheless gave a buoyant account, details of which the directors reproduced in the aforementioned memorandum to their shareholders. The underground operations were ready for the coal cutter to commence on a face over 450 metres long.[108] The surface plant

[108] Muschamp, P. (1907) *Letter to W. Clegg, 5 August.* William Olssons Archives 1899-1922 F2:20, National Museum of Science and Technology Archives, Stockholm.

included a 300-metre-long self-acting incline from the pit to the coal tip, a screen, and a self-acting incline from the coal tip to the pier, which enabled tipping directly into lighters. The five lighters could be loaded and got away at any state of the tide. The engine house and one of the three suction gas plants had been erected (Fig. 5.17). There was accommodation for a hundred men in addition to the necessary storerooms (Fig. 5.18). He therefore believed it to be a mistake to abandon the enterprise. Instead, he proposed that the company find the working capital to develop the mine in winter 1907/8 and the following summer. The only large cost would be for a tug boat to tow the lighters from the pier to the ship and back. Muschamp asserted, 'if I did not see a good prospect based upon my two years' experience in Spitzbergen, I would not recommend any further expenditure.'[109]

5.18 Advent City in summer 1907. Note an electricity pole in the centre. (Source: Photo Library np002342, Norwegian Polar Institute, Tromsø.

By now, Black was in charge on site. He immediately took two men from Campbell's to the engine house. One engine bed was in a bad state, so they decided to fill it in with concrete while building another one. It meant much extra work and the concrete took ten days to set. In the meantime, they took apart the engine and overhauled every part of it. Once it could be started, Black envisaged the coal cutter to get going and extract more coal than before. They had already produced good gas, but the man charged with erecting the gas engines during the winter knew nothing about them. Black felt that he was even too generous paying him half his wage before letting him go, but he reasoned that if the men had been able to start the engines, the whole plant would have ended up in the sea due to the rotten engine bed. In the time available, Black would only get one engine going,

[109] Muschamp, P. (1907) *Letter to W. Clegg, 5 August.* William Olssons Archives 1899-1922 F2:20, National Museum of Science and Technology Archives, Stockholm.

which would suffice for the electric lighting, the fan, and the coal cutter. Black strongly criticised the choice to send three engines to Advent City, writing that 'one at least is absolutely unnecessary and I propose asking Campbells to exchange one of the engines for an oil engine, which would be much more useful in a climate like this. The third engine we could sell next summer.'[110]

Sources concerning the closing stages of the development are again scarce. In June 1908, the American steamer delivered a cargo of four tons of anthracite coal and about 900 pounds of carbide to Advent City.[111] The coal was probably destined to fuel the gas producer plant; the carbide may have been intended for lighting the workings and probably the settlement. Both are an indication that the plant was still not working properly. Without electricity, it is doubtful that the coal cutter was ever taken into commission.

5.4.3 Employees

Evidence relating to the company's employees and the relations between the management and the workforce is anecdotal and incomplete. It most likely contains considerable bias. The expeditions of the Bergen Co. were small. In 1901, it dispatched seven Norwegians to Spitsbergen, in 1903 15.[112] Drawing on English capital, the expeditions then increased in size. In 1904, 26 men, including the first five English miners, embarked in Tromsø.[113] In the following summer, the company mustered approximately 80 men, mostly Norwegians and Swedes but also 14 Englishmen.[114] These included eight miners, two foremen, two machinists, and two company representatives.

Since 1903, the expeditions had been led by the Norwegian mining engineer Stener August Fangen, of whom Munroe was so critical. In 1905, he was assisted by the Norwegian engineer Hovde. Arthur Mangham and Frank Matthews were probably the two foremen. At the end of August 1905, Fangen left on the *Heloise* to obtain provisions for the imminent wintering. In Tromsø, a message from Sheffield reached him that he had been fired due to the unfortunate circumstances during the summer.[115] Another source maintains that Fangen and the company had a disagreement over irregular finances, and the engineer chose to leave his position.[116] The board now asked Aagaard, the vice-consul in Tromsø, to send a

---

[110] Black, E. D. (1907) *Letter to F. Davenport, 30 July.* William Olssons Archives 1899-1922 F2:20, National Museum of Science and Technology Archives, Stockholm.
[111] Dole (1922a) p. 359.
[112] Hoel (1967a) pp. 553-84.
[113] Hoel (1967a) pp. 557-8.
[114] Hoel (1967a) p. 559.
[115] Hoel (1967a) p. 560.
[116] Braastad (nd).

small steamer to Spitsbergen with intermediate instructions for Mangham. Fangen took the occasion to send a letter to Hovde, which he then read to the workers. It stated that if they wanted to have their wages paid, they would have to report in Tromsø; otherwise their wages could not be guaranteed. Consequently, the whole workforce demanded to be sent home immediately. The Scandinavians, and presumably the English, left Advent City on board the *Hørda* on September 23, 1905. Three guards remained on site.

In 1905/6, the mining engineer Percival Muschamp[117], a family relation of Bainbridge, was the first winter superintendent.[118] He was again assisted by Hovde. Muschamp and the winter workforce of not 50 but merely 24 men did not arrive on site until the end of October. On board were also two English foremen (presumably Mangham and Matthews), Dr Ulstad, and the engineer Hans L. Daae.[119]

Ludvig Hansen was a Norwegian skipper and hunter, who had been recruited by Fangen for the pioneering winter work.[120] Following Fangen's exit, he kept his employment. He later recounts there may have been 28 men, mainly Norwegians but also Swedes and Finns. Hansen was in charge of loading while that was possible and then of hunting for fresh food. The men were labouring outside to build the incline and loading facilities, except when the weather was bad, and their shifts lasted from 6 am to 5.30 pm. Hoel states that a miner at the time earned 4 kroner per day, of which 1 ½ kroner went back to the company; Hansen maintained that the pay was 6 ½ kroner, but the above amount was indeed subtracted for board and lodging. An industrious and clever worker could make between 10 and 20 kroner a day. Miners and outdoor workers received the same, the labourers usually being from northern Norway. Of daily life, Hansen related that there was no disease in winter, but came the first cold in spring, everyone got it. There used to be no flies or mosquitoes, but now there were some. In the camp,

---

[117] Hoel (1967a) pp. 560-2.

[118] Percival Muschamp (1867-1941) was a mining engineer in Barnsley by 1885. At 21 years of age, he got his managers' certificate. Passionate about the British Army, Muschamp became a second lieutenant in the 2nd Volunteer Battalion of the York and Lancaster Regiment in 1893. His promotion to lieutenant of the rifle regiment soon followed. By 1896, Muschamp was manager at Primrose Main Colliery, which was dissolved in 1899. Muschamp may then have seen action in the Second Boer War (1899-1902). Ranked as captain, the instructor of musketry resigned his commission in 1902. After being the winter superintendent at Advent City, he found long-term employment with the New Hucknall Colliery Co. in Mansfield in the East Midlands. Bainbridge being one of its directors, Muschamp became its chief agent, a position that was later occupied by his son. A member of the Institute for Mining Engineers since 1889, Muschamp was elected as its president in 1928/29 and as its chairman in the following year. Profoundly interested in engineering problems, Muschamp and his son were assigned some patents for the improvement of machinery and methods in the 1920s and 1930s.

[119] Like Mangham, Daae would later find employment with another firm, the Arctic Coal Co.

[120] Hansen, L. (1909) *Opgjör tvistighed paa Spitzbergen.* Norwegian Foreign Department Box 5146, National Archives of Norway, Oslo.

Muschamp was trading with beer and spirits, which resulted in some unruliness. In 1905/6, the trouble later stopped, but it is unclear how.

Following the first winter, Mangham was dismayed because the miners had not been able to do their work.[121] He complained to Munroe that he could have extracted at least 6,000 tons if he and his men had been allowed inside the mine, but Muschamp had been wasting time trying to erect an incline outside instead of getting coal. Under the circumstances, they only managed about 500 tons. The gravity tram was apparently worthless as well.

5.19 A domestic scene at Advent City. Note the women standing in the open doorway of Huset. Fish is drying on the wall of another house. Horses are used for transportation. (Source: Wilse, A. B. (1906) NF.W 05522. Galleri NOR [Online]. Available at: http://www.nb.no/gallerinor/index.php (Accessed: 21 September 2011).

At the end of summer 1906, the Danish magazine *Husmoderen* ('The Housewife') published an article on daily life at Advent City.[122] It was penned by Dr Julie Kinck, who had replaced Ulstad, while her husband went to work at Longyear City. They had both applied on August 9, were appointed within two days, got married within four, and boarded a ship. In Tromsø, the couple equipped themselves with fur coats for the winter. They arrived on Spitsbergen on September 7, the wonderful summer weather soon being replaced by ferocious storms. Despite the absence of a manager, about 60 workers had been employed on site, most of whom were Norwegian. The building known as *Huset* had kept its name despite the fact that there were now 30 houses or more on Spitsbergen (Fig. 5.19). It accommodated Muschamp, the doctor, and a few others. On its wall hung the world's northernmost post box (Fig. 5.20), which played an important role in summer but was entirely

[121] Munroe, W. D. (1906) *Letter to the Arctic Coal Co., 13 July*. Arctic Coal Co. Collection, Privatarkiv 101, Box 91, Regional State Archive, Tromsø.

[122] *Husmoderen* (1906) 'Fra Advent Bay'. Available at: http://polarlitteratur.com/husmod.htm (Accessed: 20 September 2011).

unused in winter. Since all mail went to Norway, only Norwegian stamps were used. The residents of *Huset* followed English fashions and ate together, although they could otherwise live quite separately. A male chef prepared their meals, but the colony's three women were in charge of desserts and baking. There were provisions in abundance. The Norwegian photographer Anders B. Wilse, who had visited the settlement and taken numerous pictures, noticed that the workers lived on a good diet of salted reindeer and dried fish.

5.20 Manager Muschamp in front of his residence Huset. Note the post box on the wall next to him. (Source: Wilse, A. B. (1906) NF.W 05521. Galleri NOR [Online]. Available at: http://www.nb.no/gallerinor/index.php (Accessed: 21 September 2011).

Muschamp returned from England for the second wintering in 1906/7. From across the bay, Bert Mangham counted approximately 60 Scandinavians, mostly Swedes, in addition to ten English and Scottish miners.[123] That winter, Frank Matthews wrote his diary.[124] Christmas, for instance, was spent most pleasantly, with Muschamp arranging Christmas dinner. In February, the entries stopped for a few months; perhaps the pioneering spirit had been replaced by a daily routine. Into this silence fell a series of events that unfolded after April 2, 1907. The men had been receiving 50 kilograms of reindeer meat. However, provisions were running

[123] Dole (1922a) p. 294-5.

[124] Diary of Frank Matthews (1906/7). Frank Matthew's biographical details have kindly been provided by his granddaughter Aileen Thompson. Matthews was born in Thorpe Hesley near Rotherham in Yorkshire in 1877 and became a coal miner at the colliery managed by Arthur Mangham. In January 1900, he signed up with the 1st (Hallamshire) Volunteer Battalion of the York and Lancaster Regiment, and within a month, the rifle corps was sent to South Africa to fight in the Boer War. In May 1901, the lance corporal returned to England and was discharged from the army. Matthews sailed for Spitsbergen in May 1905. He stayed for over two years, presumably in the role of deputy, and finally returned in September 1907. Matthews was recalled into active service at the outbreak of the First World War. He was killed in action at Gallipoli on April 29, 1915. Following his death, his family remembers his close friend Muschamp as a kind man, who occasionally brought parcels of food, sweets, and toys for the children.

low and the shooting season was coming to a close.[125] On April 2, Muschamp had therefore decided to stretch the stores by allowing the men only 30 kilograms of meat and making the remainder up of something else. On April 3, Matthews came down from the mine in the workers' dinner hour at 1 pm. The Norwegian steward explained there was no meal for the men. When he went to tell the men, they became threatening. Although Muschamp eventually granted full rations, the workers came with other demands, which were impossible for Muschamp, who had a duty to the company, to settle. So the men said that they would not work again. Matthews recounts these events with some hindsight in his last surviving entry on May 12.[126] The remaining pages of the diary have been ripped out.

Much later, the former employee Martin Brøttum recounted his experiences during the winter 1906/7.[127] Although his memory had been affected by time, confusing Percy Muschamp and Bert Mangham, he nonetheless provides a worker's perspective. Because the Spitzbergen Coal & Trading Co. considered the islands to be a wilderness, it thought it appropriate to have a Boer officer looking after the workforce.[128] He always had his hand on his gun and was never slow to draw it if there was a problem. Inside his office, he had a fearsome collection of weapons. Muschamp, as it were, was a tyrant, and the men were afraid of him. During a conflict, he had the baker fired and replaced, as a result of which the bread was both raw and burnt at the same time. On December 23, it being the Norwegian Christmas Eve, the men gathered buckets, pots, and other items to march on *Huset* and make an awful racket to show their discontent. The manager, however, thought they had come to pay him homage and prepared to give a speech. He was quickly corrected. The situation came to a head on April 3, 1907. The food had been bad, and there was a strike. The workers stormed the barracks and the kitchen, and disarmed the policemen. Muschamp arrived, bearing his gun, and wanted to shoot the strike committee. If he shot, four Englishmen were under instructions to also open fire. It could have been a bloodbath, but the workers managed to disarm the Englishmen. It did not bring order, however. The food was still bad, the men got scurvy, and the strike continued all season.

From within his confinement, Muschamp sent word to the Americans.[129] Munroe, who had drowned in a shipping accident off the Dutch coast in February 1907, had been replaced by Kenneth L. Gilson.[130] The Scandinavian workers had

---

[125] There was no official shooting season on Spitsbergen, but the Spitzbergen Coal & Trading Co. may have imposed its own rules to sustain the reindeer population, late May to mid-June being the calving season.
[126] Diary of Frank Matthews (1906/7).
[127] Brøttum, M. (1952) 'Da Bert Mangham ville skyte streikekomiteen', *Nordlys* 4/1.
[128] Refer to footnote 118.
[129] Dole (1922a) p. 295.
[130] Hartnell (2009). Gilson would later find employment with the Northern Exploration Co.

threatened to seize the company's ship when it arrived in spring. They would take everyone at Advent City back to Norway, where they proposed to take Muschamp before a Norwegian court. They demanded full wages despite the fact that they had been on strike. It was vital, therefore, that the ship should not anchor near the site. The Americans spotted her on June 3, 1907.[131] Due to ice blocking Icefiord, she had to moor at the ice edge near Safe Harbour. Being forewarned, the directors left for Tromsø as outlined above. While the Scandinavians were on strike, only ten men had continued to work in the mine.

The incoming summer workforce appears not to have disembarked until June 26, 1907.[132] Probably in solidarity with the striking workers, they refused to take orders from Muschamp. Subsequently, Black, sixteen labourers, and two representatives of Campbell's Gas Engine Co. landed on July 25.[133] Black now relieved Muschamp of his command and reported that the summer workers were seemingly content. Based on the way they were occupied at the end of July, Black judged them to be a good lot.[134] The men said they had never eaten so well in their lives, although the company was providing them with exactly the same food as the men who went on strike, with the exception of fresh beef.

Black could see his way to producing a paying output of coal by the end of June 1908.[135] The company had accommodation for over a hundred workers. Black proposed to keep 40 men at Advent City during the winter of 1907/8, to be replaced by a summer workforce of about 80 men in the first week of June 1908, ice conditions permitting. Development thus continued at Advent City during the winter 1907/8.[136] To meet the contract with the Swedish Government, approximately 30 to 40 men stayed on the site. They produced less than expected. The sources fall silent thereafter.

5.4.4 Products

Information regarding the company's products is extremely sparse. Where available, the output of coal is expressed in rough estimates. Alternative resources are barely mentioned.

---

[131] Dole (1922a) p. 304.
[132] Dole (1922a) p. 318-9.
[133] Black, E. D. (1907) *Letter to F. Davenport, 30 July.* William Olssons Archives 1899-1922 F2:20, National Museum of Science and Technology Archives, Stockholm.
[134] Black, E. D. (1907) *Letter to F. Davenport, 30 July.* William Olssons Archives 1899-1922 F2:20, National Museum of Science and Technology Archives, Stockholm.
[135] Black, E. D. (1907) *Letter to F. Davenport, 30 July.* William Olssons Archives 1899-1922 F2:20, National Museum of Science and Technology Archives, Stockholm.
[136] Hoel (1967a) p. 565.

According to a witness from the American mine, the company extracted 400 tons of coal in summer 1904 but was unable to load more than 150 tons due to storms.[137] Hoel describes an output of 500 tons that summer that was partly sold to whalers on Spitsbergen before the *Heloise* left with a full cargo at the beginning of September, which was almost completely delivered to Bergen's gas plant.[138] Shortly after Mangham arrived at Advent City in summer 1905, he mined 200 tons, but in August, the miners were ordered to do surface work.[139] Hoel assumes that the target of 6,000 tons for winter 1905/6 was not met because the dark season and the pay caused the workers to progress slowly, extracting only 600 tons of coal.[140] Since much effort went into erecting houses and surface plant, the combined winter and summer output amounted to only 1,000 tons of coal in August 1906 and loading was still done using lighters instead of a proper dock.[141] The coal was shipped to Narvik and Trondheim.

During the strike in 1906/7, only ten men continued to work in the mine.[142] They got about 2,000 tons of coal, but it contained so much stone that cleaning was almost impossible. Some coal was shipped later that season. Its quality was supposedly so poor that it gave Spitsbergen coal a bad reputation.

In summer 1907, Black's men mainly laboured at the engine house.[143] Their tasks included fetching sand, breaking stone, and mixing concrete. They were cleaning the place, which had been left in a dismal state after the strike, unloading the company's steamer, filling tubs with coal from the stockpile, loading coal into the lighters to transfer to passing whalers, and loading coal for the Norwegian market onto the steamer. Black did not want to employ many men in the mine until the fully-loaded steamer had left. Nonetheless, the miners extracted 51 tons of coal over a five-day period: '10 tons on Saturday, 17 on Monday, 12 on Tuesday, and 12 on Wednesday.'[144] Incidentally, these numbers do not at all create the impression that the coal cutter was working. Some of the coal was sold for cash. Three whalers had also promised drafts for the outstanding amounts when they received their last lot of the season.

---

[137] Jeldness, O. (1904) *Letter to J. M. Longyear, 3 November.* Arctic Coal Co. Collection, Privatarkiv 101, Box 91, Regional State Archive, Tromsø.
[138] Hoel (1967a) pp. 557-9.
[139] Munroe, W. D. (1905) *Letter to F. Ayer and J. M. Longyear, 10 August.* Arctic Coal Co. Collection, Privatarkiv 101, Box 91, Regional State Archive, Tromsø.
[140] Hoel (1967a) pp. 560-2.
[141] Hoel (1967a) p. 562.
[142] Dole (1922a) p. 304.
[143] Black, E. D. (1907) *Letter to F. Davenport, 30 July.* William Olssons Archives 1899-1922 F2:20, National Museum of Science and Technology Archives, Stockholm.
[144] Black, E. D. (1907) *Letter to F. Davenport, 30 July.* William Olssons Archives 1899-1922 F2:20, National Museum of Science and Technology Archives, Stockholm.

Besides coal, Black had been selling the company's excess stores to whalers and other people at a good paying price. The skins got during the winter were another source of income. The company had previously sold reindeer skins to Anning & Cobb in London at 3s. Black now sent 123 reindeer skins, 7 blue fox skins, and 8 white fox skins to Muschamp. For the foxes, he expected 120 kroner and 40 to 50 kroner, respectively. To further improve the financial situation, Black suggested sending a cargo of pulp from Trondheim to England when the company's steamer left Spitsbergen in the middle of September.[145] He seemingly intended to winter and had arranged for the present doctor to remain at the reduced wage of 500 kroner. Another income could be the catching of beluga whales. The 'white fish' occasionally came into the bay in swarms, and Black had already engaged a man apt in the netting of belugas, which usually sold at £10 but sometimes for twice or three times that. Black concluded, 'the venture can be made a success if properly worked, but a lot of the extra expense must be cut down.'[146]

The development carried on in winter 1907/8, presumably under Black.[147] To meet the contract with the Swedish Government to deliver 6,000 tons of coal, up to 40 men stayed on the site. Although they produced only about 3,500 tons, this was said to be of decent quality.[148]

An output of 7,251 tons of coal between summer 1904 and spring 1908 has thus been documented. This only amounted to an average of about 1,800 tons per year or 900 tons per season. In 1908, the Swedish Government was willing to pay up to 16s 9d per ton. If the company had been able to sell its entire output at this price, it would have resulted in just over £6,000. Even with the proceeds from the skins and a generous buffer to allow for any output of coal not accounted for, it is clear that the returns from the mine did not cover the outlay.

5.5 Summary and conclusion

Environmental factors and the archaeology of the Spitzbergen Coal & Trading Co. were the focus of Chapter 3 and gave rise to the company's partial actor-network in Fig. 3.17. The results of archival research in this chapter emphasise that the strength in fieldwork lay in a better understanding of natural determinants, of the spatial arrangement of the settlement and mine, and of site formation processes following abandonment. The fieldwork was substantiated by fortunate discoveries

[145] Black, E. D. (1907) *Letter to F. Davenport, 30 July.* William Olssons Archives 1899-1922 F2:20, National Museum of Science and Technology Archives, Stockholm.
[146] Black, E. D. (1907) *Letter to F. Davenport, 30 July.* William Olssons Archives 1899-1922 F2:20, National Museum of Science and Technology Archives, Stockholm.
[147] Hoel (1967a) p. 565.
[148] Dole (1922a) p. 357.

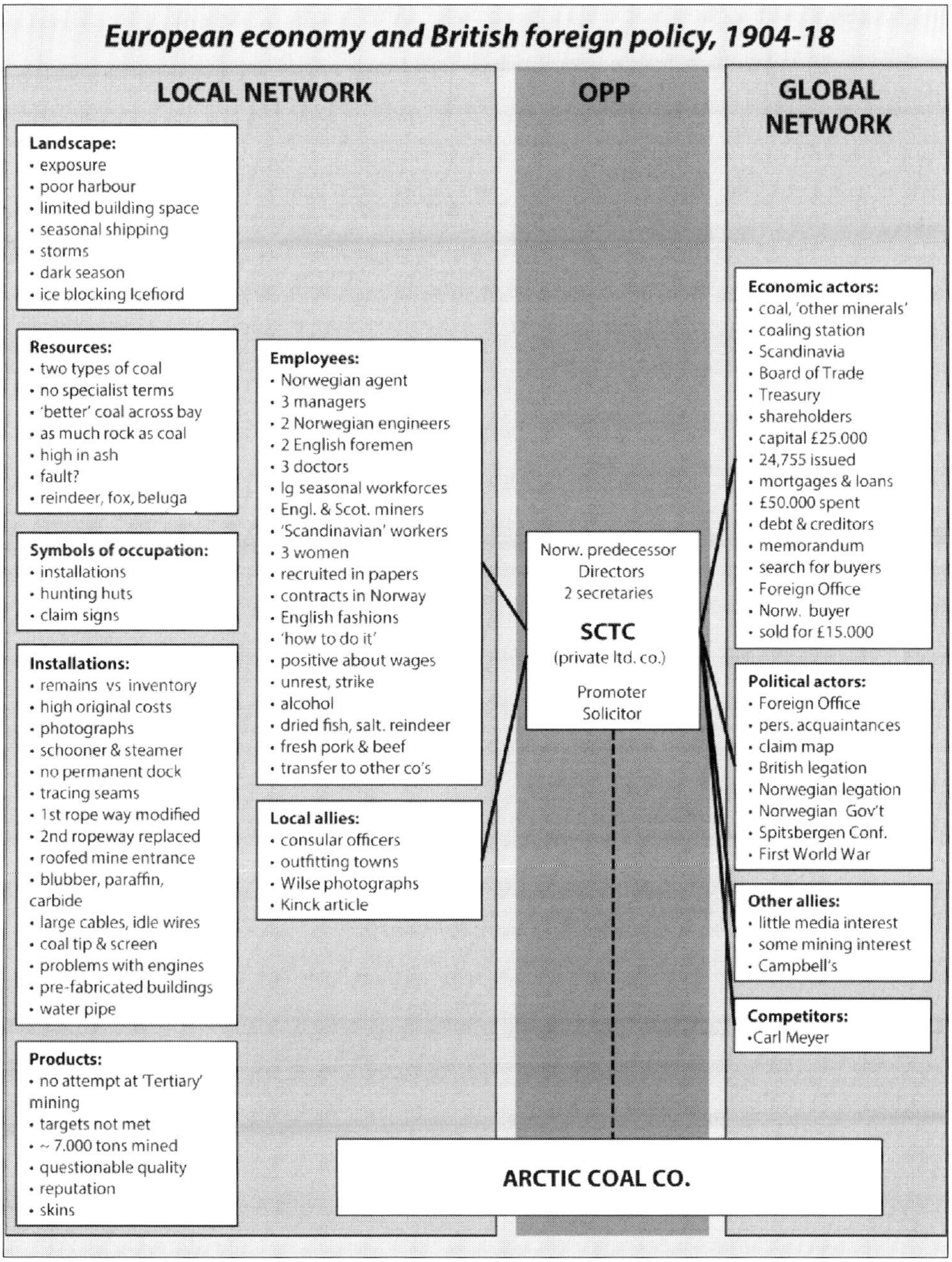

5.21 Documentary evidence completes the actor-network of the Spitzbergen Coal & Trading Co. (Chart: F. Kruse.)

such as Wilse's exceptional photographs and the inventory of original costs. In accordance with the principles of historical archaeology, the documentary evidence summarised in Fig. 5.21 completes the image of the company. Selected aspects are highlighted in this summary, which concludes with some general statements concerning the firm's motivations, strategies, and dissolution.

The composition of the head office had been practically invisible, but the documents revealed a private company limited by shares, usually restricted to 50 members. The endeavours of a predecessor were promoted by a Norwegian engineer previously known to one of the founding directors, and his information was initially accepted in good faith. The first board was assisted by a secretary, all of whom possessed in-depth knowledge of British coal mining. In addition, they enlisted in-house legal expertise. Later directors had experience in industry, too, but not in coal. Subsequent events put into question if they were as competent. Active Norwegian involvement or control was quickly phased out.

As shown in Fig. 2.2 and Fig. 5.5, British coal prices began to rise again in 1904, and it was as good a time as any to invest in the resource. Company correspondence also broadly refers to 'other minerals', but none ever played a role. The intended markets were whalers and other vessels as well as Norway and Sweden. The original claim was small as was the capital, but the financial plan included mortgages from the beginning. Black must have returned from the archipelago with an extremely positive evaluation for the capital to be increased tenfold in autumn 1904, but the members already showed reservations, and the shares were never issued fully. Only when £50,000 had been spent and another £25,000 was desperately needed did the board issue a memorandum with the suggestion to convert to a public company or sell. The lack of funds dampened the spirits, and when the unresolved financial problems caused particularly bad press, the Foreign Office got involved. The bargaining power of the company decreased relentlessly, and it was arguably a lucky circumstance that the war intervened and increased the Norwegian presence on the islands, through which a buyer came forward, and a reasonable purchase price was agreed. Debt made the company inconsistent, unreliable, and unable to deliver the goods. This only served to weaken the connections across its entire actor-network.

The links with the political actors had not been strong to begin with. Since British investments were at risk in a no man's land, the company appealed to the Foreign Office for immediate protection and the eventual settlement of the legal status. The case had no precedent, so the department adhered to the foreign policy of preserving the balance of power and announced that the company proceeded at its own risk. Despite the involvement of influential acquaintances and legal jargon, the Foreign Office was adamant. An important actant was the company's claim map, which was seemingly filed without hesitation. Although accountable to their respective foreign secretaries, the company may have turned to the British and Norwegian legations to have an audience better acquainted with the issues at stake, who might yet sway situation in its favour. The strike in 1906/7 made clear that no governmental help could be expected, even if a small legal victory had been won. The company had not been concerned with the break-up of the Swedish-Norwegian union, but when Norway's Arctic aspirations emerged, the possibility of a Norwegian take-over was not opposed but incorporated into the rhetoric surrounding the sale. Simultaneously, the Spitsbergen Conference called for Britain's participation, and the Foreign Office slightly adjusted its policy in order to safeguard all British interests. It was never the company's nor the Foreign Office's plan to control the islands.

Other allies were few and far between, as were competitors. There was little British media interest in the development, the Norwegian newspapers not having been evaluated. There were hardly any articles in British mining-related publications; there were practically none when a profitable output was not forthcoming. The manufacturers identified were merely suppliers. Only Campbell's played a further role in resuscitating the gas engines. Still, there is no certainty they ever worked. Instead, fieldwork proved that an exchange of the surplus engines never occurred, and the machines corroded on site. Resistance to the company arose as a reaction to outstanding debts. After 1908, its property was contested by two rivalling parties. The Norwegian Meyer was accused of claim-jumping, but he was under the impression that the company had ceased to exist.[149] His fellow countryman Hiorth opted for official negotiations with the board.

The American sources must be approached with caution. They nonetheless portray the Arctic Coal Co. as an actor in the global network as well as the local network. Its role as either an ally or a competitor is not clear-cut. It easily obtained intimate knowledge of Advent City and was quick to point out the company's disadvantages and incompetence. The English correspondence with the Americans, in turn, contains prime examples of the rhetoric it used to influence the overall context in its favour and handle this particular actor. Ultimately, it failed to establish itself at the obligatory point of passage, and the Arctic Coal Co. was free to use its knowledge as it saw fit. The damaging effect is evident, for instance, in the American visit to the mine. The link between the firms was always weak, and the Spitzbergen Coal & Trading Co. additionally conceded the autonomy over its network.

Using a range of documents, it has been possible to identify several employees by name and function. As such, Bruvik was the Norwegian agent; Fangen, Muschamp, and Black consecutive general managers; the Norwegian

[149] *Memorandum by Hertslet* (1913).

engineers Hovde and Daae likely deputies; and Mangham and Matthews probable mine foremen. Of the three doctors, Ulstad had been a pioneer on Spitsbergen and Kinck his female replacement. The workforces were large and seasonal. According to one witness, the few British miners also included some Scots. At the time, hewers in Britain received more 35s per ton mined (Fig. 2.4) and were an expensive commodity. This might have been the reasons why mostly Scandinavian workers were employed. Sources agree that Norwegians dominated, followed by Swedes, but Finns and Danes also find mention. Besides Kinck, there were two women. Perhaps Ida Jackson, after whom a street was named, was one of them. The workers were recruited through Norwegian newspapers, although other means may have existed. Whether they really had to sign English contracts without translation can no longer be proven. The management lived in English style and instructed its inexperienced miners with the help of illustrations underground. Wages may not have been a cause of unrest as one Norwegian stated that an industrious worker could make between 10 and 20 kroner a day. The strike may instead have erupted over cultural clashes, not least where the Scandinavians' consumption of alcohol was concerned, and the workers' dissatisfaction with the food. Provisions included bread, dried fish, salted reindeer, fresh pork, and eventually fresh beef, but it not known how this was rationed. That some employees harboured a continued personal interest in Spitsbergen showed their subsequent transfer to other companies on the island.

The consular officers of Bergen, Trondheim, and Tromsø were among the local allies, but they were, of course, not exclusively in the service of the company. What form their support took has not been traced, only that the company soon owed them considerable amounts of money. Hull, Haugesund, Bergen, Trondheim, and Tromsø were outfitting towns of the expeditions. Hammerfest appears to have been the closest seat of Norway's legal system. Less vague are the voluntary contributions of Kinck and the photographer Wilse. They created an exotic yet sophisticated image of the world's most northerly town, which initially served to enhance its reputation.

Archival research added little to the impression of the landscape generated during fieldwork, yet the frequent citation of the non-existent harbour and poor landing conditions highlights and quantifies this disadvantage. An employee referred to two coal seams on the company's claim, but if the British had also discerned the two different types of coal known to the Americans could not be shown. Scientific study was not at the forefront of their trials, and the contemporary actors did not use specialist terms such as 'Cretaceous' and 'Tertiary' as is the norm today. If they used or produced maps, they were probably topographic rather than geological. What mattered to the Americans was that their seam was of better quality than that at Advent City, which contained as much rock as coal, the coal also being high in ash. A vague map suggested a structural fault, which could not be substantiated. Reindeer, foxes, and belugas were listed as alternative resources.

It is unclear if the British took over the Norwegian strategy of using metal wire along the claim boundary. Although the claim was small by later British standards, wire along the entire perimeter seems unlikely. That the extension of the claim in 1904 was not properly communicated to the board hinted at the directors' limited knowledge and control of the local network. The identified symbols of occupation included the installations at Advent City as well as two hunting huts that sported large claim signs. Having been first in the field, the company was perhaps a little lax about defending its property. In any case, international competition only stiffened in later years.

The archaeological record at Advent City provided a thorough account of the installations. Although documentary evidence did not change the site interpretation, finds like the photographs and the inventory added interesting details and sharpen the overall conclusion. Notably, the items that remained on site were also among the most costly. Which ships were used could not be known by the archaeologists. A temporary floating pier had existed and may most likely have been removed rather than destroyed. Prospecting activities were at a minimum even in the documents. The very first ropeway had not been detected because the modified version used the same towers. The roof over the adit may have been taken away whole. Blubber, paraffin, and carbide had all lit the workings, and although cables had been laid, there was no indication that electrification was ever achieved. The archaeological signature of the coal tip and the screen had been missed. A report highlighted the ongoing problems with the engines, and as said above, an exchange was not effected. Wilse's pictures establish the sequence of settlement construction, but they fail to give an insight into the pre-fabricated interiors. One picture showed a water pipe, but whatever the source of the water, it will have frozen in winter.

In view of the diversity of actors and actants, the question remains how much the company produced. It never made an attempt to mine the Tertiary seam above the 600-metre contour, concentrating instead on the lower, Cretaceous seam. The firm repeatedly failed to meet its targets, and it appears that extensive surface works over consecutive seasons rather than poor loading conditions or the strike during one winter were to blame. In all, an output of 7,000 tons of coal is accounted for. The company never reached sustained production. Furthermore, the inferior quality of the coal had the detrimental effect of damaging the company's reputation and that of Spitsbergen coal in general. Any proceeds from skins will have been insignificant. Belugas were seemingly not exploited. The firm never recovered the original outlay, let alone produced a profit. It was unable to keep its promises to the global network.

In conclusion, the Spitzbergen Coal & Trading Co. was motivated by distinctly economic goals. It aimed for the profitable extraction of coal on an overseeable claim. It would not have engaged in political activities if it had not been necessary to secure the property in the no man's land in order to pacify its members and other stakeholders. Its operating strategy hinged on the knowledge of its directors and the combined mining experience of mainland Britain. Nonetheless, a conservative financial approach was ignored. Overspending and managerial incompetence were naturally not among the range of reasons the directors themselves gave for wanting to sell, nor were poor landing and the quality of the coal an issue. Instead, they listed difficulties with the Scandinavian workers, language problems, the inconvenient distance from Sheffield, and inexperience of prevailing conditions. They later added the death of their managing director and prime mover, without whose energy and optimism the stakeholders lost heart. All of the above played a part, but at the root of the company's failure essentially lay in ill-advised optimism and overconfidence in the supremacy of British capital.

6 The Spitzbergen Mining & Exploration Syndicate, Ltd. (1906-11)

6.1 Introduction

This chapter investigates how the limited archaeological record of Camp Morton and Camp Bell (Chapter 3) compares with the historical sources concerning the Spitzbergen Mining & Exploration Syndicate. Two private expeditions preceded the formation of the syndicate and witnessed the arrival of the adventurer Ernest Mansfield on the archipelago. While Mansfield staked out claims in Bell Sound, his friends initiated negotiations with economic and political actors in Britain. These occurrences must be viewed as an attempt to construct a global network, of which the imminent syndicate would merely be another actor. Rather than follow the networking activities of Mansfield, however, this chapter concentrates on the connections of the Spitzbergen Mining & Exploration Syndicate. Details of its formation in 1906 are accompanied by a timeline of subsequent events. Its global and local networks are analysed within the research framework of this study.

6.2 Early claims and company promotion

The origin of the Spitzbergen Mining & Exploration Syndicate lies in the chance meeting of three men. One was the doctor John Henry Salter (1841-1932), resident in Tolleshunt D'Arcy in Essex. The second was Mansfield (1862-1924), who had recently moved to nearby Goldhanger. The third was Frederick Thomas Gardner (1864-1936), rector of the Anglican Church at Goldhanger and Little Totham. This work benefits immensely from the recent publication of Mansfield's biography, which also elucidates the involvement of Salter and Gardner.[1] It is, however, useful to give emphasis to Mansfield's experience in British Columbia after 1898.[2] He became the manager of a French firm and constructed Camp Mansfield.[3] At an altitude of more than 2,200 metres, it was one of two camps with the distinction of being the highest in the region. A journalist explained, 'There are many miners who believe that the richest minerals are contained in the highest regions. When a trace of gold or silver is found in the highest part of a mining country, the chances are

[1] Barr, S., Newman, D., and Nesteroff, G. (2012) *Ernest Mansfield (1862-1924). "Gold, or I'm a Dutchman."*, Trondheim: Akademika Publishing.

[2] *Wanganui Chronicle* (1899) 'Mr Mansfield in British Columbia', 9 June, p. 3.

[3] BC Geological Survey and BC Ministry of Energy, Mines & Petroleum Resources (1997) *MINFILE 082FNW115 (Joker, L.3891)* Available at: http://minfile.gov.bc.ca/Summary.aspx?minfilno=082FNW115 (Accessed 1 December 2011).

that there is a lot of it there [...].'[4] This experience inadvertently prepared Mansfield for the Arctic.

Salter met Gardner when Mansfield brought the rector along to a dinner invitation.[5] The doctor prescribed him some rest, which in those days often meant travel, and the party discussed suitable destinations. Paramount was the idea of a voyage to the Land of the Midnight Sun. Mansfield not only whole-heartedly supported this idea; he proposed that Gardner use his time to prospect for gold. Mansfield himself would supply and instruct him. He believed that the world's undiscovered gold lay around the poles, and he was certain that there was as much ore in the European Arctic as there had been at Klondike. He therefore equipped Gardner with tools that the rector could carry on his back. Drawing on his familiarity with British Columbia, he instructed,

> When you get up there in Spitzbergen [...], the snow will have melted, the sun will have bared the land, you will find a carpet of flowers everywhere, and the rivers, which have been roaring torrents during the late spring and early summer, will have discharged into the sea and will be fordable and comparatively empty. There is no end of these rivers, but you must remember this – that they have brought bits of rock long distances, which you will find at the corners. At the corners also you will find mud, which is disintegrated rock. That has been brought some distance, perhaps not far. At any rate, make your way up some of these rivers, and into your sack put portions of this gritty stuff and mud and bits of rock, and when the other people are careering about eating cakes and dancing jigs under the Midnight Sun you will be using your time to much greater advantage. Bring your specimens home with memoranda as to where you got them, and we will see what they contain.[6]

In August 1904, Gardner and his wife travelled northward aboard the RMS *Ophir*.[7] He later recounted, 'I went out to [Spitsbergen] ostensibly for sport, but, finding the land during my travels there indicated so much mineral wealth in certain parts previously untrodden and unmapped, I became deeply interested in the infinite possibilities of that country.'[8] During his time ashore, he gathered a few promising samples of rock and mud. Upon his return to Goldhanger, Mansfield sent three of these for analysis. The analysis was undertaken by the principal assayers of the Bank of England. They found small quantities of gold in each sample.[9]

---

[4] *Wanganui Chronicle* (1899) 'Rapid developments in Camp Mansfield', 22 December, p. 2.

[5] Thomson, J. O. (ed.) (1933) *Dr Salter of Tolleshunt D'Arcy: diary and reminiscences: 1849-1932*, London: The Bodley Head, p. 335.

[6] Thompson (1933) p. 335.

[7] Reilly, J. T. (2009) *Greetings from Spitsbergen*, Trondheim: Tapir Academic Press.

[8] Gardner, F. T., 'Statement by Rev. F. T. Gardner', *The Northern Exploration Company, Ltd.* (1911) The Northern Exploration Company, Norwegian Polar Institute, Tromsø, pp. 11-8.

[9] Isaac Newton fixed Britain's gold price at 84s 11 ½d in 1700. This lasted for over 200 years. Britain suspended the strict gold standard during the First World War. (National Mining Association, '*The history of gold*', available at: http://www.nma.org/pdf/gold/gold_history.pdf, accessed: 12 July 2013.)

The small amounts of gold discovered in 1904 inspired the expedition in summer 1905. Gardner personally paid all expenses and was accompanied by Mansfield but not by Salter, who was hindered professionally.[10] They intended to travel by steamer to Norway and then board a whaler to Spitsbergen, but a whaler seems not to have been available. Gardner and Mansfield were compelled to travel in an old and unseaworthy sailing boat and were subjected to a horrendous journey.[11] It is safe to assume that Mansfield and not Gardner took charge of subsequent operations on shore. They had hired a couple of Norwegians, one of them being Oluf Martinsen from Tromsø, who acted as Mansfield's translator in 1905 and again in 1906. He later spoke very highly of the Englishman.[12] The party prospected mainly in Lowe Sound, where they supposedly found gold as well as coal, bitumen and oil, iron ore, gypsum, and phosphorus.[13] They marked these discoveries with stone cairns, buried pieces of parchment, and carefully mapped them. To further secure their claims, they attached the names of influential friends to each. The official date of discovery for the mineral areas in Lowe Sound was recorded as July 11, 1905.[14] Salter later admitted, 'we thus felt we had possession of something of world-wide value.'[15]

The resulting map (Fig. 6.1) is a testimony to their activities on Spitsbergen and the composition of their early network. The plan produced by Mansfield in July 1905 showed six large claims. The boundaries were remarkably straight, which suggests that they were arbitrarily drawn and not determined by natural features. Five claims lay to the south of Lowe Sound. According to Gardner, he and Salter had obtained 150 square miles of coal-bearing land each.[16] Their areas comprised

---

[10] Thompson (1933) p. 335; Hoel (1966) p. 420.

[11] A journalist later recounted, '[The boat's] sails were rotten as tinder, she was full of bilge water, and it was impossible to insure her. Such was the craft in which a 500 mile trip across the Arctic Ocean was attempted! Fortunately the weather was not rough, but the voyagers had frequent adventures in the ice, though they escaped being nipped. They were lost for three weeks, and nobody aboard could read the sun. Mr. Gardner was full of blood poisoning all the time from drinking polluted water and the horrible odour of the bilge. The skipper was 400 miles out of his reckoning when picked up by a whaler, who agreed to tow the boat to Bell Sound for £5, but after bringing the party in sight of the coast, cut them adrift in order to kill a whale. The currents, however, were favourable, the boat drifted shoreward, and in two days Mr. Gardner (who had eaten nothing for a fortnight) and Mr. Mansfield landed.' (*Spitsbergen's mineral wealth* (1918) MTU MS 631 LY B2/14, Michigan Technological University Archives and Copper County Historical Collections, Houghton/MI, p. 40.)

[12] Gardner, F. T., 'Statement by Rev. F. T. Gardner', *The Northern Exploration Company, Ltd.* (1911) The Northern Exploration Company, Norwegian Polar Institute, Tromsø, p. 12; Martinsen, O. (1910) 'Letter to H. Williamson, 21 October', *The Northern Exploration Company, Ltd.* (1911) The Northern Exploration Company, Norwegian Polar Institute, p. 79.

[13] Gardner, F. T., 'Statement by Rev. F. T. Gardner', *The Northern Exploration Company, Ltd.* (1911) The Northern Exploration Company, Norwegian Polar Institute, Tromsø, p. 12-6.

[14] *Memorandum by Hertslet* (1910).

[15] Thompson (1933) p. 336.

[16] Gardner, F. T. (1905) *Letter to Lord Lansdowne, 18 September.* Spitzbergen 1897-1905 FO 83/2147, The National Archives, Kew.

considerable sea frontage in Lowe Sound. Further south still lay three more claims totalling 300 square miles and presumably also comprising coal.[17] The westernmost was designated to Roderick Douglas and included a long shoreline in Van Keulen Bay. To the east of it lay the areas of Frederick Charles Byron and the Earl of Morton. These claims had no access to the sea.

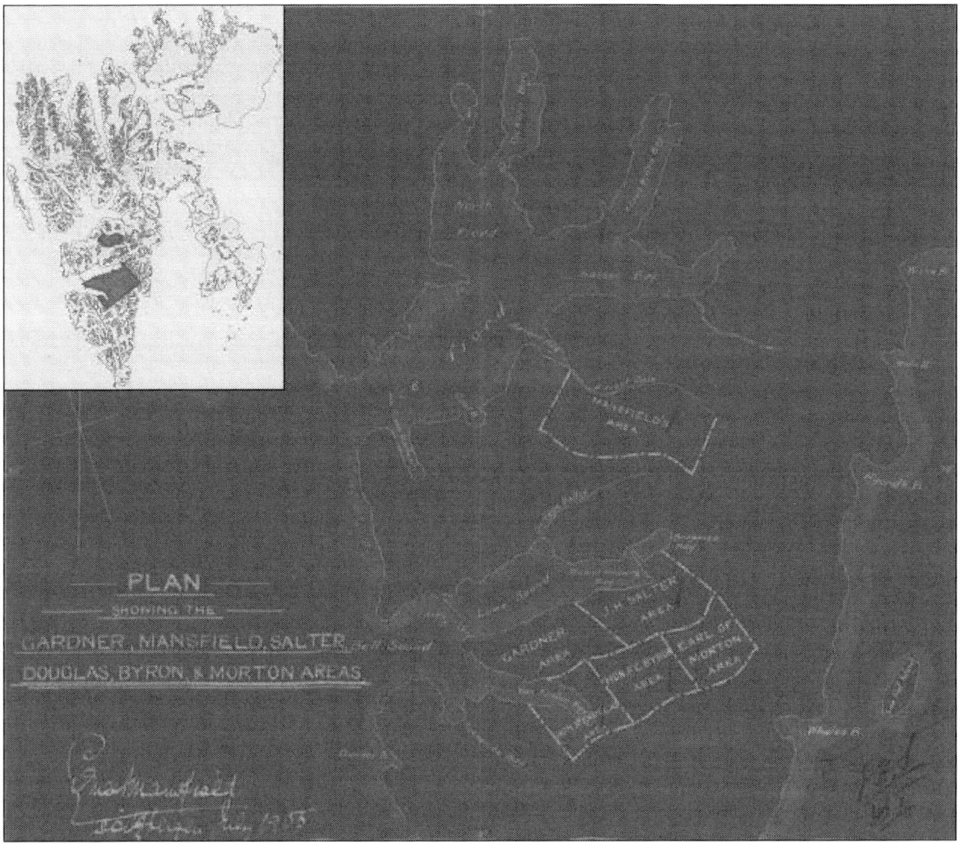

6.1 Map of six separate claims by Ernest Mansfield in July 1906. (Source: Plan showing the Gardner, Mansfield, Salter, Douglas, Byron & Morton areas (1905) Spitzbergen Mining & Exploration Syndicate BT 31/11526/88833, The National Archives, Kew.)

Gardner and Mansfield only spent a short but seemingly productive time on Spitsbergen. On July 17, 1905, Gardner, understandably not wanting to repeat the dreadful inward journey, left the islands on board the German passenger ship

[17] *An agreement made the twenty first day of May [...]* (1906) Spitzbergen Mining & Exploration Syndicate BT 31/11526/88833, The National Archives, Kew.

Moltke.[18] Mansfield remained for a time with the American mining engineer Munroe in Advent Bay. He was seemingly gathering what information he could about coal mining on Spitsbergen. During this time, he occupied the sixth claim shown on the map. This coal claim in Advent Valley measured 200 square miles, the official date of discovery being given as July 21.[19] He then boarded the German cruise ship *Prinzessin Victoria Luise* on July 29 to return via Norway to England.[20]

6.3 Formation and chronological overview

Salter's explanation as to why Gardner and Mansfield were able to make their discoveries on Spitsbergen ahead of others favoured the apparent hardiness of British explorers, '[Norwegians] only go to cut up whales and are frightened to death of going over the mountainsides for fear of seeing hobgoblins; they just go and cut up their whales and get back as soon as they can; they have no enterprise.'[21] The reality of British enterprise, on the other hand, was evident in the samples Gardner had brought back and the claim map Mansfield had produced. Gardner now spoke of a syndicate involving over 70 investors interested in gold, while Salter and possibly Earl Morton, Douglas, and Byron intended to develop the coal.[22] The need to consolidate and effectively represent the varied interests of an increasing number of stakeholders would best be dealt with by the formation of a public limited company.

The process of incorporation of the Spitzbergen Mining & Exploration Syndicate began with merging the existing claims into a single property. For this purpose, five contracts were drawn up on May 18, 1906.[23] By means of these contracts, Mansfield, Salter, Earl Morton, Douglas, and Byron sold their rights to Gardner. Respectively, they would receive 1,000, 250, 550, 500, and 250 fully paid deferred shares of 1s each in the company being formed. The payments were presumably indicative of each claimant's previous and continued involvement in the scheme. Additional applications and contracts followed in quick succession. On May 19, eight original subscribers, largely clerks and secretaries, applied for a

[18] Hoel (1966) p. 421.

[19] Gardner, F. T. (1905) *Letter to Lord Lansdowne, 18 September.* Spitzbergen 1897-1905 FO 83/2147, The National Archives, Kew.

[20] Orth, W. (1919) *Letter to Adolf Hoel, 2 June.* Norwegian Foreign Department Box 5173, National Archives of Norway, Oslo; Hoel (1966) p. 421.

[21] Thompson (1933) p. 337.

[22] Gardner, F. T. (1905) *Letter to Lord Lansdowne, 18 September.* Spitzbergen 1897-1905 FO 83/2147, The National Archives, Kew; *Memorandum by Hertslet* (1910); Salter, J. H. 'Report of Interview, 6 February 1911', *The Northern Exploration Company, Ltd.* (1911) The Northern Exploration Company, Norwegian Polar Institute, Tromsø, p. 21-8.

[23] *Prospectus of the Spitzbergen Mining & Exploration Syndicate* (1906) Spitzbergen Mining & Exploration Syndicate BT 31/11526/88833, The National Archives, Kew.

director in July 1906.[36] That month also witnessed 400 shares being divided among three other nominees before a financially quiet time followed.[37] It was presumably possible to raise a small amount of funds in October and November in view of the planned wintering at Camp Morton in 1906/7. Similarly, the expedition in summer 1907 found new sponsors in the months preceding it. An increase in nominal capital by £10,100 coincided with the peak of 1,442 ordinary shares being sold in May 1907. After the excitement again subsided, subscriptions for the rest of the year were considerably fewer. The spike in March 1908 may be an accumulation of shares allotted in the previous months and is not thought to represent a particular occurrence that stimulated potential investors. It is not possible to find a specific reason for the last shares sold in August that year.

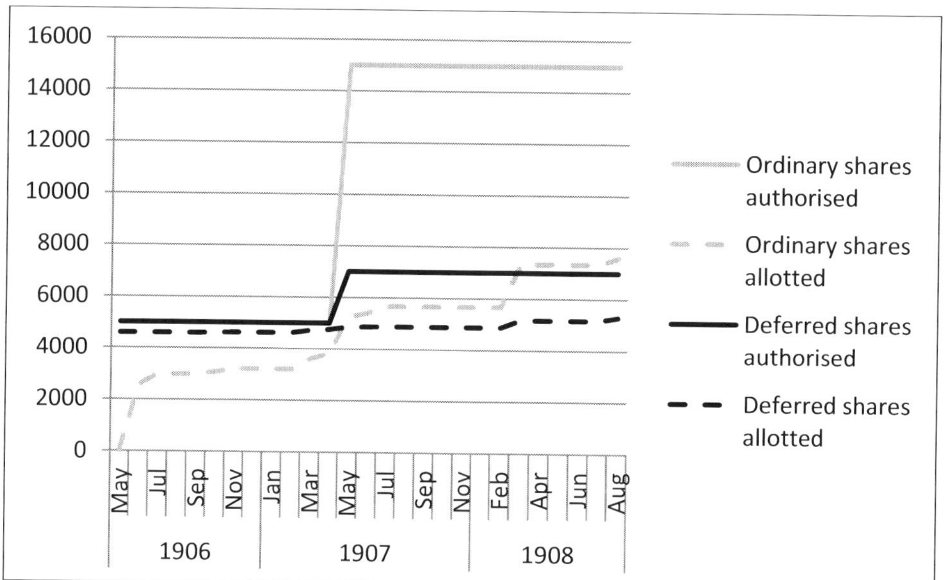

6.4 Graph indicating an overall lack of interest in the syndicate's shares. (Data: Spitzbergen Mining & Exploration Syndicate BT 31/11526/88833, The National Archives, Kew; Graph: F. Kruse.)

Despite the relative success with the Foreign Office, activities in the global network of the Spitzbergen Mining & Exploration Syndicate quickly stagnated. This may have been due to a lack of interest in its shares (Fig. 6.4). A balance sheet made up until May 31, 1908 indicated that only half its capital had been allotted after two

[36] Spitzbergen Mining & Exploration Syndicate BT 31/11526/88833, The National Archives, Kew.
[37] *Return of allotments to 2 July* (1906) Spitzbergen Mining & Exploration Syndicate BT 31/11526/ 88833, The National Archives, Kew.

years and that it owed roughly £515 to sundry creditors.[38] Among its assets, it counted £200 paid for the claims on Spitsbergen and £6,010 worth of prospecting and the development of the mine. Other expenditures included £220 for the registration of the company and £770 for office and administration, while the firm retained a cash reserve of £880. A total of £8,090 had therefore been spent. If the company were wound up now, its 98 shareholders would lose their investments, but any outstanding debt could be repaid using the cash reserves. Instead, another expedition was fitted out, the early reports and promises of which may have led to last lot of ordinary shares raising an additional £243 in August 1908. Thereafter, no financial transactions are known.

Following the final expedition, archival sources concerning the Spitzbergen Mining & Exploration Syndicate become sparse. In May 1909, the syndicate's offices had been relocated to Chapel House in Broad Street. A year went by. On April 19, 1910, the Company Registration Office addressed a letter to Broad Street to remind the firm to forward its annual members' list or be liable to a fine. The letter was returned unopened. It was then sent to Birkbeck Bank Chambers, from where it was also returned. On May 15, the Company Registrar enquired if the syndicate was still in business. A follow-up letter dated August 15 explained that, 'if a reply to [the] enquiry be not received within a month of the date of this letter, a Notice will be published in the London Gazette, with a view to striking the name of your Company off the register.'[39] As before, the letter was sent to Broad Street but returned. It was then sent to Tollemache in Wandsworth, who refused to take it in. So the Spitzbergen Mining & Exploration Syndicate was unceremoniously 'gazetted' on February 24, 1911.[40] It had lasted for less than five years.

6.4.2 Political actors

Following Gardner's original journey in 1904, the pioneering three contemplated the formality of occupying territory in the European Arctic. Based on Mansfield's experiences in the British colonies, they expected to need mining concessions to be able to work any mineral deposits. For the colonies, they could have applied to the Colonial Office; Spitsbergen being outside the British Empire, however, they identified the Foreign Office to obtain such concessions. On August 23, 1904, Gardner informed the department of the existence of promising mineral resources

[38] *Balance sheet, 31st May 1908* (1908) Spitzbergen Mining & Exploration Syndicate BT 31/11526/88833, The National Archives, Kew.
[39] Spitzbergen Mining & Exploration Syndicate BT 31/11526/88833, The National Archives, Kew.
[40] *The London Gazette* (1911) 'Joint stock companies', p. 1478.

up north.[41] Before he could consider any serious development, he enquired whether the Government would protect him if his rights were questioned; and if not, whom he should apply to for concessions.

Gardner was advised to appeal to the Foreign Secretary directly. In a letter to Lord Lansdowne, he disclosed his accidental discovery of gold and requested some security before further developments could take place.[42] His request was made official, albeit with an added note that he did not want the matter to become common knowledge.[43] Having first clarified the position of Sweden, the Foreign Office informed Gardner that 'Spitzbergen being under no particular flag, any mining operations there must be carried on at the risk of the promoters of the undertaking.'[44] Gardner probably hoped for a better verdict, but it did not hinder him and Mansfield to embark on their voyage in 1905.

Upon his return, Gardner again contacted the Foreign Office, emphasising the existence of valuable gold up north.[45] He and 71 others were now interested in developing an area of about 100 square miles, and he once more asked to be granted concessions. In addition, Mansfield, Gardner, and Salter had occupied almost 500 square miles of coal-bearing land and sought facilities to open these up, too. Gardner hoped that Lansdowne would afford them the same recognition that the claims of Lockyer[46] and Boney and of the Spitzbergen Coal & Trading Co. had received. The Foreign Office was not aware of a firm called Lockyer and Boney.[47] It did recall, however, that Bainbridge and his solicitors had previously asked for a guarantee of their property. Like Gardner, they had received the reply that Spitsbergen was under no particular flag and that the department could not give any concessions. The Foreign Office responded that Lansdowne was not aware of any facilities having been granted and that there was nothing to add to the letter sent on October 13, 1904.[48]

---

[41] *Memorandum of inquiry, 23 August* (1904) Spitzbergen 1897-1905 FO 83/2147, The National Archives, Kew.

[42] Gardner, F. T. (1904) *Letter to Lord Lansdowne, 19 September.* Spitzbergen 1897-1905 FO 83/2147, The National Archives, Kew.

[43] Barrington, E. (1904) *Internal memorandum, 21 September.* Spitzbergen 1897-1905 FO 83/2147, The National Archives, Kew.

[44] Foreign Office (1904) *Draft letter to Gardner, 13 October.* Spitzbergen 1897-1905, FO 83/2147, The National Archives, Kew.

[45] Gardner, F. T. (1905) *Letter to Lord Lansdowne, 18 September.* Spitzbergen 1897-1905 FO 83/2147, The National Archives, Kew; *Memorandum by Hertslet* (1910).

[46] Gardner may have been referring to Sir Norman Lockyer, a British astronomer and the founder of *Nature*, who undertook a cruise to Spitsbergen in 1896 in order to observe the total eclipse of the sun (Reilly, 2009, pp. 45 & 54). It is not known if Lockyer made any territorial claims.

[47] Perhaps Gardner had misheard the names Longyear and Munroe, who were in any case American and therefore fell outside the control of the British Crown.

[48] Sanderson, T. H. (1905) *Internal memorandum, 21 September.* Spitzbergen 1897-1905 FO 83/2147, The National Archives, Kew.

Salter now got involved to see what he could do.[49] He had little interest in the gold, but the occurrence of coal had swayed him.[50] Via a personal contact, he was introduced to Earl Percy, Lansdowne's right-hand man, at his Downing Street office on November 9, 1905.[51] Prior to the interview, Salter had sought legal advice to clarify the rights of a British enterprise in a no man's land. His advisers must have thought the British system, whereby minerals usually belonged to the landowner, to be unique. They concluded that any place outside Britain must be subject to European conventions, whereby individuals extracting minerals had no right to dispose of them. Salter had thus been advised that the minerals on Spitsbergen belonged to the Crown. He therefore reasoned that the Foreign Office should either give him the desired concessions or renounce any intention to confiscate any output. In addition, he suggested that the department take charge of the maps and documents, 'on which they relied to prove, in the event of rival claims being put forward either by private individuals or foreign governments, the fact that on a particular date they had explored and pegged out their claims to certain areas.'[52] Percy, however, could only submit the requests to Lansdowne and was unable to say what view the Foreign Secretary would take.

The Foreign Office feared that by indirectly supporting the claims of Gardner's syndicate, it would be seen to take responsibility for Spitsbergen, which it did not intend to do.[53] The ministry again searched for precedents as to how similar applications had been dealt with in the past. In addition to the case of Kerguelen Land in Antarctica in the previous chapter, the search revealed the examples of guano and coconut islands in the Pacific. Commonly very small, a number of these islands had been occupied by British subjects under license from the British Government. They had no native population and were thereafter only inhabited by the British and their servants. 'They are regarded as territory acquired by settlement and as forming part of the British dominions.'[54] Yet, the precedents

---

[49] Salter, J. H. 'Report of Interview, 6 February 1911', *The Northern Exploration Company, Ltd.* (1911) The Northern Exploration Company, Norwegian Polar Institute, Tromsø, p. 21-8.

[50] 'I attached more importance to the coal [...]. Here, I thought, is something substantial and certain, and, being so near the water, so easy to get, and in such demand all over the world, I considered this – the coal – the greatest find and told Mansfield so. My interest in the venture was greatly increased by the discovery of coal in the vast areas which bordered upon navigable waters. The coal must be of great worth, for there is all Northern Europe to supply.' (Salter, J. H. 'Report of Interview, 6 February 1911', *The Northern Exploration Company, Ltd.* (1911) The Northern Exploration Company, Norwegian Polar Institute, Tromsø, p. 22.)

[51] Percy, H. (1905) *Internal memorandum to T. H. Sanderson, 9 November.* Spitzbergen 1897-1905 FO 83/2147, The National Archives, Kew.

[52] Percy, H. (1905) *Internal memorandum to T. H. Sanderson, 9 November.* Spitzbergen 1897-1905 FO 83/2147, The National Archives, Kew.

[53] Sanderson, T. H. (1905) *Internal memorandum, 9 November.* Spitzbergen 1897-1905 FO 83/2147, The National Archives, Kew.

[54] Foreign Office (1905) *Internal memorandum, 21 November.* Spitzbergen 1897-1905 FO 83/2147, The National Archives, Kew.

were too few to effectively guide foreign policy. By November 28, 1905, Lansdowne was ready to ask the Treasury for assurance that the Crown would not confiscate any minerals extracted on Spitsbergen, but he would take no further action in the matter. The syndicate was free to apply to any other Government that might be a stakeholder on the archipelago. Lastly, the Foreign Office would not take the title deeds into custody. 'It would be quite contrary to precedent that the Secretary of State of Foreign Affairs should take charge of such documents.'[55] Salter should instead deposit them with a bank or other reputed financial establishment.

Gardner, Mansfield, and Salter hence applied to other Governments, whom they believed to be stakeholders on Spitsbergen, but every nation represented on the archipelago at the time denounced responsibility and could not be relied on to grant any mining rights.[56] The trio therefore set out to prove that they had been the first discoverers of their claims and held the primary title. They drew up a statement on March 6, 1906, which read,

> The intension of this document is to place on record (for purpose of identification and claim) the fact that I, Rev. Frederick Gardner, Rector of Goldhanger, Witham, Essex, and Ernest Mansfield, Mining Engineer, of 19-21, Queen Victoria Street, London, discovered minerals (gold and coal to wit) in the Island of [Spitsbergen] in the month of July, 1905. We endeavoured subsequently to ascertain the sovereign owner of [Spitsbergen], in order to procure a concession of the mineral area in accordance with the rights and customs of the country owning it, but our efforts were unsuccessful. [...] By virtue of our being the first Discoverers we claim the sole right of possession of the above-mentioned areas, with the sole right of access and to work and carry away minerals and generally to use and deal with the property for our sole use and benefit in such way and at such times as we shall consider best.[57]

On April 5, 1906, Gardner and Salter lodged the maps and documents with Salter's bank in Colchester. The title deeds could not remain in private hands, where they were at risk of falsification; they had to be locked up in a reliable institution to assure an unbroken chain of custody. 'If at any time in the future it might be required to establish the proof of the discoveries, they would be concrete

[55] Gorst, E., 'Letter to C. Strutt, 28 November 1905', *The Northern Exploration Company, Ltd.* (1911) The Northern Exploration Company, Norwegian Polar Institute, Tromsø, p. 29.

[56] Afterwards, Salter reflected, 'Now how were we to tackle this business? To take possession of a huge continent almost, like Spitzbergen, was a mighty big order. Did no country own it? Where was Norway? Where was Sweden? Where was Denmark? Where was England? Where was anybody? Surely we could get somebody on proof of value to say, "Yes, we'll own it." We went to the Ambassador of the different countries and of Russia – of every country we could – asking them if they would own this. They all said No, it was of no use to them at all. We even went to America, but no country would have anything to do with it.' (Thompson, 1933, p. 337.)

[57] Mansfield, E. and Gardner, F. T., 'A statement relative to an important discovery of minerals in "N", 9 March 1906', *The Northern Exploration Company, Ltd.* (1911) The Northern Exploration Company, Norwegian Polar Institute, Tromsø, p. 38-40.

evidence.'[58] The men could presently do no more to secure their claims on Spitsbergen. It would have to suffice if they were to attract investors and other backers to the scheme.

Simultaneously, Campbell tackled the task of securing the properties.[59] On January 20, 1906, he sent an application for Douglas' coal claim to the Foreign Office together with a description and a plan of the area. He also sent an application for Earl Morton's claim. Both were returned on January 24 with a reminder that Spitsbergen was not under British jurisdiction. On May 11, Campbell forwarded the details of Mansfield's claim; they were returned. In June, he submitted Byron's and Salter's claims. Although they, too, were returned, the Foreign Office added that 'Sir E. Grey was very much surprised that he should continue to submit these plans after it had been fully explained to him that they could not be entertained.'[60] Campbell replied that 'he had been instructed to forward the applications to Sir Edward Grey because the applicants knew of no other person to whom they might properly send them.'[61]

Meanwhile, Lansdowne's private secretary had spoken to Nansen. Norway's minister to London did not believe that any power would intervene with those who intended to exploit the natural resources of the islands. This may have been a turning point. In June 1906, Campbell succeeded in depositing the first lot of claims at the Foreign Office.[62] Although the planning of the next expedition to Spitsbergen had not hinged on this, more ordinary shares were sold in that month than in any thereafter. The new claims staked out in summer 1906 (Fig. 6.5) were correspondingly easy to lodge. In November, the syndicate's solicitors informed the Foreign Office of a second claim made on behalf of Earl Morton. The Foreign Office briefly acknowledged this and filed the enclosed map with the Spitsbergen memorandum in the Norwegian Dossier in the Western Department. On April 16, 1907, Campbell sent the particulars pertaining to the Earl's second claim as well as claims by the Countess of Morton and by Campbell himself and Edwin Ponsonby. These, too, were duly acknowledged.

Later records involving the political actors are sporadic. In November 1908, the Board of Trade informed the Foreign Office that the syndicate had not yet made a success of its coal property in Bell Sound, but it appeared not have given up hope as Mansfield was still prospecting on their behalf.[63] In October 1909, the British minister to Kristiania updated the Foreign Office that another mine was said

58 Salter, J. H. 'Report of Interview, 6 February 1911', *The Northern Exploration Company, Ltd.* (1911) The Northern Exploration Company, Norwegian Polar Institute, Tromsø, p. 24.
59 *Memorandum by Hertslet* (1910).
60 *Memorandum by Hertslet* (1910).
61 *Memorandum by Hertslet* (1910).
62 *Memorandum by Hertslet* (1910).
63 *Memorandum by Hertslet* (1910).

to be opening in Bell Sound, where Mansfield had found gold, albeit not in sufficient quantities to be worked commercially.[64] However, the history of Camp Millar, as it were, already belongs in the next chapter.

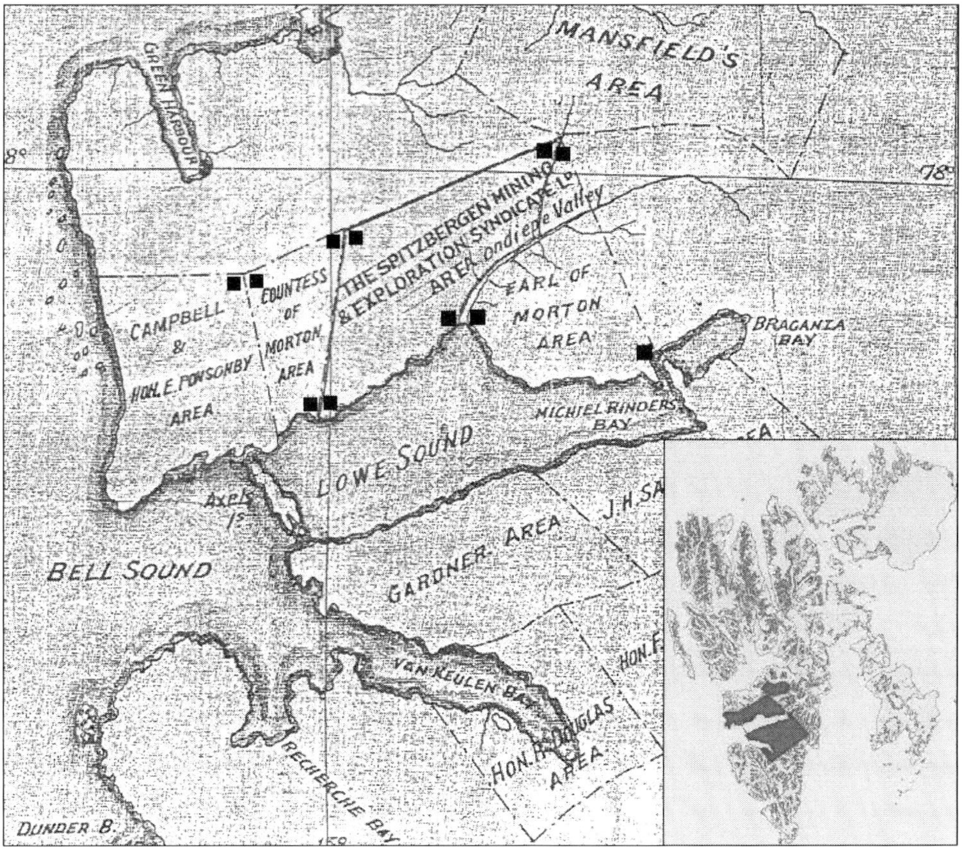

6.5 Detail of map showing four new claims occupied in July 1906. The squares allegedly mark the locations of claim posts. (Source: Plan showing area belonging to the Spitzbergen Mining and Exploration Syndicate (1906) Spitzbergen Mining & Exploration Syndicate, BT 31/11526/88833, The National Archives, Kew.)

6.4.3 Competitors

Mansfield had staked out claims to the north of Bell Sound in 1906, which encompassed *Michelsenhuset* built in 1901. Michelsen had arguably let the claim lapse by the time the sealing captain Johan Hagerup arrived on site to protest

[64] *Memorandum by Hertslet* (1910).

against Mansfield's use of the house and the area.[65] This competitor, however, was quickly won over when the men agreed to jointly pay for two trappers to winter and guard the camp. Mansfield and Hagerup would divide the catch between them.

Other rivalry involving the syndicate's claims developed over time. Barr *et al.* explain that Adolf Hoel, a Norwegian geologist, first visited Spitsbergen aboard a small mapping expedition in 1907, which fuelled his belief in Norway's natural rights in the Arctic.[66] To attain sovereignty of Spitsbergen, it was essential that as much land as possible should end up in Norwegian hands. Hoel viewed Mansfield as an opponent, and his efforts to discredit the prospector tainted later accounts of Mansfield's activities. Whether or not Mansfield had claimed coal in Bell Sound and, more importantly, in Kings Bay became crucial to Hoel's work leading up to the ratification of the Spitsbergen Treaty in 1925. He concluded that although Mansfield had prospected for gold on the south side of Kings Bay in 1906, he had not been interested in coal. Despite two photographs, one taken by Earl Morton's son William Douglas, that documented the trial works on the southern shore of Kings Bay that year, Hoel's conclusion holds firm.[67] However, this strong opposition was not aimed at the Spitzbergen Mining & Exploration Syndicate in its day but at the Northern Exploration Co. in later years, which habitually based itself on Mansfield's early claims.

6.5 The local network

6.5.1 *Claims and natural resources*

The incorporation of the Spitzbergen Mining & Exploration Syndicate coincided with the fitting out of the expedition in 1906. It was no longer only paid by Gardner but benefitted from external funds and better organisation. The primary aim was said to be proving the coal. Sponsored by many of Salter's personal friends, Mansfield and Gardner departed England in June.[68] In July, workers arrived on board the steamer *Mylingen*.[69] At the same time, Earl Morton and Lord Balfour of Burleigh undertook a cruise to Spitsbergen aboard the RYS *Cresseda* and staked out claims in Lowe

---

[65] Barr, Newman, and Nesteroff (2012) p. 87.
[66] Barr, Newman, and Nesteroff (2012) pp. 77-86.
[67] Mansfield, E., 'Report on low grade quartz gold mines in "N", 2 March 1911', *The Northern Exploration Company, Ltd.* (1911) The Northern Exploration Company, Norwegian Polar Institute, Tromsø, p. 124; http://www.douglashistory.co.uk (2011) (Accessed: 13 May 2011).
[68] Salter, J. H., 'Report of Interview, 6 February 1911', *The Northern Exploration Company, Ltd.* (1911) The Northern Exploration Company, Norwegian Polar Institute, Tromsø, p. 24.
[69] Mann, C., 'General report from Mr. Charles Mann, 21 January 1911', *The Northern Exploration Company, Ltd.* (1911) The Northern Exploration Company, Norwegian Polar Institute, Tromsø, p. 112-4.

Sound, Coles Bay, at Cape Thordsen, and in Kings Bay.[70] Shortly afterwards, other members of the Douglas family sailed the bays aboard the SY *Latona*.[71] Photographs show them in Bell Sound, in Icefiord, and in Kings Bay. Gardner travelled back on the *Latona*, dining with Salter at Goldhanger Rectory on July 18. Salter's corresponding diary entry read, 'His account is wonderful, and there seems to be a great and extraordinary result looming over all the participants, myself among the number. We shall see.'[72]

Munroe did not allow the *Mylingen* to be discharged on American territory.[73] The incoming workers, however, disembarked in Advent Bay and walked through Advent Valley to stake out several new claims. Mansfield had supposedly already discovered minerals on the north side of Bell Sound and Lowe Sound in 1905.[74] He now occupied the whole of the northern shore under different names (Fig. 6.5). Mansfield officially made the claim on behalf of the Countess of Morton on June 29, 1906. He erected posts in each of the four corners, although only three are shown on the map, which identified the owner, delineated the boundaries, and provided the dates of location and staking off. That day, he also marked out a tract of land for the syndicate, which besides the claim boards comprised the coal mine at Camp Morton in the southernmost extent. Unbeknown to Mansfield, Edwin Ponsonby had claimed an area on behalf of Campbell and himself, which Mansfield erroneously counterclaimed on July 27.[75] The mistake was quickly rectified. July 28 went on record as being the official date, when Mansfield took possession of a second, almost triangular claim designated to Earl Morton. It constituted three claim boards. All new claims had a considerable water front. Each of them was recognised by the Foreign Office.

Miners were left at Camp Morton while a party took a boat northward to Kings Bay.[76] In Kings Bay, Mansfield was so impressed by the aptly named Marble Island that he kept its discovery a secret from the foreigners in his service and from the crew.[77] Eager to prevent any suspicion, he did not blast the marble. Instead, he and Mann collected loose samples for analysis in England. While the two men ascended the summit of the island and erected a post stating their names,

---

[70] Førisdal, L. and Lien, H., (nd) *No man's land: the Scottish Spitsbergen Syndicate Ltd – SSS*. Available at: http://www.svalbardmuseum.no/skilt/index.php?skiltselskap=18&lang= (Accessed: 1 July 2011); Hoel (1967a).

[71] http://www.douglashistory.co.uk (2011) (Accessed: 13 May 2011).

[72] Thompson (1933) p. 136.

[73] Barr, Newman, and Nesteroff (2012) pp. 81-2.

[74] *Memorandum by Hertslet* (1910).

[75] An almost illegible claim board survives in the Svalbard Museum.

[76] This boat may well have been the *Latona*, but Mansfield was eager to conceal any connection with Earl Morton and the Douglas family in later sources.

[77] Mansfield, E., 'Report upon Marble Island, 2 March 1911', *The Northern Exploration Company, Ltd.* (1911) The Northern Exploration Company, Norwegian Polar Institute, Tromsø, p. 59-65.

Alexander stayed behind to keep an eye on their two sailors, who had remained in the boat, fearing that they could collect samples themselves.[78]

After leaving Marble Island, the men crossed Kings Bay and landed on its southern shore. This area had been claimed by the Bergen Co. in 1901 but abandoned in favour of Advent City. Mansfield and his men now investigated the drift and supposedly discovered traces of gold.[79] They then returned to Camp Morton, from where Mann and four Norwegians rowed to No. 10 Valley, as of yet unidentified, to stake out claims in the valley and along the coast of Braganza Bay. Yet, the north of Braganza Bay appears to be unclaimed in the map above. Mann returned to Camp Morton, where the expedition stayed for six weeks until September 1906, when they temporarily left two Norwegians in charge.

The syndicate intended to develop the coal during winter 1906/7. For this purpose, Arthur Mangham, formerly of the Spitzbergen Coal & Trading Co., arrived on October 5, 1906. He seemed to have remained until August 14, 1907. Under his supervision, a group of miners drove the headings further into the coal. His report on the work carried out and on the nature of the coal was uncharacteristically brief, 'Splendid coal all the way, it only requires further proceeding into the hill.'[80]

The syndicate sent an expedition in summer 1907. Little is known about it, but it was marked by two occurrences: the preparation of a professional report on the coal at Camp Morton and the discovery of a lucrative quantity of gold. Gardner had undertaken his fourth and final journey to the archipelago when Tollemache and the Scottish mine inspector William Galloway arrived in Lowe Sound on August 26, 1907.[81] Their steamer had been delayed at the ice edge (Fig. 6.6), which gave them a first impression of local conditions. Galloway additionally noted that all crafts destined for Lowe Sound needed to pass by Axel Island. On one hand, the island provided protection against swell and storms, but on the other hand, the narrowest part of the passage had yet to be fathomed and was presently a threat to larger ships. At Camp Morton, he found the sea floor to slope away gradually for a distance of 600 feet to where the steamer was anchored in 20 feet of water. Other aspects of the report are covered in the relevant sections below.

---

[78] Alexander, G., 'Letter to H. Williamson, 23 January 1911', *The Northern Exploration Company, Ltd.* (1911) The Northern Exploration Company, Norwegian Polar Institute, Tromsø, p. 73-5.

[79] Mann, C., 'General report from Mr. Charles Mann, 21 January 1911', *The Northern Exploration Company, Ltd.* (1911) The Northern Exploration Company, Norwegian Polar Institute, Tromsø, p. 112.

[80] Mangham, A., 'Letter to H. Williamson, 10 August 1910', *The Northern Exploration Company, Ltd.* (1911) The Northern Exploration Company, Norwegian Polar Institute, Tromsø, p. 109.

[81] Gardner, F. T., 'Statement by Rev. F. T. Gardner', *The Northern Exploration Company, Ltd.* (1911) The Northern Exploration Company, Norwegian Polar Institute, Tromsø, p. 11; Galloway, W. (1907) *Report on the mines at Camp Moreton, Spitzbergen, 20 September* YF622.33 (984) GAL, British Geological Society, London.

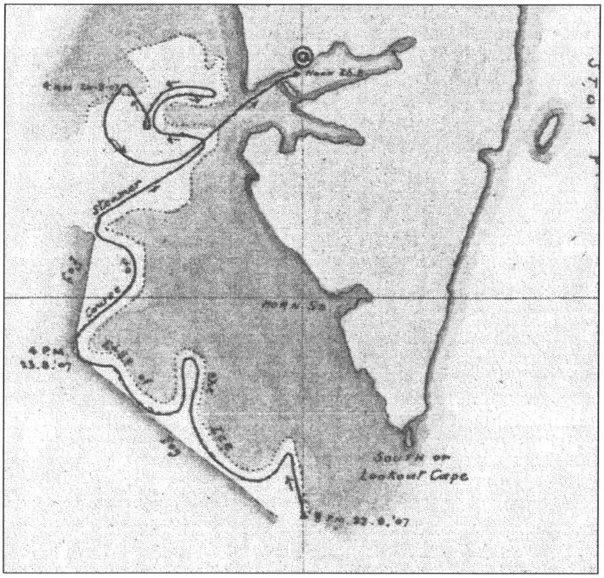

6.6 Galloway recorded the steamer's efforts to break through the ice between August 22 and 26, 1907. (Source: Galloway, W. (1907) Report on the mines at Camp Moreton, Spitzbergen, 20 September YF622.33 (984) GAL, British Geological Society, London.)

Mann returned to Spitsbergen with the expedition in summer 1908.[82] While he constructed Camp Bell, he noticed conglomerate as a possible source rock for gold. He also referred to oil and coal. Despite the possibility of gold, it was seemingly the presence of anthracite that warranted the effort being put into the construction of the camp. As many as nine workable seams were supposedly proven in the vicinity (Fig. 6.7).[83]

While the official focus of the syndicate was stated to be coal on numerous occasions, Mansfield continued the search for gold.[84] In 1907, he returned to England with a few specimens of conglomerate. One of Salter's acquaintances took a sample to Scotland and had it assayed at Edinburgh University. It contained lucrative six pennyweights (dwt) of gold. Equal to 120 grains or 9.3 grams of gold in a ton of rock, this amount excited further exploration. Throughout summer 1908, Mansfield searched for the source of the gold. He sent back another seven samples. On August 20, Salter received a telegram that the assay result was a disappointing 16 grains of gold to the ton and therefore useless. Yet, gold had been found in every sample.[85] Mansfield vowed to remain on Spitsbergen until he

[82] Mann, C., 'General report from Mr. Charles Mann, 21 January 1911', *The Northern Exploration Company, Ltd.* (1911) The Northern Exploration Company, Norwegian Polar Institute, Tromsø, p. 112-4.
[83] Gardner, F. T., 'Statement by Rev. F. T. Gardner', *The Northern Exploration Company, Ltd.* (1911) The Northern Exploration Company, Norwegian Polar Institute, Tromsø, p. 14.
[84] Salter, J. H. 'Report of Interview, 6 February 1911', *The Northern Exploration Company, Ltd.* (1911) The Northern Exploration Company, Norwegian Polar Institute, Tromsø, p. 25.
[85] Thompson (1933) p. 140.

had found the source rock. He wintered at Camp Bell in 1908/9, writing *The Icemaiden*, and returned, allegedly successful, to England in July 1909. Mansfield was back on the archipelago in summer 1910. That journey, however, marks the beginnings of the Northern Exploration Co.

6.7 This photograph of George Alexander supposedly documents the discovery of an anthracite seam. (Source: The Northern Exploration Company, Ltd. (1911) The Northern Exploration Company, Norwegian Polar Institute, Tromsø, p. 102.)

6.5.2 Manifestations

Documentary details concerning the installations on the syndicate's property are rare prior to Galloway's arrival at Camp Morton. In summer 1906, Mann had timbered up the headings that the miners had driven into the coal on either side of a mountain gully.[86] In the following winter, Mangham's men drove the headings another 400 yards.[87] Galloway went to work as he would have done for any employer in any coalfield, affording Spitsbergen no special status. He recorded that the landscape lacked vegetation, so that geological features were easily traced in mountains of stratified rock that rose sharply from near the shore to 2,000 feet or more. Galloway could see no faulting on Coal Mountain, but he suspected some in the valleys and gullies. He noted that the strata dipped gently eastward, while the drift mines had been driven north to northwest at a dip of four to five degrees. He thus concluded that, 'provided coal seams of suitable thickness and quality are

[86] Mann, C., 'General report from Mr. Charles Mann, 21 January 1911', *The Northern Exploration Company, Ltd.* (1911) The Northern Exploration Company, Norwegian Polar Institute, Tromsø, p. 112.
[87] Mangham, A., 'Letter to H. Williamson, 10 August 1910', *The Northern Exploration Company, Ltd.* (1911) The Northern Exploration Company, Norwegian Polar Institute, Tromsø, p. 109.

found anywhere along the margin of Lowe Sound the geological conditions are very favourable to their economic development and working.'[88]

In fact, a lower and an upper coal seam had been found at Camp Morton (Fig. 6.8). Layers of stone divided each seam into two or three coal beds. The miners had investigated the lower seam in drift M and the upper seam in headings N and O. In addition, they had made other openings of about a yard or two, which had been allowed to fall in and could no longer be assessed.

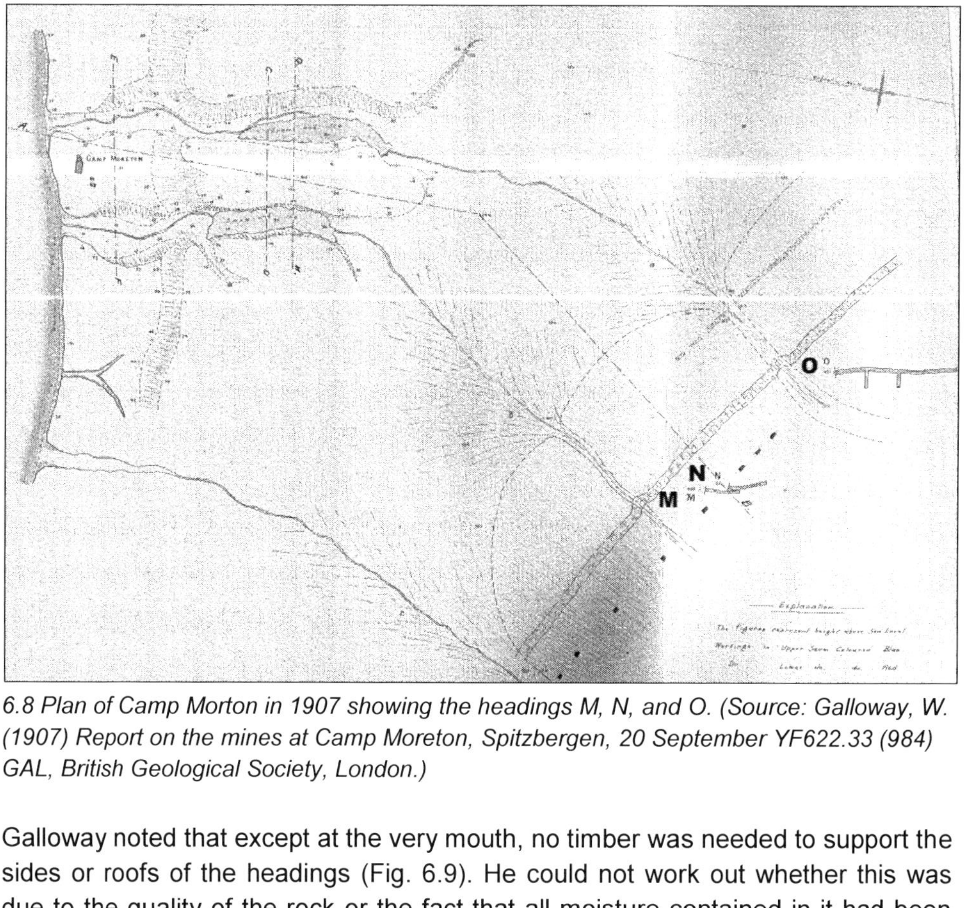

6.8 Plan of Camp Morton in 1907 showing the headings M, N, and O. (Source: Galloway, W. (1907) Report on the mines at Camp Moreton, Spitzbergen, 20 September YF622.33 (984) GAL, British Geological Society, London.)

Galloway noted that except at the very mouth, no timber was needed to support the sides or roofs of the headings (Fig. 6.9). He could not work out whether this was due to the quality of the rock or the fact that all moisture contained in it had been frozen into ice, acting as an adhesive. The headings were generally five feet wide and up to six feet high, and no coal had been taken except that obtained by driving the heading. Ventilation of the mines was not a problem, being facilitated either by

[88] Galloway (1907) p. 3.

the differences in temperature inside and outside the mines, the heat from the miners' bodies, or the burning of candles. Neither was there any trouble with water as it had been frozen. The natural conditions were altogether favourable for coal mining at Camp Morton. In the mines, wooden bars were attached to wooden sleepers, which acted as rails for small wooden tubs with steel wheels and axles that fitted a narrow gauge (Fig. 6.10).

6.9 Timber at the mouth of a mine at Camp Morton. (Source: The Northern Exploration Company, Ltd. (1911) The Northern Exploration Company, Norwegian Polar Institute, Tromsø, p. 96.)

6.10 Wooden tubs used in a mine at Camp Morton. Note Ernest Mansfield in the light jumper on the far left. (Source: The Northern Exploration Co., Ltd. (1911) The Northern Exploration Company, Norwegian Polar Institute, Tromsø, p. 94.)

In the mines, the sides and the roof were encrusted with hoarfrost. Mine M penetrated the lower coal seam at approximately 456 feet above sea level and had been driven for about 121 feet. The coal rapidly got thinner and the stone thicker until the lower bed, the stone, and the upper bed were 8 inches, 4 feet, and 1 foot, respectively (Fig. 6.11).

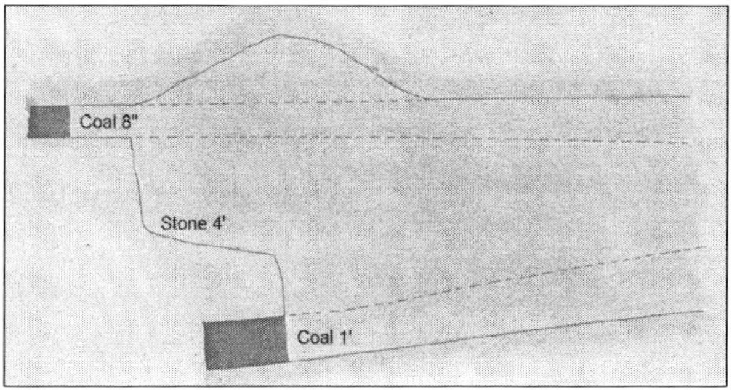

6.11 Section of the lower coal seam at the face in heading M. (Source: Galloway, W. (1907) Report on the mines at Camp Moreton, Spitzbergen, 20 September YF622.33 (984) GAL, British Geological Society, London.)

Heading N in the upper coal seam, also called No. 1 Seam, had been driven 105 feet. At the face, the lower bed comprised 1 foot of coal and stone while 1 foot 9 inches of stone separated it from an upper bed of nearly 2 foot 2 ½ inches (Fig. 6.12).

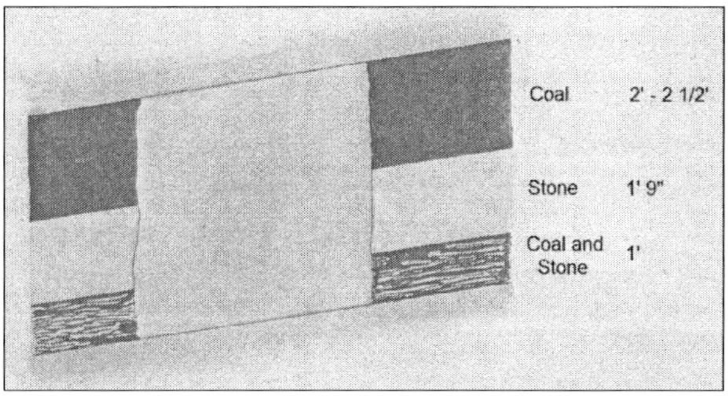

6.12 Section of No. 1 Seam at the face in heading N. (Source: Galloway, W. (1907) Report on the mines at Camp Moreton, Spitzbergen, 20 September YF622.33 (984) GAL, British Geological Society, London.)

Heading O in the upper seam was fairly variable, being level for the first 80 feet, then rising 8 feet over the next 150 feet before being level again until the face located at 350 feet. At 170 feet from the mouth, the lower bed was 1 foot 8 inches thick while the stone and the upper bed measured 2 feet 2 inches each (Fig. 6.13).

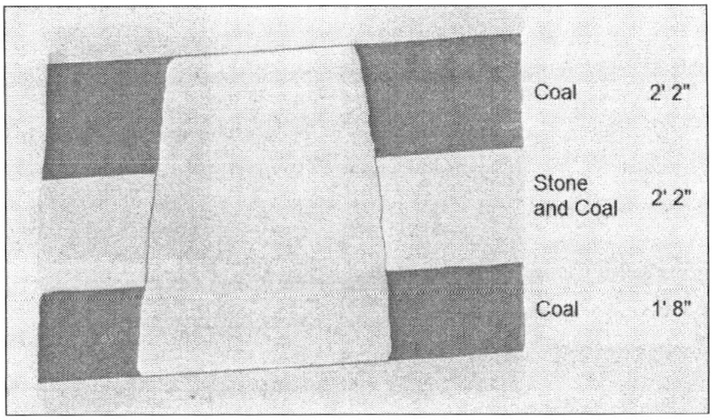

6.13 Section of No. 1 Seam 170 feet from the mouth of heading O. (Source: Galloway, W. (1907) Report on the mines at Camp Moreton, Spitzbergen, 20 September YF622.33 (984) GAL, British Geological Society, London.)

In a last section taken 35 feet metres from the face, Galloway demonstrated that the lower bed got thinner as the stone got thicker and the upper bed split into two (Fig. 6.14). Galloway concluded that 'the seams not only vary in thickness, but are also liable to become depreciated at various points by the intercalation of thinner and thicker scales and beds of shale and stone.'[89] Furthermore, 'both seams are of an erratic nature, a feature which detracts much from their value as objects of mining.'[90] Whether the seams were better at a different location was naturally difficult for him to say.

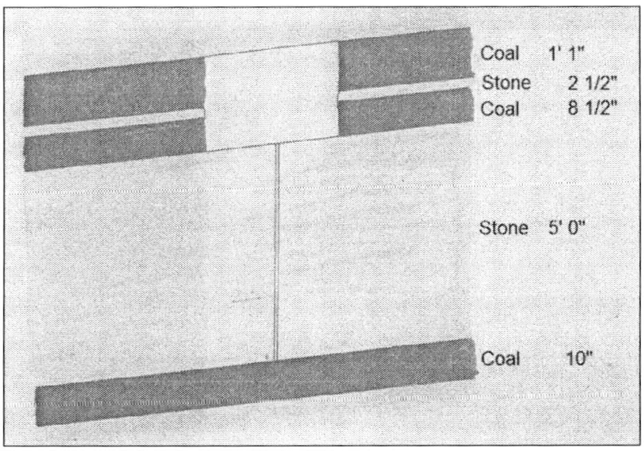

6.14 Section of No. 1 Seam 35 feet from the face in heading O. (Source: Galloway, W. (1907) Report on the mines at Camp Moreton, Spitzbergen, 20 September YF622.33 (984) GAL, British Geological Society, London.)

In addition to the workings, Galloway described the small settlement that stood on a terrace about 70 feet from the cliff at a height of 34 feet above sea level. At the time, there were three buildings at Camp Morton, two made of wood and one made of canvas. Galloway may not have known that the large wooden house was

[89] Galloway (1907) p. 5.
[90] Galloway (1907) p. 6.

Michelsenhuset. The building was in need of repair before it could be taken into use, presumably as a barrack for the Norwegian workers (Fig. 6.15). The smaller wooden house had been constructed by Mann in 1906 and accommodated the management, while the canvas structure was possibly a store (Fig. 6.16).

6.15 Michelsenhuset under repair in 1906. Note the independent Norwegian flag. (Source: The Northern Exploration Company, Ltd. (1911) The Northern Exploration Company, Norwegian Polar Institute, Tromsø, p. 122.)

6.16 Two smaller houses at Camp Morton in 1907. (Source: Galloway, W. (1907) Report on the mines at Camp Moreton, Spitzbergen, 20 September YF622.33 (984) GAL, British Geological Society, London.)

Mann also built Camp Bell in 1908 (Fig. 6.17).[91] Timber and all else had been brought from the nearest port, presumably Tromsø, and the task took him seven weeks to accomplish. He described Camp Bell as a strong and roomy building, made of thick timbers, felted both inside and out, and match-boarded over the felt. On the outside was also a bank of stones and earth. There were portable and reliable stoves in the store room and in the living room, each standing on a sheet of iron. An airspace prevented that an overheated chimney should set fire to the hut. Furthermore, the living room had two windows with inside and outside shutters.

6.17 Charles Mann at Camp Bell in 1908. Note the Union Jack above the door and the decorative whale bones. (Source: The Northern Exploration Company, Ltd. (1911) The Northern Exploration Company, Norwegian Polar Institute, Tromsø, p. 100.)

6.5.3 Employees

With new investments, the 1906 expedition was able to afford a considerable workforce. Gardner and Mansfield made their way to Spitsbergen in advance of the workers. In July, they were joined by Charles Mann and George Alexander, both from Gardner's parish, and thirteen Norwegians, all of whom arrived on board the *Mylingen*.[92] Some sources recount as many as 20 men, who had supposedly come from Hammerfest.[93] Yet, the actual number may have been closer to the 12 men, from Tromsø, who subsequently received two shares each in the syndicate.[94] How many miners were employed at Camp Morton while others assisted with the prospecting is not known. Rare photographs suggest that the completion of the

---

[91] Mann, C., 'General report from Mr. Charles Mann, 21 January 1911', *The Northern Exploration Company, Ltd.* (1911) The Northern Exploration Company, Norwegian Polar Institute, Tromsø, p. 112-4.
[92] Mann, C., 'General report from Mr. Charles Mann, 21 January 1911', *The Northern Exploration Company, Ltd.* (1911) The Northern Exploration Company, Norwegian Polar Institute, Tromsø, p. 112-4.
[93] Barr, Newman, and Nesteroff (2012) p. 81.
[94] *An agreement made the nineteenth day of September* (1906) Spitzbergen Mining & Exploration Syndicate BT 31/11526/88833, The National Archives, Kew.

buildings on site provided a reason for social gatherings, if not celebrations (Fig. 6.18 & Fig. 6.19). When the expedition left in September, two Norwegians were temporarily left in charge.[95]

6.18 Naming the canvass house. (Source: The Northern Exploration Company, Ltd. (1911) The Northern Exploration Company, Norwegian Polar Institute, Tromsø, p. 20.)

6.19 The manager's house. Note Charles Mann in the light jumper, the teapot, and the decorative reindeer head. (Source: The Northern Exploration Company, Ltd. (1911) The Northern Exploration Company, Norwegian Polar Institute, Tromsø, p. 111.)

[95] Barr, Newman, and Nesteroff (2012) p. 87.

At the beginning of October 1906, the men were joined by six English miners under the leadership of Arthur Mangham, who planned to develop the mine during the winter.[96] Mangham had previously been employed by the Spitzbergen Coal & Trading Co. Clem Burgin was one of the miners who came with him.[97] Burgin lived in the same village as Mangham and was probably his brother-in-law.[98] He stayed on for almost two years until August 1908. During this time, he mainly worked the coal at Camp Morton, as opposed to being employed in exploration, echoing the opinion that driving further into the mountain would improve its quality. Another miner was called Speight, but it is not known if he was also from Thorpe Hesley or if he had previously been at Advent City.

Before he returned to Camp Morton in summer 1907, Mansfield sent two Norwegians with provisions to the mine.[99] The wintering trappers then departed for Norway. Whatever the season's programme may have been, it was carried out by Mansfield, the Norwegians, and the English winterers. Mangham then left on August 14 and may have taken two or three Englishmen with him. Mansfield and the Norwegians workers also departed, but it is not known if this happened at the same time. Burgin and Speight were still on site when Tollemache and Galloway arrived later that month. Hoel probably meant these two miners when he stated that two or three Englishmen also wintered in 1907/8.

In summer 1908, Mansfield was back again. Mann undertook his second voyage.[100] He had been employed to build Camp Bell. Mansfield additionally named five Norwegian workers and loosely referred to several others, who were with him.[101] The prospector then wintered at Camp Bell in the company of four men, but this no longer took place on behalf of the syndicate.

6.5.4 Products

The coal extracted at Camp Morton was tipped down the slope, and pieces from mine O were occasionally collected into sacks and used in the houses. Galloway made no reference to any having been sold. The mine inspector conducted some field experiments to assess the quality of the coal. Three samples from the upper seam averaged at 31.08% volatile matter, 11.95% carbon, and 26.97% ash. Two samples from the lower seam comprised 27.75% volatile matter, 38.38% carbon,

[96] Barr, Newman, and Nesteroff (2012) p. 87.
[97] Burgan, C., 'Mr. Clem Burgan's testimony, 11 August 1910', *The Northern Exploration Company, Ltd.* (1911) The Northern Exploration Company, Norwegian Polar Institute, Tromsø, p. 110-1.
[98] Mangham, G. (2008) Email to Frigga Kruse, 29 July.
[99] Hoel (1966) p. 427-9.
[100] Mann, C., 'General report from Mr. Charles Mann, 21 January 1911', *The Northern Exploration Company, Ltd.* (1911) The Northern Exploration Company, Norwegian Polar Institute, Tromsø, p. 112-4.
[101] Hoel (1966) p. 429.

and 33.87% ash. Based on the volatile matter, these were not valued anthracite or steam coals but medium volatile coals.[102] Galloway's analysis excluded two samples, which had been collected by Burgin. Whilst volatile matter compared to Galloway's results, Burgin's carbon was much higher, while ash was lower.

By August 1907, the mines had been worked for three years. Galloway thus calculated an average advance of 100 yards per year or one foot per day. That was much less than the work of an ordinary British miner. The seams were stiff and hard, and difficult to work, which would affect competing commercially. In addition, irregular shale and stone, and high ash content were a defect in the mines' constitution. In effect, two tons of Camp Morton coal equalled one ton of Welsh steam coal. So Galloway predicted that 'the coal would not be taken by anyone who could produce better coal even at double the price.'[103] The mine inspector went on to outline problems with the pack ice and climate. Burgin and Speight had related how winter weather could be severe and how they had occasionally not been able to reach the mine. Galloway's solution was to erect houses near the mine entrance and sheds for the coal, despite the fact that frost had apparently nothing to do with its quality.

Following his thorough assessment of Camp Morton, Galloway concluded that 'there does not appear to be much object in continuing to drive headings in the two known seams as their quality is not sufficiently good to warrant the belief that they will ever be of sufficient value to mine on a commercial scale.'[104] However, he did not suggest abandoning the site. Instead, he proposed that future operations should be undertaken in view of discovering more valuable seams by boring and trenching. 'There is ample scope for prospecting operations to a great height above the known seams and also to some distance below them.'[105]

6.6 Summary and conclusion

The strength of archaeological fieldwork in Bell Sound and Lowe Sound had again been a much deeper appreciation for the environmental circumstances under which mining occurred and for the distances that were covered by early prospectors than would have been possible through brief historical descriptions and occasional black and white photographs only. Unfortunately, the material remains of the Spitzbergen Mining & Exploration Syndicate had been masked by

[102] The Engineering Toolbox (nd) *Classification of coal.* Available at: http://www.engineeringtoolbox.com/classification-coal-d_164.html (Accessed: 3 December 2010).
[103] Galloway (1907) p. 17.
[104] Galloway (1907) p. 19.
[105] Galloway (1907) p. 20.

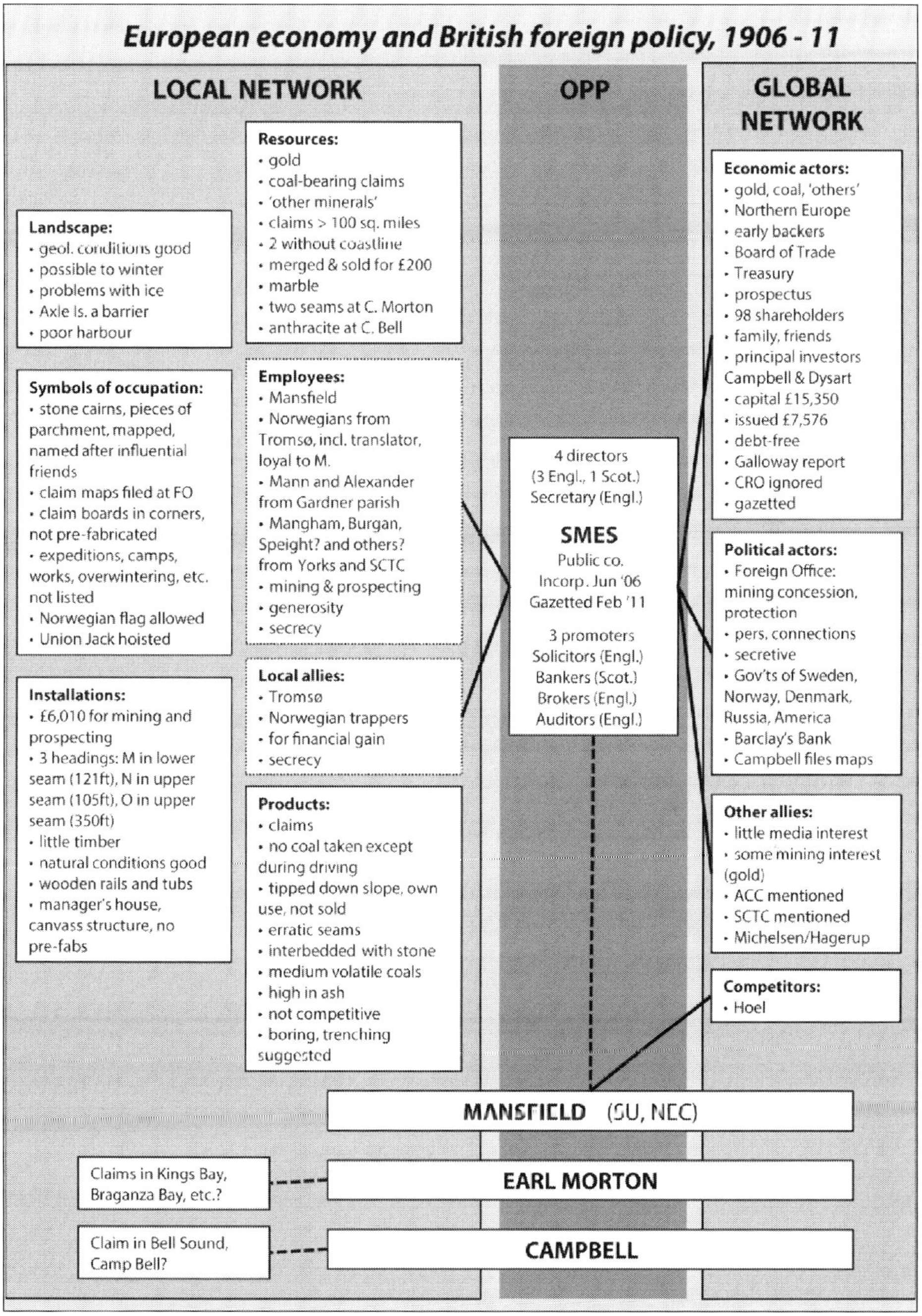

6.20 Archival documents gave rise to a very different actor-network of the Spitzbergen Mining & Exploration Syndicate than expected. (Chart: F. Kruse.)

the later developments of the Northern Exploration Co., leading to an imperfect actor-network (Fig. 3.38). Archival research gave rise to a very different actor-network than was expected (Fig. 6.20). Many of the sources consulted, however, had been prepared several years after the events and contained obvious rhetoric aimed at the Northern Exploration Co. Bias undermined their value and made it difficult to reconstruct the historical reality.

Mansfield, Gardner, and Salter were instrumental to the formation of the syndicate, which was a public company from the beginning and headed by four directors. It is noteworthy that the directors resided in London, Tolleshunt D'Arcy, Leicester, and Edinburgh. No board meeting has ever been recorded. Tollemache in London did not hold any shares; it is unlikely that the London secretary did. Although Campbell in Edinburgh was a solicitor, there were seemingly no in-house experts. The Bank of Scotland may have had a London office, which may have eased transactions and auditing.

Documentation concerning the economic actors barely referred to the prevailing commercial context at the time. Gold was the primary objective, closely followed by coal. That the syndicate may have been interested in other minerals is only ever mentioned on hindsight. None were named. Salter broadly identified a large demand and all of Northern Europe to be supplied with coal. The pioneering trio had early influential backers and no problems obliging to the prerequisites of the Board of Trade and the Treasury during incorporation. The syndicate at once issued a prospectus, a legal document not be confused with a memorandum, to the most likely investors. Although it did not mention gold, the syndicate mustered 98 shareholders. Many were family and friends, who did not contribute to the operational capital. The principal investors were Campbell and the Earl of Dysart, who invested over £1,000 each. Despite the confidence the director's investment should have installed in others, the shares were not popular, and only about half were ever allotted. Coupled with Galloway's disappointing report, enthusiasm in the endeavour slumped. Nonetheless, the syndicate practically remained debt-free. When further obligations to the Company Registration Office were repeatedly ignored, the syndicate was crossed off the register.

Neither was the political context stated, let alone British foreign policy. The pioneers contacted the Foreign Office to obtain mining concessions and protection. They utilised personal connections but were mainly secretive about their purpose. Gardner first mentioned gold in direct correspondence with the Foreign Secretary. The British Government was non-committal. According to Salter, so were the Governments of Sweden, Norway, Denmark, Russia, and America. That the first discoverers lodged their claim maps with Salter's bank, not with the syndicate's bank, was important for subsequent events. These title deeds were never retrieved during negotiations with the syndicate. Instead, Campbell simultaneously succeeded in filing other claim documents with the Foreign Office. In the future, Mansfield and his partners would be able to refer back to their deeds and bypass the syndicate as if it had never existed.

In letters to the Foreign Office, Gardner alludes to the Americans and the Spitzbergen Coal & Trading Co. as allies. Later, the American engineer prohibited the syndicate's ship to be discharged on his territory, and Advent City only proved to be an ally in that dissatisfied English miners took up employment at Camp Morton. Allies became rivals and vice versa. Although Michelsen did not complain about the use of his cabin in person, Hagerup took it upon himself to preserve Norwegian rights to the area. That this was not politically but financially motivated showed the ensuing agreement between Hagerup and Mansfield regarding wintering trappers. The patriotic opposition staged by Hoel had its roots in Mansfield's later network.

In the local network, an assessment of the landscape did not add significantly to that which was archaeologically known. While coal had been discerned as a natural resource at Camp Bell, gold and other minerals had not been documented in the field. It is unclear if the existence of marble in Kings Bay was ever reported to the global network at the time. Although geological maps were seemingly not produced, presumably in pursuit of secrecy, it is unlikely that any resource warranted the large sizes of the resultant claims. None were smaller than 100 square miles. Two did not even have access to the sea and appear not to have played a role in subsequent prospecting. While two coal seams have been thoroughly assessed at Camp Morton, the anthracite at Camp Bell is unsubstantiated.

The large claims needed to be secured. Early symbols of occupation included stone cairns, buried pieces of parchment, mapping, and naming after influential friends. Thereafter attempts were made to lodge the claim maps with the Foreign Office, which initially failed. The maps were powerful actants, which indicated the location of claim boards, named and dated. However, there were very few claim boards indeed, and they were seemingly not pre-fabricated, which gave the whole undertaking an unintentional, haphazard feel. Contemporary actors did not refer to yearly expeditions, camps, works, and wintering as acts of effective occupation, although they may have understood these as such. The importance of geopolitics was reduced when the Norwegian flag flew above Michelsenhuset at the same time as the Englishmen hoisted the Union Jack at Camp Morton and Camp Bell. If anything, the Norwegians may have roused Swedish passers-by.

The syndicate spent approximately £6,000 on mining and prospecting, although it is not known, which activities were most costly. Three headings of a total of 576 feet had been driven at Camp Morton, and the mining technology was extremely basic. It is likely, therefore, that more money had gone into prospecting –

for gold. The syndicate had erected the manager's house and the canvas structure at the mine. The house was not pre-fabricated. While this may have enhanced the building's quality, its construction may have wasted funds and time. The fieldwork had not recorded headings M and N, but it had shown that mine O was continued by the Northern Exploration Co., who also opened two mines in the seam below. All other technology, the two barracks, and the stockpile dated from this time.

The product of Mansfield's efforts was the claims. Some of these claims were made over to the syndicate. The syndicate hoped to produce coal and gold. At Camp Morton, no coal was taken, except that from driving the headings. It was tipped down the slope, and some was gathered for use in the camp; none was sold. Galloway judged the seams to be erratic in nature and interbedded with stone. They comprised only medium volatile coals with a high ash content. The coal would not be compatible according to British standards, but perhaps Galloway misunderstood the North European markets. He did not suggest to abandon the site but to search for better seams by boring and trenching instead.

Mansfield was presumably employed by the syndicate, but it is not known under which terms. He appears to have valued his independence. A varying number of Norwegians took part in the expeditions each year. From the first, this included a translator. Several of the men from Tromsø were loyal to Mansfield over time; they probably benefitted from the many acts of generosity related by Barr *et al.*[106] The Englishmen in the syndicate's service were either connected with Mansfield and Gardner or had seemingly come across from Advent City with Mangham. The generosity and the social cohesion evident in rare group photographs were most likely aimed at safeguarding Mansfield's and the syndicate's secrets, such as where the search for gold was taking place and where it seemed to be promising. As far as local allies were concerned, Tromsø was the most likely outfitting town on route, and Norwegian trappers were enlisted to guard the camps and generate alternative income. Yet, Mansfield appears not to have capitalised on this.[107]

To conclude, the syndicate was therefore entirely motivated by commercial goals, but it was dysfunctional. Reasons can be found at global and local level. Firstly, the distance between the directors was too great, and the lack of in-house expertise undermined effective management. Then there will have been discord between those who had invested in gold and those who were interested in coal. This discord prevented a functional operational strategy. To top it off, any strategy was hindered by the syndicate's failure to become indispensable. The obligatory point of passage had been breached several times.

[106] Barr, Newman, and Nesteroff (2012).
[107] Barr, Newman, and Nesteroff (2012).

The syndicate relied wholly on Mansfield, while he maintained his own network as shown by the existence of Spitzbergen United. He neglected that firm, however, because the syndicate appeared more lucrative. Although he later stressed his activities in Bell Sound, he purposefully refrained from naming the syndicate or any of its global actors, none of whom played a part in the Northern Exploration Co. Similarly, Earl Morton diffused the boundaries between the networks when he and Lord Balfour of Burleigh undertook their cruise. Morton's network has not been evaluated, suffice to say that he may have supported the elusive Spitzbergen Coal & Mineral Ltd. in 1905, and he definitely invested in the unidentified Arctic, Ltd. in 1908.[108] During the cruise, he supposedly claimed widely, including Kings Bay and possibly Braganza Bay, but he was highly unlikely to uphold these claims. Despite being a director of the syndicate, Campbell independently occupied the northern shore of Bell Sound. The question arises whether Mann built Camp Bell, a play on words, on behalf of Campbell, who may have persisted when the syndicate was already disintegrating in 1908. Mansfield, savouring his autonomy, outlasted them all. He seemingly decided that the original claims in Bell Sound and Kings Bay had lapsed and assimilated them by the time he enrolled the first actors of the Northern Exploration Co.

The disadvantages and downsides of the syndicate could have been counteracted if the local network had delivered the expected resources. Gold, however, had not been found in workable quantities, and the coal had been professionally proven to be of no value. Whether the debt-free syndicate actively decided against the investigation of the coalfield by boring and trenching is not known. It is more likely that disagreements, as indicated by the non-compliance with the Company Registrar, put a stop to everything.

[108] *Allotment letter, 5 August* (1908) GD150/3878/1 (1 & 2), National Archives of Scotland, Edinburgh.

7 The Northern Exploration Co., Ltd. (1910-34)

7.1 Introduction

The Northern Exploration Co. was founded in 1910. It existed for 24 years before it was dissolved in 1934. Although its roots lay in Ernest Mansfield's activities on Spitsbergen since 1905, it must not be mistaken for the successor of the Spitzbergen Mining & Exploration Syndicate. The company's history is a history in two parts. It began as a private company, but the First World War prompted the conversion to a public company, which subsequently experienced rapid changes in the global context. This chapter therefore elucidates its actor-networks before and after the conversion as well as its responses to the post-war depression and the ratification of the Spitsbergen Treaty. In closing, a review of the defining features of the Northern Exploration Co. offers probable answers as to its motivations, its operational choices, and reasons behind its termination.

7.2 Formation and chronological overview

In 1908/9, Mansfield wintered at Camp Bell.[1] After his return to England, he raised the funds for an independent expedition in summer 1910. Arthur Mangham led the works at Camp Bell while Mansfield visited Kings Bay and erected several huts in Bell Sound and in Braganza Bay.[2] He named Camp Millar and Camp Williamson after the London merchant Henry Edward Millar and the London secretary Henry Williamson, respectively, who were the main sponsors of the expedition. These future directors of the imminent company then requested an exhaustive account of Spitsbergen and Mansfield's previous endeavours, which resulted in an imposing 134-page brochure.[3] Testimonial evidence regarding the rightfulness and value of Mansfield's claims rang in the new company.

The Northern Exploration Co. was incorporated on November 17, 1910.[4] The number of its shareholders was restricted to 50, and its shares and debentures were not offered for public subscription; it was a private company. As such, it was

---

[1] Hoel (1966) pp. 429-32.

[2] A historical photograph (Photo Library, np003738, Norwegian Polar Institute, Tromsø) shows that the original Camp Millar was not one of the huts that occupy the site today. The picture gives the impression that the huts, which included Camp Williamson and probably the elusive Camp Margaret, were by now pre-fabricated.

[3] *The Northern Exploration Company, Ltd.* (1911) The Northern Exploration Company (N.E.C.), Norwegian Polar Institute, Tromsø.

[4] *Certificate of incorporation* (1910) Northern Exploration Company, Limited BT 31/32080/112730, National Archives, Kew.

ratification of the Spitsbergen Treaty in 1925, one of its aims being the settlement of the claim disputes on the archipelago. The official recognition of all claims was achieved in 1927.

7.3 The global network of the private company

7.3.1 *Economic actors*

The aforementioned brochure was presumably used for advertising purposes and generated considerable interest. Coupled with the fact that the company had successfully dispatched an expedition to Spitsbergen in spring 1911, it was able to enrol several fundamental investors. On July 7, the coal proprietor Frederick Lewis Davis and the banker Gerald Dudley Smith acquired 5,000 shares and 1,000 shares, respectively.[10] Within a week, another 23,500 shares were divided between 27 subscribers. One of these was Sidney Thomas Peirson, a chartered accountant, who obtained 1,000 shares, while Smith received another allotment of 2,000 shares. In addition to the chairman Millar and the managing director Williamson, Davis, Smith, and Peirson now sat on the board and would soon have Arctic camps named after them.

A group of company representatives immediately visited Spitsbergen to gain first-hand knowledge of the properties. Williamson and Peirson were accompanied by five shareholders and by David Hannah, an employee at Davis' collieries and a coal mining expert. In August 1911, they disembarked at Marble Island.[11] Since the expedition's arrival, the natural harbour had been named Port Peirson, and the directors were pleased to note that some of the men were housed in the commodious and comfortable Camp Peirson. They climbed the summit and posed for a photograph next to a stone cairn that supported a wooden post with a manufactured claim sign. After inspecting the Lesser Islands, the group departed for Camp Bell and other places before returning to Norway at the end of the month.

The company's efforts were initially directed towards the development of marble. William George Renwick, author of *Marble and marble workings* (1909), reported on 25 varieties among the samples he had received.[12] Five were ordinary and could be used for table tops and floor tiles. They could, however, come into competition with cheaper Italian and Belgian marbles. 14 were decorative and could meet the increasing demand of luxury building. Similar marbles were selling

---

[10] *Credit report from the Swedish Chamber of Commerce* (1911).

[11] *Marble Island. Short Account of the Discovery, Location and Products of the Property with Practical Notes and Criticisms upon the Unique Variety and Value of its Marbles* (1913) 443-A, British Geological Survey, Keyworth.

[12] *Marble Island* (1913) pp. 42-4.

between £6 and £12 per ton. The remainder were a class of their own that could not be obtained anywhere else and were decorative, which made them very valuable and should be reserved for high-grade architectural work. A total of 700,000 tons of marble was imported world-wide each year. The primary market would be America and Canada, secondly Germany and Belgium, thirdly France and Britain. Although Marble Island formed an infinite supply, Renwick stressed the importance of securing and maintaining stocks in anticipation of demands.

The company then sought out American architects and marble experts. On March 13, 1912, Mansfield, Williamson, and Peirson departed for New York to ascertain the situation in the world's leading marble market.[13] One architect rated their samples among Europe's finest. Even the few exceptions matched the ordinary commercial grades and would attract attention among New York buyers for their beauty and decorative quality.[14] Another observed that decorative marble was currently supplied from scattered deposits in Europe, north America, and Africa, and that demand was increasing in proportion to the amount of building.[15] Prices for interior marble were much higher than at European ports. If the company were to introduce less expensive and newer material to the market, it would encounter a large demand. The editor of *Stone*, a magazine devoted to quarrying, stressed that the greatest demand would be for lighter colours and more decorative patterns.[16] In this, the company could expect rivalry from Tunisia, Italy, and France, yet Spitsbergen marbles were attractive enough to compete. When the company enquired if it could use the experts' opinions in their propaganda, one contractor at least agreed that his letter may be referred to but not published.[17] *Marble Island* was eventually issued in 1913. This extraordinary booklet was undoubtedly aimed at mobilising fresh capital. It is surprising, therefore, that Marble Island lay deserted in 1914.

Incidentally, the Swedish industrialist and diplomat Herman Ludvig Fabian Lagercrantz was also staying at New York's Waldorf-Astoria Hotel.[18] He was a director of the AB Isfjorden-Belsund founded in April 1911.[19] That the men met was evident in an agreement dated May 11, 1912, whereby the company surrendered parts of its claims in Braganza Bay to the Swedes.[20] Following Millar's resignation due to illness, Lagercrantz was elected onto the board in June, although he did not

---

13 http://www.titanic1.org/articles/pdfs/Crossing9-March13-20-1912-V2.pdf (no date) (Accessed: 11 February 2011).

14 Fenner, B. L. (1912) 'Letter to Henry Williamson, Esq., 27 March', cited in *Marble Island* (1913).

15 Brainard, O. (1912) 'Letter to Henry Williamson, Esq., 27 March', cited in *Marble Island* (1913).

16 Hoyt, F. W. (1912) 'Letter to Henry Williamson, Esq., 1 April', cited in *Marble Island* (1913).

17 Roberts, C. H. (1912) 'Letter to Sidney T. Peirson, Esq., 2 April', cited in *Marble Island* (1913).

18 http://www.sovereignorderofsaintjohnofjerusalemknightsofmalta.org/archpdf/May_1912.pdf (1912) (Accessed: 16 February 2011).

19 *Mining Journal* (1911) 'Mining affairs in Norway: coal mining on Spitzbergen', 12 August.

20 *Memorandum by Hertslet* (1913).

yet qualify as a shareholder.[21] Another significant addition was the London gentleman Oscar Emanuel Warburg.

Meanwhile, the company had dispatched another expedition and continued the search for coal in Bell Sound. This resource was reported on less frequently and in less detail. It fell to David Hannah to examine the various developments on Spitsbergen, including those not in the company's possession.[22] After his second visit in 1912, Hannah suggested erecting a by-product plant in Norway. Since the cost of coal from Spitsbergen was low, presumably in comparison with British coal, and land and labour were available cheaply in Scandinavia, coke ovens in the vicinity of existing iron and steel works would make correspondingly large profits.

A second brochure issued after the expedition addressed existing shareholders and potential investors alike.[23] The company now owned between 1,500 and 2,000 square miles. Although two Spitsbergen conferences had failed to secure the titles beyond all doubt, the Foreign Office had alerted the affected Governments to several British claims, which it expected to be considered in any settlement arrived at. Readers were reminded of the experts' opinions concerning marble and coal, and a recent assay of three samples of iron ore had provided 60 per cent of iron. Samples of graphite and asbestos had also been collected, but the gold-bearing conglomerate had not yet produced payable amounts. The company appealed to the shareholders to enlarge the operational capital to £150,000. It intended to place the new capital among large industrial and banking firms likely to pursue a programme of development. This programme firstly entailed quarrying the marble and shipping it to market; secondly, developing the most suitable coal seams; thirdly, investigating known mineral deposits; and lastly, prospecting further. Predicting the enterprise to become much larger, the directors foresaw the formation of a subsidiary firm to take over the properties and assets and to engage in practical mining. The shareholders supported the notion after Marble Island proved safe to winter in 1912/13 and, following the instatement of Charles Edward Evans on the board in May 1913, the capital was raised in July.

As in previous years, the bond between the global and the local networks was reinforced by a group of esteemed persons inspecting Spitsbergen in summer 1913. Commanders Victor L. A. Campbell and Wilfred M. Bruce accompanied them to Recherche Bay, where the Norwegian geologist Birger Jacobsen had been prospecting for iron ore on behalf of the company.[24] As before, samples were

---

[21] *Directors* (1912) The Northern Exploration Company Limited BT 31/32080/112730, National Archives, Kew.

[22] *The Northern Exploration Company., Ltd.* (1912) Northern Exploration Company (N.E.C.), Norwegian Polar Institute, Tromsø, pp. 5-9.

[23] *The Northern Exploration Company* (1912).

[24] *Some particulars* (1914) pp. 20-1.

dispatched to renowned experts for analysis. The excellent results prompted the comparison between Gellivare and Spitsbergen.[25] Spitsbergen had a number of unique advantages that favoured its immediate development.[26] Its iron would find a ready market as demonstrated by a recent Board of Trade memorandum on the situation of iron and steel. It outlined that 'while the entire output of iron ore from leading sources of supply has only increased 10 per cent during a period covering five years, the demand, as evidenced by the quantities imported into leading iron and steel producing countries (excess of imports over exports) had increased by 72.2 per cent during the past four years.'[27] The iron ore supplied varied in chemical composition and yielded iron in the range of 20 per cent to 65 per cent of their weight. Since the costs of reducing low-grade ores were higher, high-grade ores were in greater demand. The company's product would therefore be well placed to compete.

It was the second time on Spitsbergen for the shareholder Edward J. Bullard, a brewer from Norwich. He sailed from Recherche Bay to Bell Sound to observe the company's coal seams and thereafter continued to Marble Island.[28] The company later sent coal samples for testing in Britain as well as abroad. A London laboratory compared two samples to Scotch Boghead coal, although they contained more sulphur and were non-coking.[29] They thought the coal to be valuable, especially for gas-making, and imagined that oil shale similar to that of the Linlithgow district in Scotland existed in the vicinity and should be searched for. Coal samples were also sent to the Petroleum Institute in Russia.[30] According to its president, 100 tons could produce 20 tons of petroleum products, 60 tons of coke, and the remaining 20 per cent would provide 21,000 cubic metres of gas. The commercial value would approximate £170. After subtracting the costs, a profit of about £100 could be made. If the coke were turned into briquettes, which were in large demand in the North Sea, the profit would rise.

Bertrand Mangham, since 1906 winter superintendent of the Arctic Coal Co., suggested wasting no time in getting the coal to the European markets, where

[25] *Some particulars* (1914) p. 17. Gellivare Iron Mountain in Lapland in northern Sweden was thought to be the largest and richest deposit of magnetite known. It was 17,000 feet long and 525 feet high with an average assay of 63 per cent of iron and 0.08 per cent of phosphorus. Iron Mountain in Recherche Bay was about 12 miles long, three miles wide, and 1,400 feet high. According to the company, the average assay was 64.44 per cent of iron and 0.02 per cent of phosphorus.

[26] As mentioned in other chapters, these advantages included that it lay on the seaboard. Its ore was seemingly higher in quality and more massive in quantity. Working expenses and extraction charges were lower. There were better natural facilities for cheap power and land transport. Due to its unsettled status, there were currently no taxes, duties, or customs, and there was no governmental interference.

[27] *Some particulars* (1914) p. 22.

[28] *Some particulars* (1914) p. 28.

[29] *Some particulars* (1914) p. 29.

[30] *Some particulars* (1914) pp. 35-6.

300,000 tons were currently being consumed in northern Norway alone. The Northern Exploration Co. was already saving money by not having to sink deep shafts or paying dead rents and royalties. Steamers could be chartered until the company had its own ship. Mangham knew from experience that even an output of 200 tons per day could be shipped reasonably cheaply.[31] He also recommended a by-product plant. If anthracitic coal, which had so far only been found in the company's area, could be developed, it would fetch a much higher price than ordinary Advent Bay coal. It would not only be usable as steam coal; it would also be suitable for Scandinavian households.

Those stakeholders not able to travel to the Arctic benefitted from the company's third and fourth elaborate brochures. The preface of *Marble Island* stated that it was 'simply a collection of facts, pictorial and critical, crudely put together for the private purposes of a private company.'[32] It made no apologies for the generous use of photographs and colour plates. Yet, behind the picture-perfect scenes, the company had apparently used up its capital during the expedition in 1913 and was in financial difficulties.[33] This was probably the reason for Warburg's exit from the board in October 1913.[34] The company most likely secured the funds for a follow-up expedition on the back of *Some particulars respecting the Company's properties in Spitsbergen*. It mainly entailed reprints and repetitions from 1911 and 1912. In addition, it cited Robert Neal Rudmose Brown, an academic who had been to Spitsbergen twice (with the Scottish Spitsbergen Syndicate). The geographer deconstructed the perception of the islands being negatively peripheral. In fact, their accessibility was not sufficiently realised. He believed that it might be possible to reach Marble Island throughout the year. Even the ice floes that drifted around the South Cape and into Bell Sound could be circumvented, so that ice was not normally a hindrance to exploiting Spitsbergen's minerals. Only a fraction of the natural resources had been found, and more was yet to come. The distance to the market was much less than commonly believed since the average map on Mercator's projection grossly exaggerated the remoteness of Spitsbergen.

The company's brochures may have been available to the British firms whose agents accompanied the expedition in July 1914. The group disembarked at Iron Mountain to value the iron ore. Commander Michael Barne related, 'From what I gathered of the general opinion, the iron ore was considered excellent quality on the surface, but the value of the claim was entirely problematic, as no one could

[31] At 5s 6d to 6s 6p per ton.
[32] *Marble Island* (1913).
[33] Jacobsen, B. (1919) *Letter to the Norwegian Foreign Department, 28 April*, Norwegian Foreign Department Box 5373, National Archives of Norway, Oslo.
[34] Barr, Newman, and Nesteroff (2012) p. 125.

say, in the absence of a trial boring, which the company had not seen fit to undertake, how far below the surface the iron extended.'[35] The expedition was cut short by the outbreak of war.

7.3.2 Political actors

The foremost political actor continued to be the Foreign Office. The perseverance of the Spitzbergen Mining & Exploration Syndicate had led to it recognising the registered British claims in 1906. In 1909, the British minister to Norway reported that the coalfields had an excellent future ahead of them but that claim disputes were rife.[36] The disputes had been noticed by other Governments, too, and a first international conference was held in Kristiania in 1910 to settle the islands' legal status. Although Britain did not participate, the Foreign Office simultaneously conducted an investigation into the British claims. The Northern Exploration Co. had not yet been founded, but following formation, it no longer needed to battle for the recognition of its claims. It restricted its contact with the Foreign Office to the notification that Mansfield's properties had changed owners.

The first Spitsbergen Conference issued a draft convention. This prompted the British Government to instruct its ministers in St. Petersburg, Stockholm, and Kristiania to inform the respective Governments of the British suggestion that the powers affected by claims on the islands should devise a method to determine their validity. The British ambassador to America was to 'make it clear to the State Department in Washington that contrary to what they believed to be the case, there were a number of British claims of which some were believed to be of considerable importance.'[37] It seemed that while the Foreign Office had not shown much interest in the British companies in the past, it was not going to be left out of the diplomatic discourse concerning their future. The second Spitsbergen Conference commenced in January 1912. The *Mining Journal* reported on the proceedings, which were attended by Norwegian, Russian, and Swedish delegates.[38] The passing of the meeting was also fleetingly noted in the House of Commons. Another draft convention had been prepared, and several Governments were now examining the provisions.[39]

[35] Erskine A. B. (1994) 'Victor Campbell and Michael Barne in Svalbard: the 1914 voyage of *Willem Barents*', *Polar Record*, 30 (173), pp. 117-22.
[36] *Memorandum by Hertslet* (1910).
[37] *Memorandum by Hertslet* (1913).
[38] *Mining Journal* (1912) 'Scandinavia. The Spitzbergen Conference', 3 February, p. 108.
[39] HC Deb 21 May 1912 vol 38 c1726.

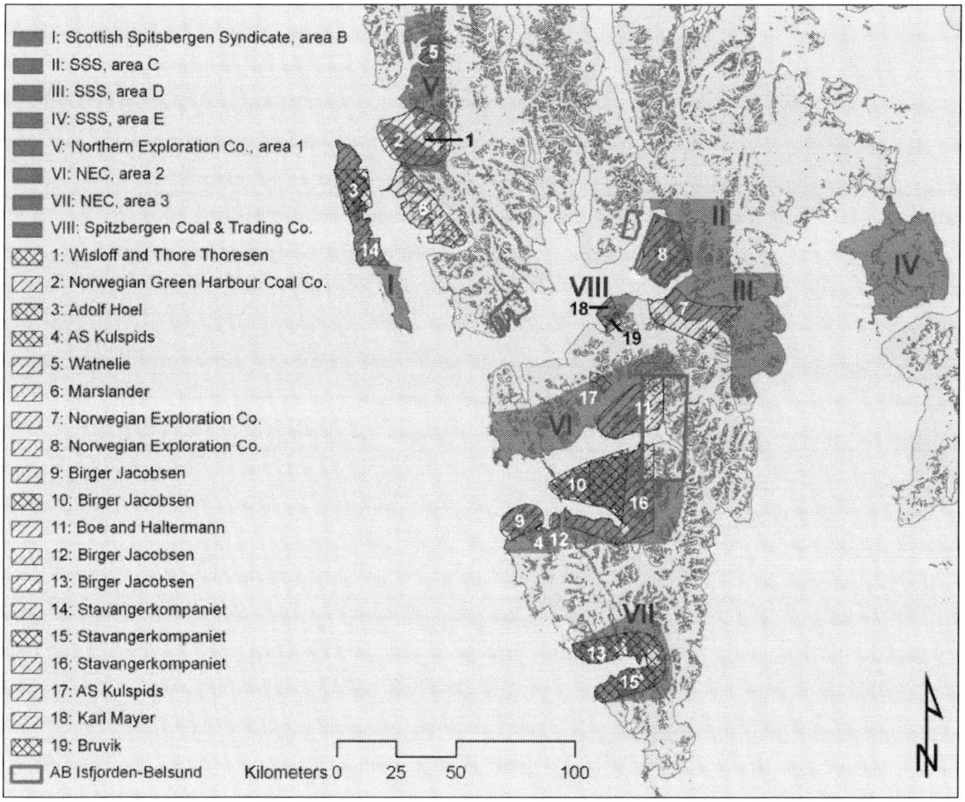

7.2 Composite map showing British, Norwegian, and Swedish claims in 1913. (Sources: Oversiktskart (1913) Norwegian Foreign Department Box 5372, National Archives of Norway, Oslo; Some particulars respecting the Co.'s properties in Spitsbergen with reports and extracts relating thereto (1914) Northern Exploration Company (N.E.C.), Norwegian Polar Institute, Tromsø; Map: F. Kruse.)

Russia had in the meantime taken an active interest in Spitsbergen, and Rusanov's *Herkules* expedition had netted 38 coal claims, many of which had already been staked out by others. This naturally increased the anxiety surrounding the islands and raised the question if Russia aimed to be independent of British coal.[40] In view of the rapidly changing situation, the Foreign Office repeated its investigation of the British claims in 1913.[41] The department was aware that the Northern Exploration Co. had come to an agreement with the AB Isfjorden-Belsund in May 1912. In November, however, the company objected to Norwegians trespassing on the

[40] *Mining Journal* (1912) 'Scandinavia. Spitzbergen. Russian claims', 2 November, p. 1061; *Mining Journal* (1912) 'The Spitzbergen question', 7 December, p. 1204.
[41] *Memorandum by Hertslet* (1913).

territory made over to the Swedes in Braganza Bay. British and Norwegian foreign officials tried to mediate between the parties, but the problem was still awaiting resolution in June 1913. The Foreign Office also knew of Christian Anker's claim in southern Kings Bay. In June 1912, the company asserted that its work on Marble Island had not been interfered with, which suggests that the Norwegian accepted British rights there while the company did not contest his. In April 1912, the Foreign Office had received details of the occupation of Horn Sound, and the claim between Dunder Bay and Recherche Bay had been supported by the Scottish Spitsbergen Syndicate in April 1913. The Foreign Office now asked the company to provide a written definition of its areas, 'so as to permit of a proper comparison with the Norwegian claims.'[42] The Norwegian Government in turn compiled a draft map, which highlighted the areas of contention (Fig. 7.2).

The outcome of the third Spitsbergen Conference in 1914 was eagerly awaited. A preliminary session between Russia, Sweden, and Norway to draw up 'proposals for regulating lawlessness by means of international police supervision'[43] was planned to take place on June 4. From June 16, Britain, France, the Netherlands, Germany, Denmark, and America were also part of the proceedings. 'The international conference [...] has been sitting, practically, from day to day with closed doors. It is impossible, at present, to form any idea as to when the deliberations of the delegates will have culminated in any general decision. Many of the subjects to be discussed are, to say the least, of a delicate, if not contentious nature.'[44] The outbreak of war, however, delayed a decisive outcome indefinitely.

7.3.3 Other allies

Among other actors that bore on the company's global network was the British press. After the expedition's return in 1911, it ran stories pertaining to rumours of gold on Spitsbergen.[45] A Norwegian tourist steamer returned from the islands without the verification that payable gold had been found. 'Spitzbergen, it must be remembered, has during the last two or three years been fairly well prospected. The secret, if secret it be, of Mr. Mansfield's goldfields has therefore been well kept, and he is to be congratulated on the loyalty of his Norwegian miners.'[46] Nor

42 *Memorandum by Hertslet* (1913).
43 *The Times* (1914) 'Status of Spitzbergen', 18 February, p. 7.
44 *Mining Journal* (1914) 'Scandinavia. Spitzbergen', 11 July, p. 645.
45 *Mining Journal* (1911) 'Norway: Spitzbergen', 26 August, p. 863; *The Times* (1911) 'Gold in Spitzbergen', 13 October, p. 13; *Mining Journal* (1911) 'Spitzbergen: rumoured gold finds', 21 October, p. 1017; *Mining Journal* (1911) 'Scandinavia: Spitzbergen marble and gold', 4 November, p. 1068; *Mining Journal* (1912) 'Scandinavia: Spitzbergen marble', 19 October, p. 1032.
46 *Mining Journal* (1911) 'Norway: Spitzbergen', 26 August, p. 863.

could other homebound vessels that pulled in at Norwegian ports confirm the discoveries. Although it would, of course, be unwise to announce such a thing in as lawless a place as Spitsbergen, British mining circles remained sceptical.

Given the right incentives, enterprising individuals could be enrolled to advance the endeavour. Hence, a deal was made with the aforementioned Birger Jacobsen, who had first come to Spitsbergen aboard a Norwegian expedition in 1909. In 1911, he discovered asbestos, coal, lead, and graphite in Recherche Bay and Van Keulen Bay as well as zinc, nickel, and gold in Horn Sound. As some of the discoveries lay within the company's territory, he was willing to avoid disputes by means of an agreement. Jacobsen would receive £50 once the Horn Sound discoveries had been proven.[47] He would also be employed until mid-February 1913, during which time he was to reside on Spitsbergen when the company saw fit in order to explore and develop the property. He was prohibited from acquiring any options for himself or others, neither was he allowed to pass on any knowledge gained in the service of the company. He would receive 3,500 Norwegian kroner for the year and ten per cent of the net profits.

In summer 1912, the company tested Spitsbergen as a health report. Two guests stayed on Marble Island to recuperate. Guy Sherris had contracted malaria in West Africa. He was 'glad to say that [the] prophesy concerning the curative power of the wonderful germless air in Spitsbergen has been completely fulfilled.'[48] He resided on the island for two and a half months until mid-September without a malarial attack and felt healthy enough to walk and work throughout. George H. Slater had also returned from Africa with malaria. He travelled to Spitsbergen in bad health, expecting harsh weather, and returned 'absolutely fit.'[49] There is no mention of any health tourists in later years.

The company further enlisted the help of other companies and experts. The trip to New York, for example, had procured American specialists to inspect Marble Island in 1912. R. Gibson of the Ingersoll-Rand Co. had 30 years' experience.[50] He was enthusiastic about the best surface marble he had ever seen and the best top-rock he had ever channeled. He was convinced that the marble at depth would prove sound, solid, and marketable. Marble Island was more accessible than other quarries, and the cost of transit-rail, dues, and royalties could be saved. Quarrying and shipping was all that was needed. John Kelly of the Sullivan Machinery Co. shared these views after having explored the island for ten

[47] *An Agreement between the Northern Exploration Co. and Birger Jacobsen* (1912) Norwegian Foreign Department Box 5373, National Archives of Norway, Oslo.

[48] *Some particulars respecting the Co.'s properties in Spitsbergen with reports and extracts relating thereto* (1914). Northern Exploration Company (N.E.C.), Norwegian Polar Institute, Tromsø, p. 45.

[49] *Some particulars* (1914) p. 45.

[50] *Some particulars* (1914) p. 47.

weeks.[51] Kelly looked back on 20 years' experience and had seen nothing like it. Gibson and Kelly were not entirely impartial. If the marble proved saleable, either firm could expect a contract to provide expertise and machinery.

The company may have employed Victor Campbell, Wilfred Bruce, and Michael Barne to take advantage of their status as polar celebrities. The three mariners were Antarctic veterans. Campbell had been the leader of the Northern Party of Scott's ill-fated *Terra Nova* expedition and returned from the Antarctic in April 1913. He retired from the Royal Navy and entered into the services of the company in July.[52] Bruce was Scott's brother-in-law and had been the officer in charge of zoological work in the Antarctic. He accompanied Campbell to Spitsbergen in 1913. Barne had originally been on Scott's National Antarctic Expedition in 1901-4 and had helped to organise the *Terra Nova* mission, after which he, too, left the Navy. He sailed to Spitsbergen in 1914. When Britain entered the war, Campbell, Barne, and the sailor McCarthy, also of *Terra Nova* fame, returned home immediately to report for active duty.

7.3.4 Competitors

The areas that Jacobsen had made over to the Northern Exploration Co. in 1912 caused a conflict with the A/S Kulspids. The Norwegian firm maintained that Jacobsen had been on its payroll when it had claimed Recherche Bay in 1909.[53] The prospector therefore had no right to make the deal concerning the asbestos deposit there (4 on the map above). Meanwhile, Anker had an option on the land for 30,000 kroner. The company could get it for the same price, about £1,660, payable by October 1914. It is not known if the directors sent a reply.

By March 1913, the company was immersed in another controversy. This time, Britain's minister to Norway argued in its favour. Mansfeldt de Cardonnel Findlay explained to his Norwegian counterpart that Mansfield had formally occupied the north of Lowe Sound, Braganza Bay, and Michiel Rinder's Bay in 1906. This included the disputed Blaahuken (11 on the map). The posts erected to mark the claim were observed by Captain Arve Staxrud in 1907 and 1908 and plotted on a map, which he later presented to the Norwegian Government. Findlay proceeded to outline the continuous activities at Camp Morton and maintained that the north side of Lowe Sound and Braganza Bay, including Blaahuken, had been effectively exploited. The Norwegian Government had been notified in 1909.

[51] *Some particulars* (1914) pp. 47-8.
[52] Erskine (1994) pp. 117-22.
[53] A/S Kulspids (1912) *Letter to Ernest Mansfield, 4 December*. Norwegian Foreign Department Box 5372, National Archives of Norway, Oslo.

Hence, the counterclaim of Messrs. Boe and Holtermann filed at the end of 1912 could have no effect.

In addition to claims being disputed, the worth of the company's resources was being challenged. Despite thorough British and American analysis of the marble in Kings Bay, some Norwegians drew a different conclusion. They were probably Staxrud and Hoel, who had charted the terraces of Cross Bay and Kings Bay among several other locations in summer 1911.[54] 'The occurrences and the district itself have been very carefully explored by Norwegians, who, as a result of their investigations, have decided that the marble was not payable. They arrived at this conclusion partly from the appearance of the fissures, and partly owing to the unfavourable situations of the proposed quarries. It is pointed out that in Norway itself there are a number of marble occurrences that have not been opened up to drift because they have been judged to be non-payable, and that no better results can be possibly expected from the Kings Bay undertaking [...].'[55] Profitable quarries were, in fact, being established in Nordland at the time.[56] Yet, the workability of Marble Island remains unresolved to this day.

7.4 The local network of the private company

7.4.1 Claims and natural resources

Mansfield led the maiden expedition of the Northern Exploration Co. in summer 1911. The chartered SS *Repertor*, meaning discoverer or explorer, made eight successive trips to and from Norway, during which the sea was mostly calm and easily navigable in perpetual daylight.[57] On July 9, the *Repertor* arrived in Kings Bay and anchored outside an inlet on Marble Island.[58] Upon sounding the natural and sheltered harbour, which was christened Port Peirson, the ship was moored at the cliff and discharged while a raft made from lifeboats landed stores on the beach. Workers, who had been on site since June 3, had erected some housing (Fig. 7.3) and commenced quarrying.[59]

The Lesser Islands in Kings Bay also attracted attention that year.[60] They were absorbed into the already poorly defined claim. No. 2 Marble Island (probably

---

[54] *Mining Journal* (1911) 'Spitzbergen: rumoured gold finds', 21 October, p. 1017.
[55] *Mining Journal* (1911) 'Scandinavia: Spitzbergen marble and gold', 4 November, p. 1068.
[56] Heldal, T. and Neeb, P. R. (2000) 'Natural stone in Norway: production, deposits and developments', *NGU-Bulletin*, 436.
[57] *Marble Island* (1913) pp. 22-3.
[58] *Marble Island* (1913) p. 31.
[59] Leech, H. W. (1920) *Report on various marble properties in Spitsbergen.* Northern Exploration Company (N.E.C.), Norwegian Polar Institute, Tromsø, pp. 1-2.
[60] *Marble Island* (1913) pp. 32-3.

Storholmen) comprised a cliff that also permitted ships to be brought alongside, although a historical photograph shows that the house for this location was brought on a raft. The contemporary account that two smaller islands existed to the north of it, one made of breccia and the other of breccia and marble, has led to some confusion, as all neighbouring islands are situated to the north-east and east.[61] The islands were adorned with obvious claim markers. Where no boulders could be found, the wooden posts were supported with turf.

7.3 Workers flying the Union Jack at both ends of a pre-fabricated barrack on Marble Island in 1911. (Source: Marble Island (1913). 443-A, British Geological Survey, Keyworth.)

In September 1911, the Norwegian Johannes H. Giæver sold his claim and hut in Recherche Bay to Mansfield for £300.[62] The hut became known as Camp Smith. Arthur Mangham and some men were landed at Camp Millar for the winter 1911/12.[63] A Norwegian caretaker was left on Marble Island, while two trappers had erected Camp Zoe for Mansfield and used it as a base. Other trappers wintered at Camp Margaret on the south side of Braganza Bay.[64]

Ingvald Svendsen sold his claim in Van Keulen Bay to Mansfield, who was en route to Spitsbergen, on July 6, 1912.[65] The latter again led the expedition, which was assisted by Jacobsen and had secured the services of Captain Th.

[61] Consequently, the Northern Exploration Co. itself appears to have confused the contemporary names of the islands (Breccia Is., Davis Is., and Maples Is.), which the author has not been able to assign with absolute certainty.
[62] Barr, Newman, and Nesteroff (2012) p. 98.
[63] Barr, Newman, and Nesteroff (2012) p. 99.
[64] Barr, Newman, and Nesteroff (2012) p. 105.
[65] Barr, Newman, and Nesteroff (2012) p. 110.

No work appears to have been done in Horn Sound. Even so, the claim was clearly marked in the first map to be included in a company brochure (Fig. 7.5). The claims in Lowe Sound had been adjusted to accommodate the AB Isfjorden-Belsund in Braganza Bay. The claims in Van Keulen Bay had crept significantly westward, probably on the back of the deals with Giæver, Svendsen, and Hagerup, and comprised the area between Recherche Bay and Dunder Bay on the West Coast. The vague claim on Marble Island extended over the whole of Kings Bay and the eastern shore of Cross Bay, where Camp Zoe was situated. Oscar II Land had been coloured in with hindsight. A hut in the Thousand Islands had not been taken into consideration, nor had the huts built by August Olsen on the company's behalf in the northerly Bjørnhamna, Magdalenefjord, and Hamburgbukta in 1912.[75]

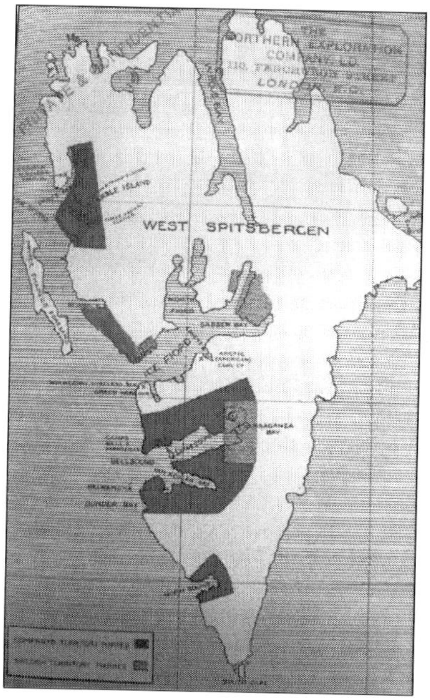

7.5 The first claim map to be included in a company brochure shows the claims in 1913. The company had already come to an arrangement with the Swedes. Oscar II Land was coloured in at a later date. (Source: Some particulars (1914) Northern Exploration Company (N.E.C.), Norwegian Polar Institute, Tromsø.)

In summer 1914, Mansfield no longer attended the expedition, which was instead led by Victor Campbell with Michael Barne as mate.[76] The company had purchased the MS *Willem Barents*, which was refitted in Norway. Barne recounted some trouble with the London office regarding the payment of bills. There was so little money that the ship left the works without being able to settle considerable debts,

[75] Barr, Newman, and Nesteroff (2012) p. 148.
[76] Erskine (1994).

and the expedition narrowly escaped being detained in Tromsø.[77] The ship sailed on May 24, Recherche Bay being the primary destination to continue the search for iron ore. Due to ice, Campbell could only land the prospectors at Renardodden, from where they walked to Camp Smith and were later fetched to Iron Mountain. The *Willem Barents* meanwhile proceeded to Camp Bell. On June 25, Campbell returned to Recherche Bay and landed some timber in a sheltered anchorage, where a hut for the prospectors was built. It was called Camp Campbell. Later, a second hut erected about a mile south of this location was named Camp Jacobsen.

At the beginning of August, the *Zoë* arrived with stores from Tromsø at the same time that war erupted in Europe. Campbell hurriedly inspected the company's camps and found Marble Island entirely unoccupied. Barne observed,

> There were thousands of pounds worth of plant left lying about including railway lines of broad gauge, leading from the place where the cutting had been made down to the sea where there is an excellent natural rock quay. There were huge derricks for working marble. However we were told that the marble apparently disintegrated when taken to a mild climate, and was valueless.[78]

Meanwhile, Jacobsen had consolidated the claims on Iron Mountain.[79] His men had placed over 200 claim markers across the range. They had also constructed houses and thus 'secured the property, so long as work is carried on, absolutely for the Company.'[80] Prospecting led to further discoveries of iron pyrites, marble, and coal. Near the end of the season, he used a boat for five days. In this period, he collected samples of gypsum in Berzelius Valley, which already belonged to the company, and estimated the deposit to contain several thousand tons. He also discovered hematite around Farm Harbour in Oscar II Land. Farm Harbour was one of the most sheltered anchorages on Spitsbergen. Jacobsen did not think that the area had been claimed by anyone yet and advised the company to do so at the earliest convenience. In Cross Bay, the geologist found an outcrop of white marble at a distance of about a mile from the sea at a good anchorage.

The plan was for Jacobsen and some men to winter on Spitsbergen. Before Campbell departed for Green Harbour, the geologist therefore requested provisions. Campbell sent 16 cases of lime juice, sauces, and pickles ashore, which escalated into an argument.[81] The commander was under instructions to take the *Willem Barents* directly to Aberdeen with only the crew. When Jacobsen

---

[77] Debts were contracted with the Harstad Mek. Verkstad, the Nordenfjeldske Steamship Co., Weltzien Holst, and others. (Jacobsen, 1919).
[78] Erskine (1994) p. 120.
[79] *Coal and iron in Spitsbergen.* Pam. (*32):622.333, Scott Polar Research Institute, Cambridge.
[80] *Coal and iron in Spitsbergen*, p. 23.
[81] Jacobsen (1919).

discovered that the workmen could not be guaranteed their wages, Campbell replied, "War breaks all contracts.'[82] He departed on August 10. The Norwegians succeeded to leave for Tromsø that same evening.

7.4.2 *Manifestations*

The archaeological landscape of exploration that had emerged during fieldwork was very much complemented by the archival material. The camps Morton, Bell, Margaret, Millar, and Williamson had been built prior to the company's formation, and the transfer of ownership had been arranged by Mansfield. In 1911, the *Repertor* brought 200 tons of cargo from Aberdeen to Marble Island.[83] This included two cranes (Fig. 7.6), tubs and rails, seven pre-fabricated buildings, and general merchandise.

7.6 One of the two cranes made by John M. Henderson & Co. of Aberdeen, which were delivered to Marble Island in 1911. (Source: pol02970, Polar Museum, Tromsø.)

The workers assembled a wooden barrack, which could accommodate twelve men, and a smaller house for three staff.[84] They then began to quarry. The holes to place the shot were drilled manually (Fig. 7.7). Roughly 40 pounds of gunpowder displaced up to 400 tons of marble, including blocks weighing up to 35 tons.[85] The blasted rock displayed few fissures and was said to be of good quality. No. 1 Quarry was situated at sea level in a little bay in the north-west corner of Port Peirson. It was approximately 70 feet wide, 20 feet high, and had been driven 25 feet before it was abandoned because the marble here was very broken and unsound. A short distance to the west, No. 2 Quarry was started. Although the

[82] Jacobsen (1919).
[83] *Marble Island* (1913) p. 30.
[84] Leech (1920) pp. 1-2.
[85] *Marble Island* (1913) p. 19.

marble appeared sound, in 1920, marble manager Leech advised not to continue at this location due to the proximity to sea level and the likelihood of water influx. Before the use of machinery, the early workings were invariably irregular in shape (Fig. 7.8). This barely set them apart from natural features.

7.7 Drilling to place a shot into marble. (Source: Marble Island (1913) 443-A, British Geological Survey, Keyworth.)

7.8 A blasted quarry on Marble Island may look like a natural feature in later years. (Source: Marble Island (1913) 443-A, British Geological Survey, Keyworth.)

No. 3 Quarry was commenced on June 1, 1912, but the delivery of necessary equipment was delayed by the dockers' strike in London in July.[86] The construction of houses and auxiliary buildings occupied the workers in the meantime. They completed another four barracks and two staff houses, raising the settlement's

[86] The docks strike in July 1912 was part of the pre-war industrial unrest as workers pressed for higher wages and better conditions. The strike wave subsided with the outbreak of war in 1914.

capacity to 70 (Fig. 7.9). The auxiliary buildings included repair shops, a forge, and stores. The stationary crane at the cliff was linked to a second one at No. 3 Quarry by a broad-gauge railway. Water wells were sunk, and water pipes supplied the houses, the quarry, and the shore.[87] After the late arrival of the equipment on August 17, its unloading and fitting took several days, during which parts were conveyed by manpower (Fig. 7.10) This included a third crane, presumably the mobile one made by Taylor & Hubbard of Leicester, as well as two channelers of seven tons and ten tons.

7.9 Five barracks and three staff houses on Marble Island in 1912. The settlement's capacity was 70. (Source: pol02984, Polar Museum, Tromsø.)

7.10 Manpower conveyed equipment on the broad-gauge railway. (Source: pol02971, Polar Museum, Tromsø.)

[87] A map reproduced in Barr, Newman, and Nesteroff (2012, p. 97) also indicates a pumping station, although this has not been verified in the field.

Channeling was a predominantly American mode of marble production on even terrain to obtain regular blocks of good size and to reduce waste.[88] The channelers were put to work at No. 3 Quarry, which lay within 500 yards of the settlement (Fig. 7.11). A space of 33 feet wide was channeled from a feather edge to a depth of four feet six inches (Fig. 7.12).

7.11 Channelers at work at No. 3 Quarry in 1912. Note the running water. (Source: pol02988, Polar Museum, Tromsø.)

7.12 Channeler producing a cut at No. 3 Quarry in 1912. (Source: pol02979, Polar Museum, Tromsø.)

Only about one-eighth of the stone was then cut out and removed. The remainder had been channeled into blocks, but their base was still attached. On September 8, quarrying was discontinued due to frost: cutting required running water and could not commence during winter. Before Renwick departed on September 19, he marked out a number of locations from which to extract cores to a depth of 100

[88] In hilly country, a wire saw would be better employed.

feet, expecting to see an improvement of the quality at lower levels. The wintering party additionally began work on a fourth quarry.

In summer 1913, a large and spacious storehouse (Fig. 7.13) was put up.[89] To include a junction branch to this storehouse, the railway was extended to a total of 550 yards. Yet, despite the flurry of activity and despite the shareholder Bullard supposedly observing large bulk samples being loaded onto the *Activ*, Leech later recounts little actual quarrying.[90] In 1914, the site lay entirely abandoned.

7.13 The storehouse built in 1913 was made of galvanised iron. Note the Union Jack. (Source: Some particulars (1914) Northern Exploration Company (N.E.C.), Norwegian Polar Institute, Tromsø.)

Elsewhere on the company's properties, pre-war installations were limited. Camp Zoe was built in 1911. In 1912, a pre-fabricated staff house was ferried across to Storholmen, while a Norwegian erected the three northerly huts on behalf of the company. A pre-fabricated barrack and staff house were also assembled at Camp Millar to facilitate the wintering of 16 men in 1912/13.[91] In 1914, Lægerneset was probably the chosen site for Camp Campbell, while Camp Jacobsen must have been the original hut at the later Iron Mountain Camp.

7.4.3 *Employees*

The expedition in 1911 was conducted aboard the *Repertor* captained by David Thomson.[92] Chief engineer John Laffray and chief officer J. S. MacGregor were part of its otherwise unnamed crew. Seasoned adventurer Ernest Mansfield naturally took the lead. He guided the visiting directors and shareholders. Besides Williamson and Peirson, the inquisitive shareholders were A. S. Bell, E. J. Bullard,

89 *Some particulars* (1914).
90 *Some particulars* (1914) p. 43; Leech (1920) pp. 3-4.
91 Barr, Newman, and Nesteroff (2012) p. 106.
92 Testimonial evidence as to the Englishmen and Scotsman in attendance from *Marble Island* (1913) pp. 18, 19-20, 22-3, 30, 31, 32-3, 34-41 and *Some particulars* (1914) pp. 47 and 53.

L. F. Davis, R. G. Davis, and M. A. Fenton. Also aboard were Arthur Mangham, by now an Arctic veteran and probably foreman in all things coal; David Hannah, who had been charged with the task of examining the coalfields; an unidentified John Topham; and the mechanic David Booth. It is likely that the quarry manager E. C. Millar arrived at the same time. 16 British men are thus accounted for, although Mansfield took about 20 quarrymen from Aberdeen alone.[93] Barr *et al.* assume the connection with Scotland being due to the origins of Mansfield's second wife. More importantly, the Scots were the only folks in Britain with any knowledge of working native marble – a connection of greater interest to investors.

During his time with the Spitzbergen Mining & Exploration Syndicate, Mansfield had built a good rapport with the Norwegians who assisted him on and off the islands. For the benefit of the Northern Exploration Co., he now affirmed that 'ample reliable labour is within easy reach. Good men are obtainable at a moderate wage. Whatever labour may from time to time be required, can be easily and speedily engaged, and there is no difficulty in securing reliable overseers.'[94] He did not explicitly refer to northern Norway, but it must have been clear that the workers were and would be coming from there. Observers believed that Mansfield's gold finds had only remained a secret due to the loyalty of his Norwegian miners.[95] In 1911, he hired several Norwegian carpenters.[96]

7.14 Ten men and a puppy in front of a barrack on Marble Island in 1911. (Source: Marble Island (1913) 443-A, British Geological Survey, Keyworth.)

[93] Barr, Newman, and Nesteroff (2012) p. 96.
[94] *Marble Island* (1913) p. 36.
[95] *Mining Journal* (1911) 'Norway: Spitzbergen', 26 August, p. 863.
[96] Barr, Newman, and Nesteroff (2012) p. 96.

Photographs attest to the workers' presence on Marble Island from June onwards. In Fig. 7.3 above, 21 men – and the photographer – were engaged in constructing the first houses on site. Ten men and a puppy later posed in front of a workers' barrack (Fig. 7.14); they were presumably the new occupants. The official group photograph (Fig. 7.15) was arranged later in the summer, when this particular barrack had been named Camp Peirson. Among the 30 men were the visiting directors and shareholders, Mansfield in a central position, the ship's uniformed officers and crew, the workers, and the cooks in white. Yet, the image merely depicted those employed in Kings Bay. Men were also engaged in Bell Sound, so the total number of participants in the expedition was somewhat higher. The men left at the close of the shipping season in September.

7.15 Group photograph taken in front of the workers' barrack 'Camp Peirson' on Marble Island in 1911. (Source: Marble Island (1913) 443-A, British Geological Survey, Keyworth.)

Mansfield found Norwegian trappers in Tromsø, who were willing to winter on Spitsbergen and guard the company's properties.[97] Who they were is largely inconsequential to this chapter, suffice to say that they appreciated the unusually generous provisions they received, maintained the huts, and occasionally assisted with mining. According to Torberg Pedersen, Mangham was responsible for 11 men to work coal and gold at Camp Millar. The men blasted about 1,000 tons of

[97] Barr, Newman, and Nesteroff (2012) pp. 99-106.

stone, which was removed by crane.[98] The Norwegian remembered Mangham as a pleasant man, although he fired two workers for complaining about the cold. There were no arrangements for a doctor, which became a serious problem when an unidentified Englishman died in spring 1912. The atmosphere was dismal by June because supplies were running low, the men had not found coal where they expected it to be, and water was seeping into the apparent gold mine. Delayed by the docks strike, Mansfield only arrived on July 11.

In summer 1912, the *Activ* was manned by Th. Tangeraas and his crew.[99] Mansfield again accommodated visitors such as the company's secretary John Richard Maples, two shareholders, and the medical officer F. G. Gardner. Gardner commented positively on the workers' health during his one-month stay on Marble Island. Meanwhile, Mangham and Jacobsen prospected in Bell Sound and Horn Sound while David Hannah scrutinised the coal deposits for a second time. William George Renwick, who had previously studied the marbles in the comforts of Britain, now kept an expert eye on the development in Kings Bay from mid-July until mid-September. He was able to consult with Gibson and Kelly from the Ingersoll-Rand Co. and the Sullivan Machinery Co., respectively. These names were only a fraction of the entire expedition said to have been 75 strong, including 40 men from England.[100] Yet, they create the notion of complex contracts, agreements, and understandings as well as a constant coming and going of various parties on the company's sites. In terms of photographs indicating the workforce, let alone actual numbers, Fig. 7.10 above shows 15 toiling men, the most captured that season.

The wintering party of 1912/13 comprised the quarry manager Millar and nine or ten men.[101] This included three experienced Norwegian trappers and Mansfield's two brothers-in-law. Originally from Glasgow, James Booth was the cook, while the aforementioned David Booth assisted with the core drilling. The latter kept a diary from September 19, 1912 until July 23, 1913, and Barr *et al.* correctly state that such a document offers a rare insight into the company's expeditions from the perspective of a British member.[102] The diarist recounted that the cores were mostly fractured, which he attributed to the nature of the marble and to possible sabotage by some of the workers. He did not, however, question the suitability of the technology in the Arctic or the correct use of it.

---

98 This crane may have been the second stationary crane originally delivered to Marble Island. Its former foundations have been recorded archaeologically both at No. 3 Quarry and at Camp Millar.

99 Testimonial evidence as to the men in attendance from *Some particulars* (1914) pp. 10, 23-7, 40-2, 44, 45, 46, 47, 47-8.

100 Barr, Newman, and Nesteroff (2012) p. 108.

101 Leech (1920) p. 3; Barr, Newman, and Nesteroff (2012) p. 112.

102 Barr, Newman, and Nesteroff (2012) p. 114-7.

In early May 1913, a Norwegian advance party headed by Jacobsen explored Martin Range.[103] Besides four crew, it entailed Peder Engen, Erling Engen, and Peder Gamager, who had been foremen at various metal mines in Norway and Sweden.[104] Meanwhile, the *Mining Journal* reported that the *Activ* had arrived in Tromsø on May 23 on her way to Spitsbergen.[105] Led by Mansfield for the third time, the main expedition comprised 46 Englishmen, about ten of whom were engineers, the rest workers.[106] The different contracts lasted six months, a year, or one and a half years. Yorkshire miners were engaged in Bell Sound. Mangham, J. Housley, W. Walker, and J. Kelly also confirmed the presence of iron ore on Iron Mountain.[107] Later in the season, esteemed persons arrived in Recherche Bay. Campbell, Bruce, and Bullard then targeted the coal mines and the marble quarries, where Renwick studied the varieties while the development progressed under Gibson, the manager, and H. Mitchell, the general foreman.[108] In a photograph, some 30 men celebrate the completed storehouse (Fig. 7.16). A beer bottle is held above the crowd; perhaps the company used liquor as an incentive. Little is known about the inspection of the American marble expert Minard, assisted by Sherer and Barnett. Barr *et al.* assume that his report may have contributed to Mansfield's discontinuation at the company.[109]

7.16 A group of men celebrate the finished storehouse on Marble Island in 1913. (Source: pol02998, Polar Museum, Tromsø.)

[103] *Some particulars* (1914) pp. 7-11.
[104] *Some particulars* (1914) p. 13.
[105] *Mining Journal* (1913) 'Scandinavia: Spitzbergen Minerals', 31 May, p. 525.
[106] *Tidens Tegn* (1913) 'To Spitsbergen after coal, marble and goal', 19 July.
[107] *Some particulars* (1914) p. 19.
[108] *Some particulars* (1914) pp. 40-4.
[109] Barr, Newman, and Nesteroff (2012) p. 125.

Ernest Mansfield gave a last statement to the company on November 21, 1913.[110] He did not lead the expedition in 1914, nor were he or Mangham part of it. In a confused article, the *Mining Journal* related that the consul Andreas Aagaard was instrumental in purchasing the *Willem Barents*. If so, it is the first mention of a Norwegian ally on the mainland. A contemporary photograph showed the staff and crew of the ship in 1914. It named the engineer A. Strachan, the expedition leader Victor Campbell, the second officer Michael Barne (holding a puppy), and the geologist Birger Jacobsen. Peder Engen stood on the far left while Peder Gamager was seated on the far right. The expedition relieved the winterers at Camp Bell, one of the two being Gainer, who had been kept company by some trappers and Norwegian visitors from Green Harbour.[111] Furthermore, the quartermaster Young found mention, and the sailor Mortimer McCarthy was among the party that departed Spitsbergen in August 1914.[112]

The company's bad financial state after summer 1913 forced Jacobsen to take some drastic measures. Wanting to make something of his original claims, he handed in his notice on January 31, 1914, resigning his position and terminating his agreement with the company.[113] Yet, he attended the expedition in 1914, after which he accused the firm of gross misconduct and flagrant breaches of the workers' contracts, which he himself had signed on the company's behalf. Gainer had not received his fee for wintering in Bell Sound in 1913/14, while other winterers' wives, who depended on the money in their husbands' prolonged absence, were also left waiting. After the expedition, Jacobsen's monthly salary to his wife was three months in arrears, and the payments to the workers' families failed to come in on the pre-arranged dates. Captain Johan Hagerup, who had been the pilot on the *Willem Barents*, was still owed 700 kroner by autumn 1918.

In an epilogue to the expedition, Jacobsen and 12 Norwegians, who had left Spitsbergen of their own accord, encountered the *Willem Barents* in Tromsø; the ship had a broken propeller. Since the company had been unable to meet its contracts, the vessel was seized by the town magistrate. In 1915, the extensive, unredeemed stock at the custom house in Tromsø was sold in a public auction to pay for the storage. The surplus also covered the company's debt to the wireless station at Green Harbour. The firm's agent at Tromsø, Captain Scheen, however, could not recover his fees and his outlays for 1914 until 1916 through the repossession of the ship.

---

[110] *Some particulars* (1914) pp. 17,
[111] Erskine (1994) p. 119.
[112] Erskine (1994) p. 121.
[113] Jacobsen (1919).

7.4.4 *Local allies*

In summer 1911, Mansfield, Williamson, and Peirson contacted the Arctic Coal Co. The Americans had experienced strikes both at Advent City and at their own coal mine. The men therefore went to see Longyear, wishing to form an offensive and defensive alliance for mutual protection in case of future trouble and uprisings, but Longyear did not think it wise to commit to such an agreement.[114] It is unclear what happened next. The Foreign Office did, in any case, receive a letter from the company that such an alliance had in fact come about and that a copy of their letter would also be forwarded to Kristiania.[115] In 1912, Hannah was able to go over the American mine, noting semi-bituminous coal being worked by longwalling. It was cut by electric coal cutters before being blasted down. Hannah thought that method would also suit the English workings. Any dealings between the two firms were probably eased by the fact that the company's foreman Arthur Mangham was the father of the Americans' winter superintendent Bertrand Mangham.

The Scottish Spitsbergen Syndicate did not cross paths with the Northern Exploration Co. until 1913. The Scots appeared to have been oblivious to Mansfield's activities, but they now wondered if these were entirely legal. Their enquiry established that Mansfield had originally explored Spitsbergen in 1905 and had then sold his claims to the Spitzbergen Mining & Exploration Syndicate in 1906.[116] When this group had ceased to exist, he had re-sold the claims to the Northern Exploration Co. in 1910. The Scots discovered that although Lord Morton's Co., as they called it, had been crossed off the company register in accordance with the law, it still had the right to appeal. The law did not say what became of the assets of a dissolved company, but it did state that the liabilities of the directors, officers, and members continued. This implied that their rights to the assets also continued. It was up to a lawyer to decide if Mansfield had acted unlawfully. The Scottish Spitsbergen Syndicate could not clarify the matter and suggested that William Speirs Bruce, their promoter, should simply ask Lord Morton next time they met. Although the Scots clearly had suspicions, their correspondence with the Foreign Office had nonetheless supported the company's claim between Recherche Bay and Dunder Bay.[117] This marked the beginning of a love-hate relationship between the two remaining British firms on Spitsbergen.

---

[114] Dole (1922a) p. 91.
[115] *Memorandum by Hertslet* (1913).
[116] Ross, J. (1913) *Letter to W. S. Bruce, 4 June.* Scottish Spitsbergen Syndicate (S.S.S.), Norwegian Polar Institute, Tromsø.
[117] *Memorandum by Hertslet* (1913).

7.4.5 Products

Factual reports such as Thomson re-shipping 120 tons in 1911, which evidently included 15 tons of marble and 50 tons of coal, are rare.[118] The company's products between 1911 and 1914 were more likely to be eyewitness testimonies, expert statements, and product descriptions bundled into one of the four lengthy brochures that have been quoted throughout this chapter.

Samples of marble were initially sent to British specialists. F. F. Mullet was asked to provide 'an independent and unbiased report [and] was surprised to find such an assortment of fancy and coloured marbles of high quality, suitable for decorative purposes, and to learn that they can be imported in practically illimitable quantity, and at prices likely to revolutionise the marble industry.'[119] He found that 'the non-absorptive properties of the stone are so considerable that contained moisture is very small, and, in consequence, that the atmospheric disintegration which ordinarily renders surface stone friable and valueless, has not here acted to any appreciable extent even during the severe weather.'[120] Despite the fact that the marble had been obtained by blasting, H. T. Bounds was able to cut blocks evenly into three-quarter inch slabs using an ordinary frame-saw. 'More severe conditions than these could not have been given to any stone. The deduction from this is that given stone a few feet beneath the surface, and the adaptation of modern channelers in quarrying, you will be able to cut sound blocks of practically any required dimension.'[121] Hayward specifically looked for faults and failings common to surface stone, but he could not discover any, indicating that the rock at depth would be exceptionally sound.[122] Where architects and builders had hitherto been reluctant to use marble due to frequent supply delays and undependable quality, Marble Island could provide reliable and high-quality service. Some marbles could even fetch monopoly prices. Renwick offered figures regarding production costs and sale prices. Extraction, transport, loading, and freight to Britain or northern Europe would not exceed 35s per ton. The lower grades would sell at £3 15s per ton, the medium grades at £6 10s, and the high grades between £10 and £15. At an output of 50,000 tons per year, the annual profit would approach £223,750 gross. Yet, the producing stage was not reached before the war.

---

[118] *Marble Island* (1913) p. 22; *The Times* (1911) 'Gold in Spitzbergen', 13 October, p. 13.

[119] Mullet, F. F. (1911) 'Letter to the Northern Exploration Co., 28 November', cited in *Marble Island* (1913) p. 16.

[120] Mullet (1911).

[121] Bounds, H. T. (1911) 'Letter to the Northern Exploration Co., 12 December', cited in *Marble Island* (1913) p. 25.

[122] Hayward, C. J. (1912) 'Letter to the Northern Exploration Co., 26 February', cited in *Marble Island* (1913) pp. 28-9.

Regarding coal, the company commonly stated the thickness of various seams on its properties before subtly digressing to the workings of other firms. Hence, a seam at Camp Bell was four feet thick, another at Camp Morton between four and five feet.[123] Although Hannah only saw a portion of the company's land, he inferred that it possessed a large area of good coal. In mines located high above sea level, drainage could occur naturally, and haulage could make the most of gravity. Good rock roofs required little timbering, and the closeness to the shore reduced expensive land transport. How to work the seams was effectively demonstrated by the Swedes in Braganza Bay and the Americans in Advent Bay. Observations and suggestions of machinery presumably served to give the impression that the company was already half way to extracting a commercial output and to beating its competitors to market. Hannah estimated the cost of coal to be between 7s 6d and 8s per ton. With an outlay of £30,000, a daily output of 1,000 tons could be achieved.

Similarly, the products of Iron Mountain were at the time restricted to samples, test results, and inferences. Thus, well-known British experts presented the promising assays summarised in Fig. 7.17. This, too, would have encouraged the company and its stakeholders to go ahead with the enterprise.

Experts		*Total iron (%)*
The Amalgams Co., Sheffield		66.98
Prof. T. Turner, Birmingham	Sample 1	59.68
	Sample 2	65.25
Pattinson & Stead, Middlesbrough	Sample 1	66.89
	Sample 2	65.39
Riley & Harbord, London	Sample 1	68.91
	Sample 2	65.30

7.17 Summary of iron ore assays. (Source: Some particulars (1914) Northern Exploration Company (N.E.C.), Norwegian Polar Institute, Tromsø, pp. 15-7.)

The company reports are notably quiet about gold. Jacobsen later recounted that Mangham had been eager to drive an adit into what he and his workers thought was a gold-bearing conglomerate at Camp Millar in winter 1911/12. He shipped several barrels to the London office at the earliest opportunity in 1912.[124] The disappointment can only be imagined when the directors recognised the mineral to

[123] *The Northern Exploration Company* (1912) p. 5-9.
[124] Jacobsen recounted three barrels, Tangeraas ten. Barr, Newman, and Nesteroff (2012) pp. 107-8.

be iron pyrite, also known as fool's gold. Work at the 'gold mine' was discontinued that season.

7.5 The First World War

7.5.1 Global changes...

From August 4, 1914 until November 11, 1918, Britain was at war with Germany, and the activities of the Northern Exploration Co. were few and far between. Firstly, there was a shortage of men, since there was pressure on every citizen to take part in the war effort. This meant departing Spitsbergen and voluntarily taking up military or supportive tasks.[125] Men like Campbell and Barne reported for naval duty and saw active service. Others like the director Blackstone supported the home front. His factory became government-controlled for the manufacture of munitions, marine engines, and spare parts for the Admiralty.[126] Secondly, there were restrictions in transport. Most ships were in war use, and vessels for private purposes were hard to come by. Even if one succeeded, insuring such a vessel was near impossible due to the threat of it being damaged or sunk by enemy action. In addition, Britain imposed a naval and economic blockade on Germany. The company did not fit out an expedition for the time being. 'As a matter of courtesy', Jacobsen drew attention to the fact that it needed to employ guards on its properties if it expected to retain the claims and protect its machinery and stores.[127] He was convinced that not the British authorities prevented any such arrangements but the firm's total lack of funds.

In May 1915, the *Colliery Guardian* published some articles of interest to Spitsbergen stakeholders. The magazine stated that 'one of the minor results of the war now being waged in Europe has been the renewed attention to the coalfields under development in Spitzbergen.'[128] It painted a sober picture of these coalfields. If anything, it downplayed the British interests: the Spitzbergen Coal & Trading Co. had possibly been unsuccessful due to mismanagement; an English firm under Lord Morton had worked a few promising openings; and a Scottish syndicate was prospecting in Tertiary deposits of secondary importance. This gave the impression that the *Colliery Guardian* was only interested in actual mining, not in speculation. The article nonetheless concluded that Spitsbergen held abundant coal reserves,

[125] Compulsory military service had hitherto been unpopular in Britain and was only introduced in 1916.
[126] Michael Kay (2003) *Official Blackstone Engine Website*. Available at: http://www.oldengine.org/members/blkstone/History2.htm (Accessed: 28 March 2011).
[127] Jacobsen (1919).
[128] *Colliery Guardian* (1915) 'The coalfields of Spitzbergen', 14 May, pp. 1013-4, 1069-70, 1119.

and while there were no technical obstacles to their exploitation, the economic conditions were somewhat strained. Most significantly,

> [...] the question may be asked as to whether there exists a sufficient market for the coal in Scandinavia, where water-power has been so largely utilised and the railways have been electrified purposely to avoid the use of fuel. In answer, it is pointed out that Norway contains important deposits of metals that as yet remain undisturbed; finally, the prospect of utilising coke for the electric smelting of iron presents itself, and although the Ice Fjord tertiary deposits do not yield coke, it is possible that others lend themselves to this application. Spitzbergen coal has so far found a ready market in Norway. As to whether it can compete with British and German coal, it is difficult to say, and the success of more ambitious trading would depend upon the better organisation of the industry. A reference has already been made to the possible introduction of further Russian capital, and since the war started renewed interest has been shown in Norway in the mineral wealth of this "No Man's Land."[129]

In April 1916, the *Colliery Guardian* reported that two Scandinavian firms intended to mine coal in the near future.[130] Firstly, the Swedish AB Isfjorden-Belsund had decided to commence on a great scale, increasing its capital from 290,000 kroner to 2,000,000 kroner.[131] Secondly, a syndicate of Norwegian bankers and ship owners had bought the Arctic Coal Co., the Anker Coal Co., and the Norwegian Spitzbergen Coal Co. This Store Norske Spitsbergen Kulkompani, or simply Store Norske, was in a position to continue, where the Americans had left off. In August 1917, the journal related that the Baltic-White Sea canal now enabled the transport of Spitsbergen coal directly to Petrograd and that Russian capital was making a move to obtain Arctic concessions or mines to accumulate coal stocks.[132] On the whole, the information suggested increasingly stiff competition on the archipelago in the absence of any British actors.

The anxious response of the Northern Exploration Co. and the Scottish Spitsbergen Syndicate was to jointly apply to the Foreign Office to consider the re-annexation of Spitsbergen in September 1916.[133] In November, the question whether the islands and their resources belonged to anyone was briefly raised in the House of Commons, which was reminded that the international conference, which assembled in June 1914, had to be adjourned.[134] When asked if the British Government would prevent German concessions, the Foreign Office could not

[129] *Colliery Guardian* (1915) 'The coalfields of Spitzbergen', 14 May, p. 1119.
[130] *Colliery Guardian* (1916) 'Coal on Spitzbergen', 14 April, p. 718.
[131] The AB Isfjorden-Belsund was reorganised into the AB Svenska Spetsbergen in 1916. The Northern Exploration Co. retained its ties to the Swedes.
[132] *Colliery Guardian* (1917) 'Russia and Spitzbergen coal', 17 August, p. 305.
[133] *The Times* (1918) 'Letter to the editor', 11 March, p. 9.
[134] HC Deb 14 November 1916 vol 87 c542.

permit such British action in a no man's land. However, 'whether any concession pegged out during the War will be recognised at the peace is of course another question.'[135] This indicated that in the mind of the British Government, war was an exceptional state of affairs, and territorial advances and material gain during these years may not be valid afterwards.

On March 3, 1918, Russia and the Central Powers, headed by Germany, signed the Treaty of Brest-Litovsk. Besides marking Russia's exit from the war, the treaty riled the British press because it was believed to signal Germany gaining a foothold in the Arctic.[136] The editor of *The Times* wrote, 'There is a curious clause which relates to the future of Spitzbergen. So far as we can judge, the clause implies that the Germans and the Bolshevists decided to share Spitzbergen between them. [...] As we cannot imagine M. Lenin and his colleagues descending upon Spitzbergen, the real meaning of the clause is that Germany proposes to claim the archipelago as well as the rest of the earth.'[137] The Treaty of Brest-Litovsk was the turning point in Britain's public opinion of the Spitsbergen issue since 'the extraordinary interest in Northern politics which is now being displayed by Germany is naturally exciting general attention.'[138] The press releases that followed in its wake were decidedly political rather than economic. Shortly afterwards, the Foreign Office was ordered to confer with the Admiralty on Spitsbergen and to report to the War Cabinet.[139]

The Treaty of Brest-Litovsk ignited public discourse surrounding the Spitsbergen Question in Britain. The Northern Exploration Co. was among the first to respond to *The Times.* Secretary Maples wrote,

> Realizing the value to the nation of the many mineral deposits we have located, and the extreme danger of their becoming other than British-owned, this company has strongly urged the British Foreign Office, both before and during the war, to annex Spitzbergen, either by reviving the claim made thereon on behalf of King James I. (which has never been annulled) or by formally taking possession of British-claimed territory, and arriving at an arrangement with other nationals respecting their holdings.[140]

---

[135] HC Deb 21 November 1916 vol 87 c1255W.

[136] The Treaty of Brest-Litovsk comprised Chapter IX titled 'Organization of the Spitsbergen Archipelago'. A translation of Article 33 read as follows, 'The contracting parties will direct their efforts to having the international organization of the Spitzbergen Archipelago, which was contemplated at the Spitzbergen Conference in 1914, carried out on a footing of equality for both parties. To this end the Governments of both parties will request the Norwegian Government to bring about the continuation of the Spitzbergen Conference as soon as possible after the conclusion of a general peace.' (*The Times* (1918) 'Spitzbergen', 11 March, p. 9.)

[137] *The Times* (1918) 'Spitzbergen', 11 March, p. 9.

[138] *The Times* (1918) 'Scandinavian suspicion. German attacks on Norway.', 11 March, p. 7.

[139] Memorandum for the War Cabinet on the Spitsbergen Question (1919), British Legation 1919 FO 609/123 Files 427/3/1 (start- pp. 17998), National Archives, Kew.

[140] *The Times* (1918) 'Letter to the editor', 11 March, p. 9.

The historical claim and the supremacy of territorial possession were promptly adopted into popular rhetoric. In turn, other correspondence with the newspaper concerned Spitsbergen's iron ore, which had been offered to the British Government two years previously, but despite recommendations from the Ministry of Munitions to accept it, the Treasury had turned the proposition down.[141] Furthermore, a letter from the Royal Geographical Society to the Foreign Office was reproduced. It had been penned by Douglas W. Freshfield, debatably the only president with a political agenda in the history of the Society.[142] He urged the Foreign Office 'to take such steps to settle the future of Spitzbergen as they may deem expedient, either by proclaiming the sovereignty of Great Britain over the islands, or over that part of them to which, since the annexation in the seventeenth century, we hold an historical claim superior to that of any other nation.'[143] The incorrect rendition of the place-name led to a public announcement by the Society that Spitsbergen, being a Dutch word, should not be spelt with a German 'z'.[144] This historical truth and anti-German sentiment was thereafter cultivated by *The Times* and others.[145] Meanwhile, a Norwegian source was quoted, saying 'that England may calmly await her hour, as, owing to her dominating maritime position, she is always sure of a decisive influence in the final settlement. The question of influence over Spitsbergen is one between Norway and Great Britain.'[146]

The spring of 1918 was a highly-strung time. Public and governmental awareness of Spitsbergen was mounting, which led to an awakening of the dormant Northern Exploration Co. To turn this favourable political climate into an economic advantage, the company increased its nominal capital to £500,000 on May 8.[147] It then took the decisive step of converting to a public company. On August 28, its shares opened at 29s 6d on the London Stock Exchange.[148] In conjunction with the conversion, the company published a fifth brochure titled *Coal and iron in Spitsbergen*. Contemporary readers would have understood this to hint at the two resources most urgently needed during the war. The company maintained that, 'The figures concerning these mineral deposits, given by experts,

---

[141] *The Times* (1918) 'Letter to the editor', 12 March, p. 5.

[142] '[He] had the courage to defy the tacit convention that the Society should have no dealings with politics. Party politics, he admitted, must be rigidly excluded; but at a time when all parties were at one on national questions he held that the Society was entitled to make its views known on pertinent facts and the wider aspects of national policy. A minor case in point arose on a paper by Sir Martin Conway on the unsettled status of Spitsbergen, and communications were made to the Foreign Office on the subject.' (Mill, H. R. (1930) *The record of the Royal Geographical Society 1830-1930*, London: Royal Geographical Society.)

[143] *The Times* (1918) 'British claims in Spitzbergen', 13 March, p. 5.

[144] *The Times* (1918) 'Letter to the editor', 14 March, p. 5.

[145] *The Times* (1918) 'April reviews. Russia in revolution.', 1 April, p. 10.

[146] *The Times* (1918) 'Future of Spitsbergen. British and Norwegian interests.', 14 March, p. 5.

[147] *Financial News* (1919) 'Spitsbergen', 27 June.

[148] *The Times* (1918) 'Stock exchange', 28 August, p. 10.

are so huge that they seem exaggerated to anyone who has not visited West Spitsbergen. Germany, however, fully appreciates the value of Spitsbergen's minerals, as was shown by the Spitsbergen clause in the Treaty of Brest-Litovsk.'[149] As many others, the company believed that economic supremacy lay with the nation that controlled the world's coal and iron. It unmistakably intended to incorporate the natural resources of Spitsbergen into the British Empire (Fig. 7.18).

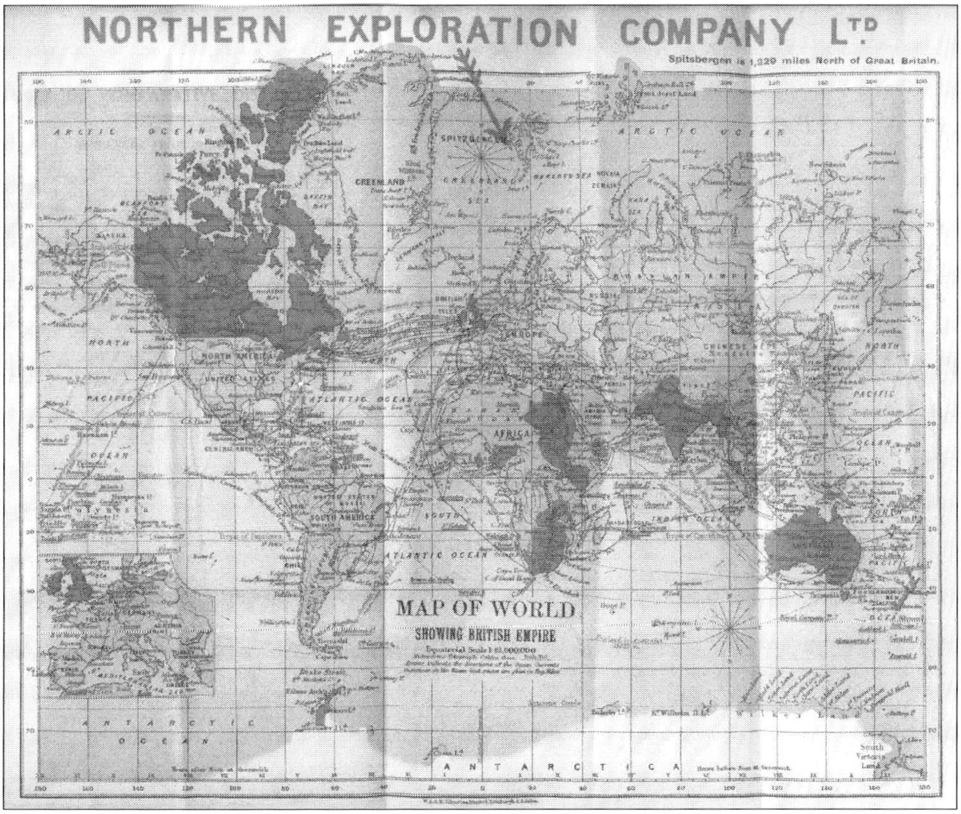

*7.18 Map hinting at the company's plan for Spitsbergen within the British Empire. (Source: Coal and iron in Spitsbergen (1918) Pam (*32):622.333, Scott Polar Research Institute, Cambridge.)*

[149] *Coal and iron in Spitsbergen* (1918) Pam (*32):622.333, Scott Polar Research Institute, Cambridge, p. 7.

7.5.2 ... local responses

During the war, an expedition to Spitsbergen needed to be officially sanctioned. Since the British Government would benefit from knowing the status of the natural resources, the Northern Exploration Co. received the backing of several departments.[150] The Foreign Office approved of Ernest Shackleton, the Antarctic explorer, to be in charge; Captain Frank Wild, also of Antarctic fame, was released by the Admiralty to be second in command; and the surgeon McIlroy could be obtained, too.[151] The Royal Navy provided the armed merchant ship SS *Ella* captained by Thomson.[152] Commander Norman Carlyle Craig accompanied the expedition in order to report to the Admiralty.[153] The company was represented by its latest managing director Frederick William Salisbury-Jones, the secretary Maples, and Noel B. Davis, son and spokesman of the chairman. The mining engineers, contracted for three years, were Arthur and Bertrand Mangham, at least until Arthur broke a rib on board and had to disembark in Tromsø. H. Rogers was the foreman of the largely Norwegian workforce. There were some Swedes. Herbert G. Ponting, photographer and cinematographer on Scott's ill-fated *Terra Nova* expedition, was probably the 'special correspondent' behind eleven lengthy articles published in the *Financier* between October 8 and November 12, soon to be reorganised into the company's sixth brochure called *Spitsbergen's mineral wealth.*[154] According to Ponting, the aims of the expedition were threefold: to spoil German intentions on Spitsbergen; to vindicate British interests against neutral aggression; and to start mining camps to develop the company's claims.[155]

The journey from England took two weeks due to the shipping shortage, submarine dangers, delayed sailings, and devious routes. In Norway, the company representatives received troubling news of Norwegian trespassing, which prompted Salisbury-Jones and Craig to travel to Kristiania to lay their complaints directly before the British minister and the Norwegian Government. They were reunited with the party in Tromsø, where local newspapers appeared to conspire against them by running stories of German U-boats lying in wait, which resulted in the Norwegian crew being unwilling to proceed, and by urging Norwegians to exploit

---

[150] *Coal and iron in Spitsbergen* (1918) p. 8.

[151] Shackleton had been on the *Discovery* expedition 1901-3, the British Antarctic or *Nimrod* expedition 1907-9, and the Imperial Trans-Antarctic or *Endurance* expedition 1914-17. Wild had been on all of these in addition to Mawson's *Aurora* expedition 1911. McIlroy had been the surgeon on the *Endurance* expedition.

[152] *The Financier* (1918) 'Spitsbergen's mineral wealth', 8 October.

[153] Copy of paper read by Mr. Herbert G. Ponting (1918) RGS/CB8/Conway, RGS/IBG Archives, London.

[154] *Spitsbergen's mineral wealth* (1919) Longyear Spitsbergen Collection MTU MS631 LY B2/14, Michigan Technological University, Houghton/MI.

[155] *The Financier* (1918) 'Spitsbergen's mineral wealth', 8 October.

the Arctic before anyone else did. Shackleton was suddenly taken ill and had to turn back.[156] Wild now assumed command.

When the *Ella* arrived in Recherche Bay, her gun was fired 'as a warning that infringements of British rights would not be permitted.'[157] A first land party inspected Iron Mountain and the depredation by Norwegian interlopers. Although 60 tons of iron ore seemed to be missing, it was made out to be nothing compared to what the claim held in store, and the claim-jumpers had, in fact, done them a service by exposing the lode. A second party departed to the south side of the bay, which they also expected to be raided, but their boat was swamped by a surge of water before they could land. A third party was dispatched in search of fresh water, a stream being found nearby. A fourth party began a survey of Iron Mountain, charged with the task of ascertaining the practicability of a ropeway, and it is possible that the construction of pre-fabricated barracks at Camp Campbell (Lægerneset) and Camp Jacobsen (Iron Mountain Camp; Fig. 7.19) commenced in 1918. Wild later recollected that 'the standard army huts they had brought with them were totally inadequate and it took a long time to make them moderately weatherproof, even when they were covered with snow.'[158]

7.19 Panorama of Iron Mountain Camp in Recherche Bay. Note the narrow-gauge railway in the front that conveyed provisions from the shore to the army barracks. (Source: pol02985, Polar Museum, Tromsø.)

The company was convinced of the economic potential of Iron Mountain, yet '[when] the heads of the expedition had visited the coal properties on the north side of Lowe Sound, they decided that, tempting as the other prospect was, it would be

[156] Huntford, R. (1985) *Shackleton*, London: Hodder and Stoughton, p. 662.
[157] *The Financier* (1918) 'Spitsbergen's mineral wealth', 15 October.
[158] Recounted in Barr, Newman, and Nesteroff (2012) p. 132.

better to defer the exploitation of the iron till the wire ropeway is installed by July 1919.'[159] The coal property at Camp Morton, despite not being mentioned by name in any of the articles, was ready for development, and coal was currently selling at £12 per ton in Norway. Owing to the gentler gradient, the company proposed a one-mile-long monocable system, which could transport 50 tons per hour, allowing for an output of 1,200 tons per day if successive shifts were worked. Archaeology showed that a 'flying fox', a zip line to a railway track, was indeed set up, but it is not known if an output of 5,000 tons from the mines was realistic.[160]

In Kings Bay, a Norwegian firm had been working an alleged British coal claim. On the way there, the *Ella* struck a reef but stayed afloat. In a letter of complaint addressed to the Kings Bay Coal Co., Salisbury-Jones and Maples stated that during the war, it had been impossible for their company to develop the claim further as men and tonnage were needed elsewhere and that it was inacceptable that men from neutral Norway had encroached on it.[161] Yet, despite the claim dispute, the ship required bunker coal from the mine. It then proceeded to Cross Bay.

7.20 Detail of an American drawing of a bird's-eye view of Spitsbergen, highlighting the hoisting of the Union Jack at Ebeltoft Haven. (Source: The Sphere (1918) The British flag hoisted at Spitsbergen, 9 November, pp. 100-1.)

On September 3, 1918, the Northern Exploration Co. arrived at the abandoned German meteorological station in Ebeltoft Haven and pulled down five enemy claim signs before erecting British ones. The actions were felt to be sufficient. 'Having formally dispossessed the Germans of the most northern territory in which they have ever had a footing we completed the act of annexation by hoisting the Union Jack; then laden with loot and proud of our achievement we re-embarked for

[159] *The Financier* (1918) 'Spitsbergen's mineral wealth', 15 October.
[160] Barr, Newman, and Nesteroff (2012) p. 132.
[161] Salisbury-Jones and Maples (1918) *Letter to the Kings Bay Coal Co, 2. September*, Norwegian Foreign Department Box 5372, National Archives of Norway, Oslo.

another destination.'[162] No photographs depicting the event have been unearthed, but other sources are equally potent (Fig. 7.20).

Subsequently, the men who had remained in Recherche Bay were picked up and conveyed to Lowe Sound, where a winter camp comprising two prefabricated army huts was under construction. The plan was to open up the proven seams and to erect a temporary gravity plant before machinery and coal-cutters would arrive in 1919. Craig also located an additional seam at the shore, on which Norwegian miners were put to work under the supervision of an English overseer (Fig. 7.21).

7.21 Coal seam discovered at sea level at Camp Morton in 1918. (Source: The gate to the northern markets (1919) Northern Exploration Company (N.E.C.), Norwegian Polar Institute, Tromsø.)

As in pre-war expeditions, a party called at the Swedish mine in Braganza Bay, in which the company had financial interests. They observed the adit, the coal dump, the rail to the jetty, and the trestles for a new ropeway, on the basis of which they credited the Swedes with good organisational skills since the workings would be sustainable for years to come. They again loaded bunker coal.

Having sent a letter of protest to the Norwegian A/S Kulspids, which declared that the land surrounding Recherche Bay belonged to the company and which otherwise read like the one above, a successful landing on the south side of the bay was now effected.[163] It showed that asbestos had been carried off from here, too, and the men lost no time to salvage the situation (Fig. 7.22).

A last act of the expedition was to name the new settlement in Lowe Sound, formerly Camp Morton, after the company's chairman, asserting 'what

[162] *The Financier* (1918) 'Spitsbergen's mineral wealth', 8 October.
[163] Salisbury-Jones and Maples (1918) *Letter to A/S Kulspids, 31 August*, Norwegian Foreign Department Box 5372, National Archives of Norway, Oslo.

Juneau is to Alaska, Davis City will be to Spitsbergen.'[164] Travelling mostly in convoy, the *Ella* reached England without further incidences. A wintering party stayed at Davis City.

7.22 Inspecting an asbestos mine in Recherche Bay. (Source: The gate to the northern markets (1919) Northern Exploration Company (N.E.C.), Norwegian Polar Institute, Tromsø.)

Following the expedition, the *Financier* went to great length to put Spitsbergen's coal and iron resources into a particular geopolitical light. It claimed that the world faced an iron and steel famine at the time when demand was at its greatest. The predominant rhetorical device was no longer the 'war effort'. It had morphed into the pertinent 'post-war reconstruction'. Thus, the seventh article in the series claimed, 'it is vital that new and abundant supplies under British control should be available.'[165] Therein lay the importance of the company's claims on Spitsbergen. If Britain had previously been at the industrial forefront, it was now being surpassed by America and Germany. While the domination of a country as vast as America was to be expected, 'it may not be so obvious why Germany, which in living memory scarcely ranked as an industrial country, has beaten [Britain] so ignominiously.'[166] British possession of Spitsbergen would enable the nation to compete in Europe and maybe regain its former advantage over Germany and America. This was facilitated by the German surrender on November 11, 1918.

[164] *Spitsbergen's mineral wealth* (1919) Michigan Technical University MTU MS631 LY B2/14, Houghton/MI, p. 14.
[165] *The Financier* (1918) 'Spitsbergen's mineral wealth', 29 October.
[166] *The Financier* (1918) 'Spitsbergen's mineral wealth', 29 October.

7.6 The global network of the public company

7.6.1 *Economic actors*

Following the war, the company reinforced the connection with its existent stakeholders and potential backers by means of a seventh brochure, which portrayed Spitsbergen as *The gate to the northern markets of Europe.*[167] The opening map delivered the essential facts: Spitsbergen lying in the Arctic; the positive influence of the Gulf Stream; the relatively short distances from Britain. An inset depicted the claims of the Northern Exploration Co. to date in the imperial pink. These now included part of Oscar II Land. The brochure recounted the events of the 1918 expedition, and rare photographs were upstaged only by extraordinary and undoubtedly pricey colour plates of the company's marble. Having been sidelined during the focus on coal and iron as key commodities during the post-war reconstruction, marble now made a tentative comeback.

Though the rhetoric primarily aimed for commercial support, a new oratory phenomenon intended to woo a more radical audience. As such, the brochure maintained that 'the Great War has brought to light the general and colossal ignorance of the rulers of the Empire regarding the internal resources of the British Commonwealth.'[168] Britain was faced with a coal famine and a serious shortage of iron ore, which could hinder its reconstruction period. The Great War Debt could only be met if Britain continued to export coal to neutral countries dependent on it, which was invariably where the company had assigned a central role to the Arctic. While this criticism may not have charmed the British Government into supporting the firm, it may have enticed fervently patriotic investors. The brochure was issued immediately before the commencement of the Paris Peace Conference, possibly in view of stoking the geopolitical debate surrounding Britain's possessions on the islands. Another provocation thus read, 'there is only one way in which the Northern European markets can be regained by Great Britain, and that is by the rapid development of the British-owned deposits in Spitsbergen. To neglect this means would be a betrayal of the national interests of the gravest kind.'[169]

In the absence of shareholders' lists, the effect of the changed political situation can be traced in the boardroom. In June 1918, the London barrister Harry Ernest Brittain became a director. He realised his political ambitions as MP for Acton from December that year until May 1929, although he only spoke of

[167] *The gate to the northern markets* (1919) Northern Exploration Company (N.E.C.), Norwegian Polar Institute, Tromsø.
[168] *The gate to the northern markets* (1919) p. 3.
[169] *The gate to the northern markets* (1919) p. 8.

Spitsbergen once before the House of Commons.[170] That summer, Frederick William Salisbury-Jones offered his services as managing director. He was contracted to take up 80,000 shares, but he, in fact, placed 350,000 shares altogether, thereby financing the 1918 expedition.[171] He was the mastermind behind most of the company's geopolitical constructs. The first half of 1920 saw several changes to the board in rapid succession. Of these, Lieutenant Arthur Drummond Borton and Major Edmund Colquhoun Pery had obvious military connections. Yet, the signing of the Spitsbergen Treaty left little room for militant expression on the islands.

The seventh brochure was also the company's last. After the conversion to a public company, its information reached the public via official channels. Subsequently, several newspapers published the intelligence, blurring the boundary between economic actors and other allies. The *Yorkshire Observer* quantified the mineral resources contained within the company's territory.[172] The decorative marble industry, for example, had been growing until the war prohibited imports, but an increase was now predicted, which would benefit from the company's unlimited supply. War had also enhanced the uses of asbestos, the versatile mineral being another one of the company's products. Consequently, the facts that the first steamer of 1919 had reached Lowe Sound and that the company was planning to install two wireless stations now caught the public eye.[173]

In 1919, Salisbury-Jones continued the tradition of inviting shareholders, investors, and Government advisers to visit the islands and the properties. The group travelled on Admiral David Beatty's yacht.[174] Beatty had risen to Commander in Chief of the Grand Fleet during the war, in which capacity he was instrumental in the surrender of the German High Seas Fleet. While Beatty himself did not inspect Spitsbergen, the association would have awed and assured many stakeholders. The company also initiated the practice of contracting engineers and geologists to produce professional reports. On the back of post-war buoyancy and increasingly professional conduct, the number of investors reached 1,200 by the end of the year.[175] *The Times* reported that the company proposed to extend its developments, which necessitated the increase of its capital to a staggering £1,000,000.[176] The decision to go ahead with the expansion was made in

[170] HC Deb 14 May 1919 vol 115 cc1570-1.
[171] *The Times* (1919) 'Company meeting. The Northern Exploration Company', 27 June, p. 22.
[172] *Yorkshire Observer* (1919) 'The marble industry in Britain', 1 May; *Yorkshire Observer* (1919) 'Natural resources of Spitsbergen', 6 May.
[173] *The Times* (1919) 'Opening up of Spitsbergen', 12 May.
[174] *The Financier* (1919) 'Spitsbergen, influential party's trip on Sir David Beatty's yacht', 5 August.
[175] Watt, D. M. (1919) *Letter to the Foreign Office, 13 November*, British Legation 1919 FO608/123 Files 427/3/1 (start – pp. 17998), National Archives, Kew.
[176] *The Times* (1919) 'Company results. Northern Exploration', 11 December, p. 22.

December 1919, and Companies House was notified on April 15, 1920.[177] This expression of commercial optimism coincided with the onslaught of the post-war depression 1920-21, which in turn rang in a decade of market stagnation.

7.6.2 Political actors

Unaware that the Foreign Office had surrendered much of its former power to other forces, the Northern Exploration Co. again tried to enrol it as the foremost political actor in the Spitsbergen Question.[178] Salisbury-Jones sought out an interview with Under-Secretary Robert Cecil in December 1918. He then posted a letter to the Foreign Office, citing his reasons for the British annexation of the archipelago, or at least its administration.[179] He believed the islands to be of political and strategic importance. Resources such as coal, iron, and petroleum were more valuable now that Britain was falling behind in production. Yet, the Foreign Office was unimpressed by his speech-making as it concentrated on the Paris Peace Conference in January 1919. In March, Salisbury-Jones informed the ministry that his board had invited a number of companies to take matters into their own hands.[180] Although a meeting between the companies never materialised, the Foreign Office received word from the Dutch minister that his Government would not be bound by any arrangement between private firms and considered these to have no effect on the eventual settlement.[181] Having communicated this to the British legation in Paris, the latter minuted, 'The Netherlands Government seem to sense danger when the Northern Exploration Co. gets on the trail. Their fears as to F. W. Salisbury-Jones are probably justified.'[182] In fact, after a conversation with him on April 8, the succeeding Under-Secretary Cecil Harmsworth was under the impression that Salisbury-Jones would not hesitate to lead Britain into a serious international dispute.[183]

---

[177] Hoel (1966) p. 459.
[178] Warman, R. M. (2003) 'The erosion of Foreign Office influence in the making of foreign policy, 1916-1918', *Historical Journal,* 15 (1), pp. 133-59; Steiner, Z. and Dockrill, M. L. (2003) 'The Foreign Office reforms, 1919-21, *Historical Journal,* 17 (1), pp. 131-56.
[179] Salisbury-Jones, F. W. (1919) *Letter to the Foreign Office, 11 January*, British Legation 1919 FO608/123 Files 427/3/1 (start – pp. 17998), National Archives, Kew.
[180] Salisbury-Jones, F. W. (1919) *Letter to the Foreign Office, 18 March*, British Legation 1919 FO608/123 Files 427/3/1 (start – pp. 17998), National Archives, Kew.
[181] *Foreign Office docket, 4 April* (1919) British Legation 1919 FO608/123 Files 427/3/1 (start – pp. 17998), National Archives, Kew.
[182] *Foreign Office minute, 4 April* (1919) British Legation 1919 FO608/123 Files 427/3/1 (start – pp. 17998), National Archives, Kew.
[183] Foreign Office (1919) *Letter to A. J. Balfour, 16 April*, British Legation 1919 FO608/123 Files 427/3/1 (start – pp. 17998), National Archives, Kew.

Salisbury-Jones proceeded to Paris in the hope of speaking with the British Prime Minister. Subsequently, the delegate Esme Howard related how Salisbury-Jones planned an expedition in summer and expected some collision between his employees and Norwegians on the islands. 'He therefore urged that some immediate settlement might be come to in order to avoid conflict and said that he hoped there would be no bloodshed. (During a previous conversation in London [...] Mr Salisbury-Jones declared there would be war in Spitzbergen.)'[184] Howard repeated that 'there was absolutely no question of the islands being annexed by Great Britain, and that he might as well give up all hope of this.'[185] The delegate nonetheless suggested that a joint Anglo-Norwegian commission might be organised to report on any disputes. The Foreign Office, however, saw no point in such a commission, and none was sent. Similarly, Salisbury-Jones' attempts to present the interests of the Northern Exploration Co. directly to the Paris Peace Conference were fruitless.

The first practical steps towards a post-war expedition saw the board apply for capital, but the response by the Committee on Fresh Issues of Capital was negative.[186] In a letter to the Treasury, the Foreign Office quickly denounced 'all responsibility for the effect which this decision may have both with respect to the possible sacrifice of British interests in Spitsbergen and on the position of his Majesty's Government with regard to the future status and administration of the island.'[187] According to the company, the Foreign Office had on the whole been supportive, with Foreign Secretary Arthur Balfour writing to the Treasury, the Admiralty, the Ministry of Shipping, and the Ministry of Munitions on its behalf.[188] Although the Ministry of Munitions had approved an expedition early in the war, the Treasury had not sanctioned the amount necessary and had practically prevented any activity until 1917. It increased the capital in May 1918 at last.

On July 9, 1919, the Inter-Allied Supreme Council at Paris eventually discussed the Spitsbergen Question. Already in spring, the Norwegian Government had put in a definite claim for sovereignty, which it hoped would be supported by its British counterpart.[189] It seemed that the Council was inclined to recognise this claim and proposed to set up a Spitsbergen Committee comprising British, French,

---

[184] Howard, E. (1919) *Letter to the Foreign Office, 16 April*, British Legation 1919 FO608/123 Files 427/3/1 (start – pp. 17998), National Archives, Kew.
[185] Howard (1919) *Letter to the Foreign Office, 16 April.*
[186] Capital Issues Department (1919) *Letter to the Foreign Office, 25 January*, British Legation 1919 FO608/123 Files 427/3/1 (start – pp. 17998), National Archives, Kew.
[187] Foreign Office (1919) *Letter to the Treasury, 31 January,* British Legation 1919 FO608/123 Files 427/3/1 (start – pp. 17998), National Archives, Kew.
[188] *The Times* (1919) 'Company meeting. The Northern Exploration Company', 27 June, p. 22.
[189] Earl Curzon (1919) *Letter to Earl Findlay, 2 April,* British Legation 1919 FO608/123 Files 427/3/1 (start – pp. 17998), National Archives, Kew.

American, and Italian representatives.[190] The company immediately contacted the Foreign Office, requesting that Salisbury-Jones be invited to put forward its claims and assist in the decision-making regarding the islands' administration.[191] The British legation took note of the request. It communicated with the Foreign Office that although the Committee would most likely hear councils from neutral states, it was unlikely to desire the opinion of private companies.[192] If it did, the Northern Exploration Co. would doubtlessly be invited. However, it was not.

On September 26, 1919, the news broke that Spitsbergen had been placed under Norwegian control, albeit with special reserves concerning the mining rights and other privileges for foreigners.[193] The directors instantly lodged a formal complaint with the Foreign Office, reminding it of its earlier verbal and written assurances.[194] They were indignant to have been kept entirely in the dark and to have been informed only by the press. They would be glad to know if the news could be confirmed by the department and what effect the decision would have on British possessions. The Foreign Office, in turn, explained that the settlement would not surrender British rights and interests. In fact, a Danish commission would soon be charged with the impartial examination of all claims.[195]

On February 9, 1920, the Spitsbergen Treaty was signed, which was duly but inconspicuously reported in Britain's papers. Although the treaty bestowed sovereignty on Norway, it contained 'the most careful provisions for the protection of British enterprises.'[196] Furthermore, it established 'equality of rights between all the nationals of the contracting Great Powers, especially as regards fishing, hunting, and mining. Arrangements are made for adjudicating upon claims in respect to rights previously acquired.'[197] If the Foreign Office had been bracing itself for a public outcry proportional to the haughty propaganda of the Northern Exploration Co. and the zealous canvassing by Salisbury-Jones, none was forthcoming. The majority of parties that had busied themselves with British supremacy on Spitsbergen before the settlement remained eerily quiet.

---

[190] *Daily Mail* (1919) 'Norway claims Spitzbergen', 9 July.
[191] Watt, J. M. (1919) *Letter to the Foreign Office, 10 July,* Peace Conference, British Legation FO608/120 File 418/1/3 on Spitsbergen, National Archives, Kew.
[192] British Legation (1919) *Letter to Earl Curzon, 21 July,* Peace Conference, British Legation FO608/120 File 418/1/3 on Spitsbergen, National Archives, Kew.
[193] *The Times* (1919) 'Future of Spitzbergen', 26 September.
[194] Watt, D. M. (1919) *Letter to the Foreign Office, 26 September,* British Legation 1919 FO608/123 Files 427/3/1 (start – pp. 17998), National Archives, Kew.
[195] Foreign Office (1919) *Letter to the Northern Exploration Company, 4 October,* British Legation 1919 FO608/123 Files 427/3/1 (start – pp. 17998), National Archives, Kew.
[196] *The Times* (1920) 'Spitsbergen Treaty signed', 10 February, p. 11.
[197] *The Times* (1920) 'The Spitsbergen Treaty', 11 February, p. 15.

7.6.3 *Other allies and competitors*

Support from others was not readily forthcoming. The precarious connection with the Royal Geographical Society, for instance, was demonstrated by the catastrophic effect of Herbert Ponting's lecture on December 9, 1918. The photographer spoke after Martin Conway on his own impressions of the economic aspects of the country. Being a shareholder of the company and having been remunerated for his appearance, however, his talk was ill-received.[198] The Society's secretary was outraged, 'I laid stress on the point that, while we should like information, we absolutely refused to have anything to do with company promoting; [...] we were particularly anxious not to have anything like exploitation of our meeting in the interests of any particular company [...].'[199] William Speirs Bruce commented, 'It is true that during recent years I have been driven into doing a certain amount of business in regard to Spitsbergen. I am glad to have done this less competently than Ponting and hope that geography may still remain my more important interest.'[200] The Norwegian minister to London regretted that throughout the evening 'the Norwegians were characterized as cunning, pushing and knowing people. It is very painful for me [...].'[201] He saw it as his duty to inform the British Foreign Secretary and the Norwegian Government of what had transpired regarding 'acts of violence and depredation committed by lawless men who claim Norwegian nationality.'[202] The consequence of the company's aggressive campaigning was commonly alienation rather than support.

Ponting attempted to redeem himself in a letter to *The Times*. His narrative concentrated on crucial resources in 'the years of reconstruction' and the national preoccupation to 'regain supremacy in iron and steel'.[203] His assertions promptly found criticism in a *Truth* article.[204] Its author stressed the absence of reliable published evidence and disapproved of the latest unconvincing brochure. Sweden's Professor de Geer had, in fact, not found iron ore worth developing, and while both Swedish metal-mining expertise and Swedish bias towards British claims could not be ignored, there was as of yet no dependable British expert

---

198 Summary of share capital (1921) Northern Exploration Company, Limited BT 31/32080/112730, National Archives, Kew; *Precis of correspondence re Ponting* (1919) RGS/CB8/Conway, RGS/IBG Archives, London.
199 Royal Geographical Society (1918) *Letter to Freshfield, 11 December*, RGS/CB8/Conway, RGS/IBG Archives, London.
200 Bruce (1918) *Letter to Hinks, 13 December*, RGS/CB8/Conway, RGS/IBG Archives, London.
201 Vogts (1918) *Letter to the Royal Geographical Society, 10 December,* RGS/CB8/Conway, RGS/IBG Archives, London.
202 Norwegian Legation (1918) *Letter to the President, RGS, 12 December*, RGS/CB8/Conway, RGS/IBG Archives, London.
203 *The Times* (1919) Letter to the editor, 9 January, p. 11.
204 *Truth* (1918) 'Is there rich iron in Spitzbergen?', 25 December.

opinion to the contrary. The author did not wish to 'discourage legitimate development of Spitzbergen, but the raising of exaggerated hopes founded on insufficient evidence is the surest way to serious disappointment and consequent difficulty in raising funds for further prospecting such as may well be justified.'[205] To support his case, he cited the company's one-time share price of £3 10s to have fallen to about 3s.

In another damaging example, Aubrey Strahan, director of the British Geological Survey, who had received Galloway's report of Camp Morton and who had visited Spitsbergen in 1910, quoted the Swedish geologist R. Högbom as having said that profitable mining on Spitsbergen was only possible within ten kilometres of the coast, greater distances being disadvantaged by difficult transport over glaciers, snow, sliding talus, and Arctic soil-flow.[206] Regarding its iron ore, Spitsbergen had not been considered a source at the International Geological Congress in Stockholm in 1910; in 1914, Högbom stated that the ore contained only about 10 to 20 per cent of total iron; and G. C. Lloyd's 1917 summary on the sources and production of iron and other sources in the iron and steel industry concluded that none of the various Spitsbergen surveys observed iron-ore in sufficient quality to warrant development. According to Strahan, 'it would appear therefore that the great mass of magnetite now reported to exist must either have escaped the notice of all previous observers, including the Swedish and Norwegian geologists, or must have been regarded as of little commercial value.'[207] The Northern Exploration Co. fought hard against disarming sarcasm on the home front, but the truth was that it indeed lacked reliable published evidence.

As seen above, the company invited a number of competitors to resolve the Spitsbergen Question themselves. Addressing Store Norske, Maples maintained that the islands' administration should receive the attention of those directly involved and asked the Norwegians as well as the Scottish Spitsbergen Syndicate and AB Svenska Spetsbergen to send their representatives to London.[208] As extra leverage, he warned that the British opportunist Cruikshank would soon interfere with Norwegian rights and that a satisfactory international settlement would take years. An arrangement between the companies would possibly be gladly accepted by the respective Governments. Store Norske opted for not attending the meeting. In the first instance, the Scots were unknown to them, while the Northern Exploration Co. had failed to include other Norwegian

[205] *Truth* (1918) 'Is there rich iron in Spitzbergen?', 25 December.

[206] Strahan, A. (1919) 'The political status of Spitsbergen: discussion', *Geographical Journal*, 53 (2), pp. 91-6.

[207] Strahan (1919).

[208] Maples, J. R. (1919) *Letter to Store Norske, 6 February*, British Legation 1919 FO608/123 Files 427/3/1 (start – pp. 17998), National Archives, Kew.

firms of relevance. Although the company viewed the Kings Bay Coal Co. and A/S Kulspids as trespassers, Store Norske would not join a meeting from which they were excluded.[209] Secondly, the location should be Kristiania as per the previous conferences and in light of the principal interests on the archipelago being Norwegian. Even if a meeting were arranged, Store Norske was not inclined to discuss politics, 'this question being wholly one concerning our State authorities.'[210] Instead, issues of mutual interest such as administration, labour, sanitation, charting and surveying, and scientific study should be on the agenda.

7.7 The local network of the public company

7.7.1 Claims and natural resources

In September 1918, Norway's prime minister Christian Michelsen transferred his rights to two houses built in 1901 and the surrounding areas to Alf Frantzen.[211] The houses stood at Camp Morton and in Calypso Bay. Frantzen was a director of the Spitsbergen Kul- & Mineral A/S. In 1919, he also acquired a number of other properties.[212] The Northern Exploration Co. was interested in a coal seam in Calypso Bay, but the Spitsbergen Kul- & Mineral A/S wrote to the company in January, objecting to its conduct.[213] Among the objections were pre-dating its claims to 1905 in 1912 and 1913 and attaining several useless Norwegian single-board huts. It protested that 'your Company can have no direct practical use of same whatsoever, you have nevertheless connected the cabins with certain ground rights'.[214] The crux of the matter was that the Norwegians would not surrender the ownership of Calypso Bay. Even so, *The Times* announced that the company had obtained a large area covering the southern peninsula of the island of Spitsbergen down to the South Cape by July 5, 1919.[215] This included the acquisition of coal deposits in Stor Fjord, where Gustav Adolf Lindquist had discovered both an

---

[209] Salisbury-Jones, F. W. (1919) *Letter to the Foreign Office, 18 March*, British Legation 1919 FO608/123 Files 427/3/1 (start – pp. 17998), National Archives, Kew.

[210] Store Norske (1919) *Letter to the Northern Exploration Co., 1 March*, British Legation 1919 FO608/123 Files 427/3/1 (start – pp. 17998), National Archives, Kew.

[211] Barr, Newman, and Nesteroff (2012) pp. 138.

[212] Barr, Newman, and Nesteroff (2012) pp. 138-9. These properties included an area around Davis Harbour, an area in south-east Spitsbergen from Kvalvågen to Edge Island, an area around Magdalene Bay, an area around Trygghamna and Daudmansøyra, and an area around Coles Bay.

[213] Spitsbergen Kul- & Mineral A/S (1919) *Letter to the Northern Exploration Co., 20. January*, Norwegian Foreign Department Box 5373, National Archives of Norway, Oslo.

[214] Spitsbergen Kul- & Mineral A/S (1919) *Letter to the Northern Exploration Co., 20. January*.

[215] *The Times* (1919) 'City news in brief', 16 April, p. 22; *The Times* (1919) 'Northern Exploration Company', 5 July, p. 21.

anchorage, which he named (Lady) Davis Harbour, and a coal seam.[216] He staked out the coal and sold his claim to Ernest Mansfield. The sale for 1,000 kroner was witnessed by Tinus Aune and Carl S. Sæther.[217] Mansfield, in turn, sold the east coast property to the company on July 8, 1919. The Foreign Office was notified, but it only tentatively rectified its Admiralty chart.

The 1919 expedition stood before the monumental task of upholding effective occupation on the company's vast territory. Estimated measurements in square kilometres indicated this to be Kings Bay and Cross Bay 950, Oscar II Land 600, Bell Sound 3,220, and Horn Sound 480, totalling 5,250.[218] Since the recent purchase on the east coast (4,000), the possessions had almost doubled in size and amounted to 14.2 per cent of the land surface of the island of Spitsbergen. Therefore, the firm's resident engineer John Alston Bevan coordinated the fieldwork of several land parties to cover more ground.

C. W. Boise led one of the land parties to undertake a detailed survey of known mineral occurrences on some of the properties.[219] Its secondary goals were a general geological study in view of future prospecting and development as well as the examination of certain coal areas belonging to other companies. Five men arrived in Recherche Bay on July 12 and conducted far-reaching fieldwork over the next 67 days. An eighth of the time (13%) was spent in Recherche Bay to study iron ore, asbestos, and coal. Thereafter, they examined Lowe and Bell Sound (16%), Van Keulen Bay and Middle Hook (8%), and spent a fraction of their time in Kings Bay and St John's Bay. Notably, they used almost half the time available to observe Store Norske. A summary of the general geology revealed no surprises besides the mention of a fine conglomerate which had been extensively searched for gold near Camp Millar. A photograph of the flooded mine suggested that mining had long since discontinued.[220] The fieldwork concluded on September 17, and the men left the archipelago three days later.

On July 13, the expedition took up work on Marble Island.[221] Although the company had protested against vandalism and theft in 1918, it had not left watchmen, and the condition of the camp had worsened still. The tasks at hand consisted of taking stock and putting the stores back in order. Throughout the period, marble manager Herbert W. Leech kept a note book and a diary, which tell

---

[216] *Avskrift. Tromsø, June 17th, 1919* (1919) Norsk Polarinstitutt 238, Regional State Archives, Tromsø.
[217] *This indenture made the eighth day of July* (1919) Norsk Polarinstitutt 238, Regional State Archives, Tromsø.
[218] Boise, C. W. (1920) *Report on the mining properties of the Northern Exploration Company Ltd. on the island of West Spitsbergen*, Northern Exploration Company (N.E.C.), Norwegian Polar Institute, Tromsø.
[219] Boise (1920).
[220] Photo Library NBR9201/07742, Norwegian Polar Institute, Tromsø.
[221] Leech (1920).

of problems with the ship, continual breakdowns and delays, and trouble with the staff and the contractors.[222] There was no quarrying due to the lack of men and machinery until the arrival of the SS *Kristoffer Ellingsen* on September 7.

In winter 1919/20, a miner, in the company of caretakers and trappers, was left at Calypso Bay to work the coal.[223] Winter workmen also stayed on Marble Island. They commenced in September under superintendent Sant.[224] Within 30 yards of No. 3 Quarry, a winter quarry was opened. Leech later criticised that the marble was the same as in No. 3 Quarry and that the trench had been driven along a natural slip. The use of explosives had given it an irregular shape, and it would be necessary to remove more rock and channel a cut to make a workable quarry floor and face. In addition, a fault ran through the trench that had been allowed to flood. Since the winter quarry was nowhere near the producing stage and the same marble could be got in No. 3 Quarry once its floor was cleared, it would be a waste of resources to continue the winter work. A more promising cut had been made to the west of the small island.

In 1920, the company dispatched an advance expedition headed by the engineer Bevan. He was charged with demonstrating the possibility of landing on Spitsbergen much earlier than post-winter communication had hitherto been re-established.[225] The advance party was to set up camps and start prospecting both at Davis Harbour on the east coast and at Copper Camp in St John's Bay. The MS *Ornen II* was chartered in Norway, and on her first attempt to leave Tromsø on April 6, the ship was damaged off Bear Island and needed to turn around. The second attempt brought the men to Recherche Bay on April 23. Some disembarked at Calypso Bay and spent considerable time putting the headquarters back in order, which involved building a new store hut. Although Recherche Bay was the focal point throughout the season, no work was done on Iron Mountain or at Camp Asbestos besides routine visits to ascertain their condition.

Bevan now tried to reach the east coast, but this early in the year the route south of Bell Sound and in the vicinity of the South Cape was blocked by ice floes.[226] Therefore, the advance party went north to St John's Bay. The inlet was frozen to within a mile of its mouth, and the ship moored on the ice edge while five men transported the stores on sledges to Copper Camp at a distance of six and a half miles away. The ship left the miners on April 30, but ice prevented her from returning to base before May 7. After another attempt to reach the east coast, the

[222] *Rare List* (2012) Available at: http://www.rarelist.co.uk/book-show.php?book=29774 (Accessed: 17 February 2012).
[223] Bevan, J. A. (1920) *Annual report on the mineral properties of the Northern Exploration Co, Ltd.'*, Northern Exploration Company (N.E.C.), Norwegian Polar Institute, Tromsø.
[224] Leech (1920).
[225] Bevan (1920).
[226] Bevan (1920).

ship was also locked out of Bell Sound and forced to sail for Kings Bay. On May 25, the winter workforce on Marble Island was relieved and returned to Norway on board the *Ornen II*. While the ship was away and the staff engaged in Calypso Bay and on Marble Island, Bevan undertook some prospecting in Cross Bay. He was accompanied by Birger Jacobsen, and the two men stayed at Camp Zoe, making their own way back to Kings Bay in mid-June.

The programme for Marble Island envisaged the expenditure of £25,000 and allowed for 50 quarrymen excluding staff to enable 1,000 tons of marble to be shipped that season.[227] At the beginning of June, 14 quarrymen, some staff, and additional machinery left Aberdeen. Headed by Leech, the group arrived on June 15, and the manager decided to continue the proposed quarry on the west side of the island. An area had already been cleared to enable the examination of marble under the weathered zone. The broken nature of the surface rock disappeared at shallow depth, where the fractures were closely cemented and the hammer rang soundly. The soundness was a reassuring feature, but there were other unsettling aspects. The type was greyish blue with white bands. Leech thought the grey to be too prominent to be valuable, and it would be too expensive to cut out the white bands. In addition, beds of the same marble protruded through the quarry floor at a different angle, which would ultimately mean a reorientation of the face and lead to much waste. To obtain information about the lower strata, Leech let boreholes be drilled to a depth of about seven feet, but the marble crumbled as the drillers struck a sub-surface fault. Not disheartened by the fault but faced with a shortage of water, the manager temporarily abandoned the area.

After consulting with Bevan and Jacobsen, Leech altered the plans for the season. The programme now included winning marble on Breccia Island (probably Juttaholmen), which commenced on June 28, and testing the white marble outcrop in Cross Bay.[228] The work on Breccia Island took place under extremely difficult conditions. The weather was unfavourable to living under canvass. Cooking in the open was a problem in strong winds. The absence of fresh water was a severe handicap and would continue to be so, unless electricity could be supplied from Marble Island.[229] The space between the cliff-edge quarry and the sea was so cramped that eventually the number of men was reduced. Due to bad weather and transport difficulties, the development on Breccia Island stopped on August 28.

Meanwhile, Leech and three others travelled to Cross Bay to examine a white outcrop near Tinayre Glacier.[230] The outcrop was about 50 feet wide and could be traced inland for roughly half a mile. It was made of crystalline limestone

[227] Leech (1920).
[228] Leech (1920).
[229] Did Marble Island have electricity?! There is no evidence, but perhaps it was planned for the future.
[230] Leech (1920).

of medium quality with interbedded schist. The men broke out a small hole in which the quality seemed to improve at a shallow depth. On a follow-up visit, they produced a map of the outcrop and selected three sites for investigative boreholes. Thereafter, the drilling crew from Marble Island was transferred to the location, which had a decent harbour, and the MS *Magdalena* came alongside the small cliff to unload. Drilling commenced on August 6 and was discontinued ten days later, although Leech believed the marble to warrant further boring in the future. For the time being, the drillers were sent to assist oil geologists in Green Harbour.

The main expedition arrived on Spitsbergen on July 12, 1920. On board were professionals, who again led land parties on geological fieldwork. The American graduate H. N. Coryell was in charge of the search for coal in Van Keulen Bay, Lowe Sound, and on the east coast.[231] On June 27, it had finally been possible to land two men and provisions for six months at Davis Harbour.[232] Their tasks were to hold down the claim, gather general information about the coal, and collect meteorological data. On August 4, Coryell's group arrived to study the potential of the deposit and to comment on problems regarding its distribution and development. They built a new hut, which could house six to eight men (Fig. 7.23), in addition to pitching three tents. During the month on site, bad weather prevented fieldwork for a total of almost two weeks. The winds shattered a rowing boat, tore the tar paper off the hut, and ripped the tents. The loss of a boat left inadequate transport to either move the camp to another location or make daily trips along the coast. The men were compelled to wait for the ship. On September 2, they returned to headquarters.

7.23 Camp at Davis Harbour under Hedgehog Mountain on the east coast, here in 1936. (Source: Photo Library np003856, Norwegian Polar Institute, Tromsø.)

[231] Coryell, H. N. (1920) *Report*, Northern Exploration Company (N.E.C.), Norwegian Polar Institute, Tromsø.
[232] Coryell (1920).

The mining engineer E. P. Crawford and his team investigated reported mineral occurrences and examined rock formations of economic significance.[233] They visited the north side of Bell Sound; the Dunder Bay region south of Bell Sound; the lead-silver occurrence west of Volage Glacier and on the slopes of Martin Range; the north side of Horn Sound; and Goose Bay on the south side of Horn Sound. On the return journey to Calypso Bay, Crawford lastly mapped the East Island of the Dun Islands prior to leaving Spitsbergen on September 6.

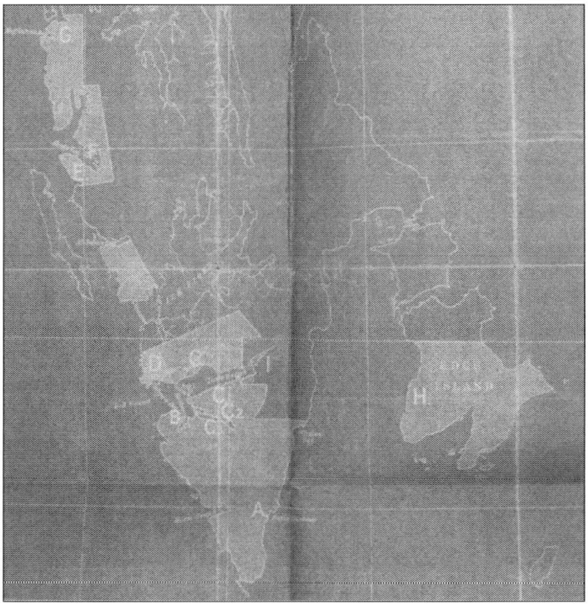

7.24 Detail of the map showing the Northern Exploration Co.'s maximum territorial extent after 1920. (Source: Map of Spitsbergen, Norwegian Foreign Department Box 5374, National Archives of Norway, Oslo.)

Before Leech departed Kings Bay on October 2, he left written instructions with the winter supervisor.[234] Steel was in charge of the wintering party comprising Allan and Spry until the first boat returned in spring 1921. His duties were to protect the company's properties and interests. He had to fill in a monthly messing report and keep a ledger. He had to take over the stores and was authorised to sell goods to the Kings Bay Coal Co. or any other buyer for cash. Steel was also responsible for keeping a weather chart and should occasionally visit the Lesser Islands. All huts needed to be cleaned out and made weather-proof after the summer workforce had left. The machinery also needed to be tended to and might occasionally require greasing. If the wintering party needed fresh meat or medical help, they

[233] Crawford, E. P. (1920) *Report on properties at Bell Sound, Dunder Bay and Horn Sound, West Spitzbergen*, Northern Exploration Company (N.E.C.), Norwegian Polar Institute, Tromsø.
[234] Leech (1920).

could get this from the Kings Bay Coal Co. The wireless operator Allan was to arrange no more than two calls per week; the London office would be in touch if necessary. Leech's letter represented the most detailed wintering instructions to have survived.

Following the expedition in 1920, the company's claims reached their maximum extent. On September 13, Alf Frantzen was made a director of the company.[235] This presumably settled all disputes with the Spitsbergen Kul- & Mineral A/S. On October 20, Frantzen transferred his various rights to the company. This included part of Edge Island. A copy of the resultant map (Fig. 7.24) also reached Norway's Foreign Department. The company made no other claims on the archipelago after the sharp financial downturn in January 1920.

7.7.2 Manifestations

The pre-war constructions were not untouched when the company returned to Spitsbergen in 1918. While Scandinavians had reused buildings and machinery during the conflict, the company itself may have recycled some between 1918 and 1920. Changes occurred during the last days of the no man's land that are barely traceable now. They perhaps account for today's 'missing' huts, for example Camp Margaret, the original Camp Millar, Camp Violet in Van Keulen Bay, and Camp Campbell at Lægerneset. Yet, their absence is eclipsed by the emergence of new structures. In 1919, the company transformed Calypso Bay into a small settlement worthy of being its headquarters. The site was the subject of several photographs, including stereoscopic images to give a three-dimensional impression. Smaller huts were erected at Copper Camp in St John's Bay and at Davis Harbour on the east coast. Less significant and not photographed, they were nonetheless part of a continuous programme of prospecting and development, being placed with the advance party of 1920 in mind. In 1920, an additional store was built at Calypso Bay, while the hut at Davis Harbour was extended.

In contrast, progress related to actual mining was very limited. In 1918, company representatives had visited Iron Mountain and Camp Morton and phrased clear goals for the sites.[236] Owing to the steepness of the former, they envisaged a double-rope-system over one and a quarter mile and postponed any development until this could be delivered in July 1919. Their attention then turned to Camp Morton, where gentler slopes would facilitate a one-mile monocable. Regarding Iron Mountain, the absence of physical remains indicates that no ropeway was ever

[235] *Directors* (1920) Northern Exploration Company, Limited BT 31/32080/112730, National Archives, Kew.

[236] *The Financier* (1918) 'Spitsbergen's mineral wealth', 15 October.

installed. Perhaps the few drilling rods recorded at Lægerneset signal a change in strategy to conduct a drilling programme first. There is no archaeological or archival evidence, however, that boreholes were ever sunk into the ore body. At Camp Morton, the 'flying fox' was set up, but it was neither written about nor photographed. Survey work has shown it to have been destroyed by melt water, while other components such as the single gravity rail to the stockpile and the double-acting inclined plane to the shore survived. Quarrying machinery had existed on Marble Island since before the war, but post-war extraction was hindered by vandalism and theft.

Investigative boreholes were drilled both in Kings Bay and in Tinayre Bay.[237] In the first hole into the white outcrop, the steel cutters were lost in broken rock at five feet six inches. In the second hole, good marble was proven to three feet six inches, but it was not reached again before the hole was abandoned at 15 feet. The third hole was placed 100 yards to the west. Broken marble was replaced by better quality marble, but the cutter passed through a variety of rock before the hole was also abandoned at 11 feet in order not to jam the drill in loose ground. Operations were discontinued after ten days.

7.25 The wireless station at Calypso Bay in 1921. (Source: Photo Library np001930, Norwegian Polar Institute, Tromsø.)

[237] Leech (1920).

The technological coup of the post-war expeditions consisted of the establishment of two wireless stations. These were purchased from Guglielmo Marconi's Wireless Telegraph & Signal Co. in Britain.[238] They were erected at Calypso Bay (Fig. 7.25) and on Marble Island and operated through the Norwegian installation in Green Harbour. Although each transmitted message came at a cost, the stations allowed for two-way communication with London all year round. The head office may have hoped to speed up investigations and development in this way, giving rapid instructions, avoiding misunderstandings, and preventing its local managers from acting too autonomously in the absence of other controlling measures. In addition, the antennae were inescapable claim markers.

Regarding shipping, the company used the SS *Kristoffer Ellingsen* and two motor cutters in 1919.[239] In 1920, no less than five ships were in service, maybe more. Both the 70-ton MS *Ornen II* and the SS *Isfjeld* were fitted with wireless telegraphy.[240] This puts the radio stations in Calypso Bay and on Marble Island into a different light because the vessels greatly enhanced communication and mobility between the properties. The MS *Magdalena*, the MS *Mina I*, and *Vesleanna* were in use, too. The *Magdalena* soon became a liability.[241] In mid-July, the motor showed the first signs of trouble. Leech contacted London with regards to wintering the ship. The answer was to send her to Tromsø before the bad weather set in. In mid-August, the engine failed, and the *Ornen II* towed the ship from Green Harbour to Kings Bay. When all efforts to get her to Tromsø failed, Leech loaded her with salted beef and blocks of marble in the hope that someone might yet tow the ship southwards and moored her at the wharf of the Kings Bay Coal Co. To make matters worse, the ship was in the Arctic without a certificate, which had expired in July. This would need to be explained if she ever reached Tromsø, but the ship's officers did not care to take the responsibility. It is not known what became of the *Magdalena*, let alone the blocks of marble.

The biggest triumph of the season should have been the fast and far-reaching reconnaissance of Spitsbergen by air. Already in December 1918, the *Daily News* had run the story that Salisbury-Jones proposed to build a runway in Lowe Sound.[242] A seaplane with a Rolls Royce engine was in fact purchased for a considerable £4,000, and aviators were on the archipelago to study the conditions for polar flight.[243] Yet, the seaplane remained in storage in Tromsø.[244] For as long

---

238 *Aftenposten* (1920) 'Spitsbergen og dets rigdommer. Fremmede selskabers virksomhed der oppe', 1 November. Available at: http://polarlitteratur.com/aspits.htm (Accessed: 25 September 2012).
239 Barr, Newman, and Nesteroff (2012) p. 133.
240 Aftenposten (1920).
241 Leech (1920).
242 *Daily News* (1918) 'By air to the pole', 30 December.
243 Aftenposten (1920).

as it did, adversaries needed no prompting to ridicule the company's wasteful post-war expeditions, especially in light of plummeting share prices.

7.7.3 Employees

The timely change from inflated narratives to professional reports was accompanied by a gradual substitution of illustrious expedition leaders with qualified personnel. From 1919 onwards, John Alston Bevan was the resident engineer, who directed the efforts across the properties. Prospecting staff seems to have been hired on an *ad hoc* basis. In 1919, Charles W. Boise was the mining geologist in charge of four British geologists and mining engineers. He prepared the first fully accountable report to be found in the archives. On the topic of labour, he recounted that in the working mines, the workforce continued to be recruited from Scandinavia, but while workers had generally been cheap and reliable before the war, they now took advantage of the urgent situation to demand higher wages for fewer tasks.[245] The geologist envisaged that labour from northern Asia or mechanisation of the mines could counteract this. Neither Boise nor his party were engaged in later seasons.

British sources about the staffing during the wintering 1919/20, which was under Wild's command, are brief. A miner was left at Calypso Bay, and Sant supervised a group on Marble Island. A Norwegian article, however, related that no less than 11 men remained in Recherche Bay, including the chef Harrison and four Norwegians.[246] Similarly, 12 men stayed in Kings Bay, including eight Norwegians. Three Swedes occupied Camp Morton. The Scandinavians may have been trappers, who sold their skins to the company. Wild is said to have been happy to leave after the unproductive winter.[247] He was replaced by the highly decorated Lieutenant Colonel Arthur Drummond Borton, VC, CMG, DSO.[248] Next to nothing is known about Borton's conduct on the islands.

In spring 1920, Bevan headed the 16-strong advance party comprising two mining engineers, an assistant mining engineer, an assayer, a wireless operator, and eleven miners and workmen.[249] Captain Kenneth L. Gilson, formerly of the

---

[244] The seaplane may have been a variant of the Fairey III reconnaissance biplane, powered by a Rolls Royce (375-hp) Eagle VII engine. The make first flew in September 1917. The Northern Exploration Co. maintained a connection with the Fairey Aviation Co., a mortgagee, in 1920.

[245] Boise (1920).

[246] *Aftenposten* (1920) 'Spitsbergen og dets rigdommer. Fremmede selskabers virksomhed der oppe', 1 November. Available at: http://polarlitteratur.com/aspits.htm (Accessed: 25 September 2012).

[247] Recounted in Barr, Newman, and Nesteroff (2012) p. 132.

[248] Victoria Cross; Companion of the Order of St Michael and St George; Companion of the Distinguished Service Order.

[249] Bevan (1920).

Arctic Coal Co., commanded the assayer, the wireless operator, a miner, and four others at Calypso Bay.[250] The advance party continued to St Johns Bay, where the assistant engineer, a foreman, two miners, and a cook were left.[251] On July 7, this group was replaced by Cornish metal miners, who discovered pockets of copper ore before they struck camp again on September 10.[252] Bevan escorted the remaining men, D. H. Strutt and C. Jacobsen, to Davis Harbour before he attended to prospecting himself, accompanied by Birger Jacobsen.

Jacobsen was an opportunist, and his loyalty to the Northern Exploration Co. was subject to the firm's changing fortunes. At the outbreak of war, it was practically bankrupt. In 1915, Jacobsen had gone to Rotterdam with view of selling the company's claims to the German firm W. D. Müller & Co.[253] He vehemently denied these allegations.[254] He emphasised that his relationship with the company was not as simple as it was made out to be since Shackleton had personally invited him back in 1918.[255] Thereafter, the Spitsbergen Kul- & Mineral A/S had employed Jacobsen and a group of men to winter in 1918/19, although they sailed too late in the year and failed to reach Spitsbergen. By 1920, Jacobsen was again in the service of the Northern Exploration Co. He presumably secured a very lucrative arrangement.

The work of the advance party was continued by two prospecting parties. One was led by the American geologist H. N. Coryell and comprised H. S. Kane, J. T. Theedom, F. C. Ryan, and A. Jacobsen.[256] The other was headed by the mining engineer E. P. Crawford, who directed the miner J. E. Stevens and the camp assistant Sidney Light.[257] The latter may have been a relation of the secretary Charles W. Light. Coryell and Crawford, and presumably most of the men, were employed for one season and were of no consequence thereafter.

The temporary nature of the prospecting parties was contrasted by the permanent plans for Marble Island. Manager Leech began his second stretch in July 1920 and brought 14 men from Aberdeen with him.[258] The Scots having

---

[250] The assayer was H. G. Elston, the wireless operator G. Lash, the miner F. Moody. The four others were the storekeeper G. Lowdnes, the blacksmith Hegg, the cook A. Ericson, and the assistant cook A. Hobbs.

[251] H. G. Elston now also seemed to be the engineer in charge at Copper Camp. The others were S. H. Light, J. T. Theedon, F. C. Ryan, and J. E. Stevens until they were replaced by the Cornish.

[252] *Mining Journal* (1920) 'Cornish miners in Spitzbergen', p. 789.

[253] Findlay, M. de C. (1919) *Letter to the Norwegian Foreign Department, 19 April,* Norwegian Foreign Department Box 5373, National Archives of Norway, Oslo.

[254] For more information regarding Dutch prospecting and mining on Spitsbergen: Haas, H. R. de (2011) 'In search of profit in the High Arctic', in Hacquebord, L. (ed.) *LASHIPA – History of large-scale resource exploitation in polar areas*, Groningen: Arctic Centre and Barkhuis, pp. 47-60.

[255] Jacobsen (1919).

[256] Coryell (1920).

[257] Crawford (1920).

[258] Leech (1920).

experience of native marbles, the company appears to have recognised skilled labour and selected the workforce accordingly. On the trip to Tinayre Bay, Leech was accompanied by Sant, Steel, and Leslie. While Sant had been the winter superintendent of 1919/20, Steel would remain in Kings Bay in 1920/21. R. Leslie was the engineer in charge of the drilling crew.

Once again, British reports named British personnel, but this time the workers were largely accounted for and left little room for Scandinavian labour. One important Norwegian who was of service to the company in 1919 was Carl S. Sæther. Sæther would be the firm's agent in Tromsø until its dissolution.

7.7.4 *Local allies*

In 1920, the company supported a group of oil geologists. Reynolds[259] and his men travelled aboard the MS *Terningen.* Their activities may be inferred from the ship's schedule.[260] On June 30, the vessel arrived at Calypso Bay to be provisioned. On July 2, she left for the east coast, from whence she returned a fortnight later. On July 17, she targeted Lowe Sound. Within nine days, the unidentified Clutterbuck and Richards received provisions here for another two weeks. On July 31, they were seemingly picked up at Davis City and conveyed to Calypso Bay. A day later, the ship sailed for Green Harbour. At one point, Reynolds began to drill for oil at Finneset, but the Northern Exploration Co. held no claims there, and the operation was fiercely contested by Store Norske (Fig. 7.26).

7.26 Drilling operations in Green Harbour being contested by Store Norske. (Source: www.geo365.no. Accessed: 24 January 2012.)

The ship anchored in Green Harbour from August 18 until August 20. Perhaps her crew backed up Reynolds during the protest. Her departure from Green Harbour on

[259] This may have been the geologist George Bernard Reynolds, whose oil discovery in Iran in 1908 stood at the beginning of the Anglo-Persian Oil Co. (and ultimately British Petroleum).
[260] Bevan (1920).

August 20 may have marked the end of the drilling, unless the prospecting party were picked up by a different boat. After again being provisioned at Calypso Bay on September 2, the *Terningen* left for Norway on September 6.

The Northern Exploration Co. had high hopes for Reynolds. He had prospected on behalf of the Northern Petroleum Syndicate, which in turn had been formed by the longstanding British shipping firm Hull, Blyth & Co.[261] If the results were promising, the company would be entitled to 25 per cent of the capital of any subsidiary company established to exploit the oil. In spring 1921, however, the shareholders learned that the syndicate did not intend to exercise its option.[262]

Towards the end of summer 1920, the Northern Exploration Co. realised that the financial situation in Europe was dire and that it should try to make money out of Spitsbergen yet. On August 11, it therefore instructed Leech on Marble Island to dispose of surplus stores, clothing, and provisions.[263] This occupied much of his time, and he criticised the storehouse as being too far from the shore, where items purchased by the Russiske Kulfelter Co. and the Kings Bay Coal Co. now needed to be reloaded. The disposal of surplus interfered with the routine work on the island. The *Magdalena* with a cargo of salted beef and marble had been moored at the wharf of the Kings Bay Coal Co. The Norwegians were willing to take a few iron barrels to Norway, but they would not load the marble or surplus stores into the space required for their coal. The winter supervisor Steel was authorised to sell any goods for cash. The London office would cable an encoded message depending on whether he should stick to the fixed price, reduce it, or increase it. If his wintering party needed assistance, they could ask the Norwegians across the bay. Despite the ongoing claim dispute between the companies, relations between the staff were amicable.

7.7.5 Products

The post-war products of the company can again barely be expressed as tonnage. The miner who remained in Calypso Bay in winter 1919/20 extracted 133 tons of coal, but mining here was discontinued in 1920 and could in any case only ever be on a small scale.[264] Although Coal Mountain in Lowe Sound had a workable thickness of coal, it contained more shale than in Braganza Bay or in Green Harbour. Bevan cited this to be the reason why mining at Camp Morton, lately Davis City, was suspended in 1919. During summer 1920, a single miner produced

---

[261] *Financial Times* (1920) 'Northern Exploration', 17 July.
[262] *Summary of proceedings* (1921) Northern Exploration Company, Limited BT 31/32080/112730, National Archives, Kew.
[263] Leech (1920).
[264] Bevan (1920).

a trial shipment of 30 tons of coal at Davis City, which realised 160 kroner in Norway. If such high prices prevailed, a market could be found for this low-quality coal, and the question of production at this location could be reconsidered, especially in conjunction with processing the coal. With hindsight, it was a big if. There had also been a marketable output of marble from Breccia Island, but its fate was sealed onboard the *Magdalena*.

Being an exploration company, it continued to produce *intelligence*. The propagandist brochures had been disposed of in favour of professional reports on the properties. Coryell concluded that neither the coal in Lowe Sound nor in Van Keulen Bay merited more work. On the east coast, however, there was a seam which he recommended for development. The severe weather was the chief agent of problems. There was no suitable area to build a camp near the proposed mine, and it would be necessary to construct a storage space for coal if loading was to take place in Dead Bear Bay, where a transport system and a loading pier would be needed. Coal boats would be unable to anchor for long in the prevailing winds. Alternatively, coal could be stored two and a quarter miles away at Davis Harbour, but transport to this location would be met by problems with steep talus and glacial ice. Loading would have to be done using lighters.

Crawford's work to the north of Bell Sound was intended to trace mineral deposits.[265] It carried him far beyond the company's territory, but he found nothing of economic interest. The samples he had taken in Dunder Bay tested negative for gold and silver. A one-day inspection of the lead-silver occurrences at Volage Glacier gave a single assay of 2.8 ounces of silver per ton, which had no commercial value. During 39 days spent in Horn Sound, his party saw no evidence of mineralisation of any importance. There were no igneous rocks on the Dun Islands, and although Goose Bay was said to have gold-bearing quartz, the assays were negative. Crawford summarised his findings in two short paragraphs: the examination of Bell Sound and Dunder Bay revealed nothing of economic interest; Horn Sound contained frequent mineral occurrences, but these were either too local or too sparse to have any commercial bearing.

Leech delivered a comprehensive report covering all phases of development on Marble Island to date.[266] Coring had proven that the Devonian deposit was solid at a shallow depth. Although there was not enough data to justify winter quarrying neither at present nor during the first couple of years of production, he suggested making arrangements for the following summer early. He was confident that 1,000 tons could be mined in 1921.

---

[265] Crawford (1920).
[266] Leech (1920).

7.8 Post-war depression

At the height of economic buoyancy, the Northern Exploration Co. converted to a public company. Its shares opened at 29s 6d in August 1918 (Fig. 7.27). After the war, however, Britain's industrial output dropped, and the need to repay the war debt was a major cause of financial instability. In April 1920, the company's offer of 100,000 shares at £1 each was met with a very poor response, owing to the unfavourable development on the London Stock Exchange.[267] From now on, a sharp downward trend characterised the movement of its shares.

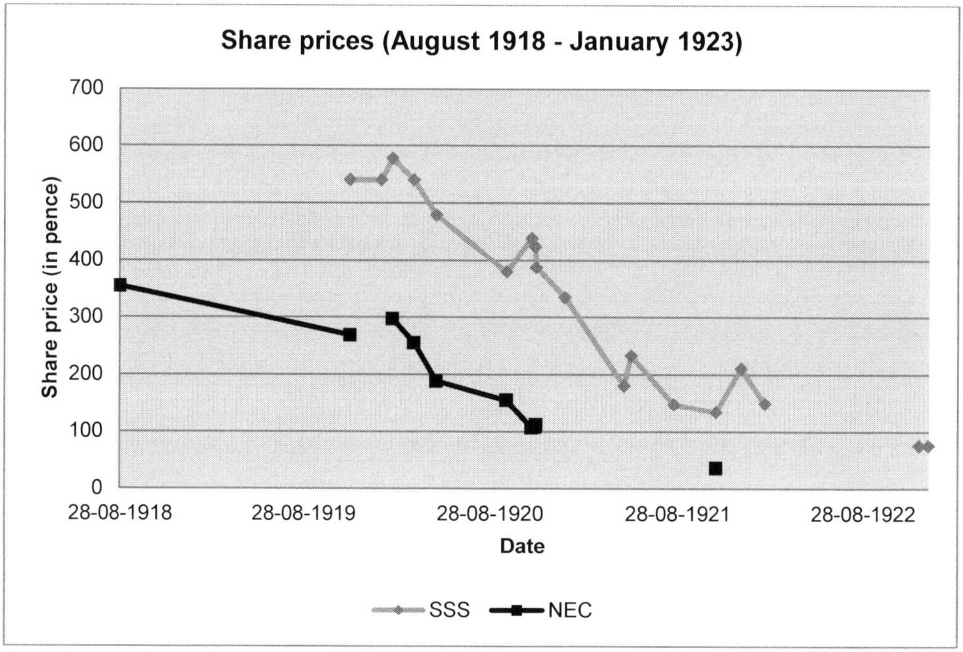

7.27 Shares of the Northern Exploration Co. (black) and the Scottish Spitsbergen Syndicate (grey) between 1918 and 1923. (Source: Financial Times; Graph: F. Kruse.)

A balance sheet made up to June 30, 1920, outlined the financial situation.[268] It included the costs of the 1919 expedition as well as the fitting out of the 1920 expedition. The liabilities section stated the authorised capital of £1,000,000, of which only 513,268 shares had been issued for the price of about £479,250. Sundry credits, payable bills in Tromsø, a loan from the London County West-

[267] *Financial Times* (1920) 'Northern Exploration', 17 July.

[268] *Balance sheet made up to the 30th day of June 1920* (1920), Northern Exploration Company, Limited BT 31/32080/112730, National Archives, Kew.

minster & Parr's Bank, and an overdraft added up to £46,760. Among the assets, the company's properties and rights on Spitsbergen were in prime position at £302,050. Houses, fittings, plant, machinery, and stores had consumed £69,600. The two wireless stations had cost almost £4,290, while the aforementioned seaplane contributed significantly to the overall transport costs of £17,920. Hospital equipment, scientific instruments, livestock, vehicles, prospectors' gear, and office furniture were listed at a total of £4,190. The balance sheet showed that the company had investments in the AB Isfjorden-Belsund to the order of £5,000. Sundry debtors (ca. £7,440) and the share commission (ca. £53,950) indicated that it engaged in other financial deals, too. General expenditure approached £61,000, its most noteworthy item being Salisbury-Jones' salary of almost £6,690. He had since terminated his contract, and it is perceivable that he took with him a fraction of the large sum, which he had previously so generously invested. His successor Major Pery only received £300, while other directors had not been paid at all.

The company headed for difficult times. This was supported by an item in the *Financial Times* on July 17, 1920.[269] The company had offered its shareholders at par £150,000 convertible debentures in order to raise the funds for another expedition. The article's subsequent reminder that investors had responded slowly to the offer of shares in April did little to increase the confidence in this proposal. At the end of July, the company registered a first £10,000 debenture.[270] More were to follow as the firm's financial manoeuvring got increasingly complicated and, in light of insufficient information, decreasingly transparent. By September, there were negotiations with a Scandinavian group who intended to provide fresh capital under the condition that it could obtain a large chunk of the issued capital of the Northern Exploration Co.[271] So the shareholders were asked to give the trustees Charles Oak Crisp and Lindsay Eric Smith a three-month option over their shares. The trustees would transfer the necessary number of shares to the group. However, the Foreign Office raised objections to the negotiations with the foreign group, and the matter was dropped.[272] The Board of Trade showed itself willing to recommend that financial assistance be given to the company, but it could not give any guarantee. So the finance was largely arranged by selling deteriorating stores and with the assistance of the directors and their friends.

In spring 1921, the Northern Exploration Co. held a general meeting, outlining the results of the 1920 expedition, which again attracted criticism. The

[269] *Financial Times* (1920) 'Northern Exploration', 17 July.
[270] *Register of mortgages and charges,* Northern Exploration Company, Limited BT 31/32080/112730, National Archives, Kew.
[271] *The Times* (1920) 'Northern Exploration. Scandinavian Group control', 27 September, p. 16.
[272] *Summary of proceedings* (1921).

Truth published an article titled 'The Spitzbergen slump'.[273] Basically, the shares had dropped dramatically after the energetic boosting of a few years ago had fallen victim to a nominal global market. The author had wanted professionals to assess the true commercial value of the claims, but the venture continued to be a highly speculative gamble. The annual report now indicated that the majority of minerals could not be developed, and it remained to be seen what could be made of the coal. There had been no revenue from the marble yet. The author commented that the board had changed almost completely from only a couple of years ago, which served to show how little trust the former directors placed in the undertaking. He could not help noticing the money with which Salisbury-Jones had departed. The article contemplated that 'the propaganda work of about a couple of years ago must have absorbed a substantial sum. Elaborate brochures cannot be prepared and printed for a few pence apiece, and I do not suppose the composition of articles in advance of those papers willing to accept and publish inspired leaders or notes was done gratis.'[274] The author considered, 'who directly benefitted it is impossible to say, but it is obvious that whatever money was paid for stimulating public interest in the Northern Exploration affairs was extracted indirectly from the pockets of those members of the public who opened their mouths and swallowed whatever was offered them'.[275]

The summary of share capital and shares released in April 1921 specified, who said members of the public had been.[276] About half a million shares had been taken up, a fifth of which had been paid for in other than cash. The accrued debt was registered at £20,000. The most prominent shareholders were the London County Westminster & Parr's Bank at 92,681 shares; former director and chairman Davis (10,290); director Smith (8,226); and both Ponting and a commander of the Royal Navy (7,000 each). The names of Bell, Borton, Frantzen, Gardner, and Mansfield were also among those with considerable investments at stake. It was clear that the Northern Exploration Co. had to succeed on Spitsbergen or come up with an alternative idea on how to recover these investments. A small expedition led by Sæther was dispatched in summer 1921, but the season had not yet come to a close when the company arranged a trust deed over £150,000 with the trustees Crisp and L. E. Smith in August.[277] This amount was substantial in light of

---

[273] *Truth* (1921) 'The Spitzbergen slump', 30 March.
[274] *Truth* (1921) 'The Spitzbergen slump', 30 March.
[275] *Truth* (1921) 'The Spitzbergen slump', 30 March.
[276] *Summary of share capital and shares* (1921) Northern Exploration Company, Limited BT 31/32080/112730, National Archives, Kew.
[277] *Particulars of a mortgage or charge* (1921) Northern Exploration Company, Limited BT 31/32080/112730, National Archives, Kew.

the shares having plummeted to 3s in *The Times'* last mention of them in November.[278] Thereafter, the company probably withdrew from the listing.

The Scottish Spitsbergen Syndicate experienced a similar downward trend (Fig. 7.27), but the British firms were not alone. The *Colliery Guardian* reported that the Swedish Spitsbergen Coal Co. had recently applied to the Swedish Government for financial assistance.[279] The application was supported by two Swedish politicians on the board. The extent of the post-war recession was becoming apparent, and at present, Spitsbergen coal could not compete with the low price at which English coal was being sold (Fig. 7.28).

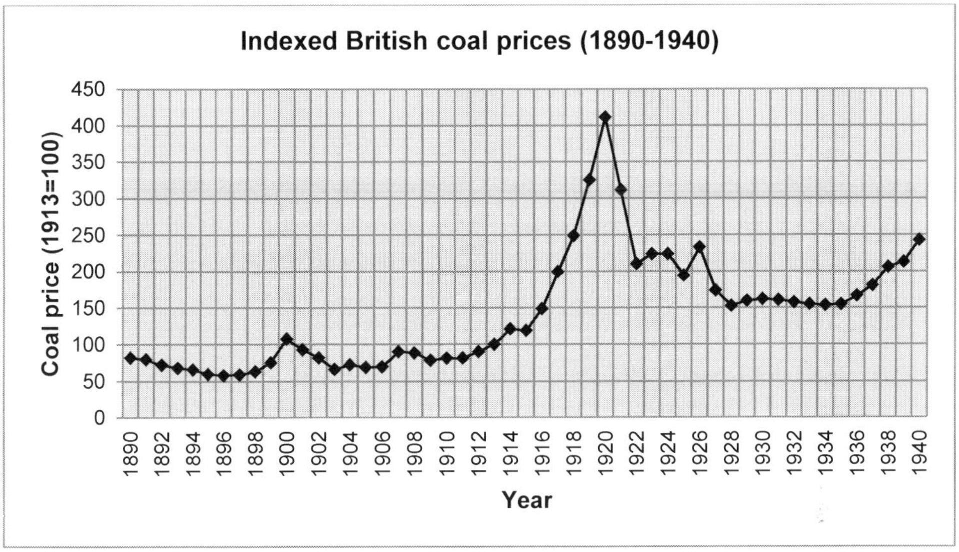

7.28 Indexed British coal prices 1890-1940. (Sources: Church, R. (1986) The history of the British coal industry: Vol. 3 1830-1913 Victorian pre-eminence. Oxford: Clarendon Press; Supple, B. (1987) The history of the British coal industry: Vol. 4 1913-1946 The political economy of decline. Oxford: Clarendon Press.; Indexing courtesy of H. R. de Haas; Chart: F. Kruse.)

The Swedes expected the cost of production and the sale price to return to a more normal situation, but for the time being, they had entered a wage agreement with the men, reducing their wages a further 20 per cent. The Swedish firm intended to suspend production in autumn until the coal market would recover. The Norwegian companies on Spitsbergen and Bear Island already enjoyed the continued support

[278] *The Times* (1921) 'Stock exchange', 10 November.
[279] *Colliery Guardian* (1922) 'Spitsbergen Coal', p. 860.

of the their Government.[280] Prematurely referring to Svalbard, Hoel maintained that the coal could in fact be worked in competition with the global market.[281] In 1921, Store Norske had been able to sell 50,000 tons of coal to the Norwegian State Railway ahead of England. The British Foreign Office was not happy with the conduct of the Norway's authorities regarding the Kings Bay Coal Co. It observed that 'the action of the Norwegian Government in openly encouraging a rival claimant company when the respective claims are about to be adjudicated upon by a properly constitutional tribunal is very regrettable and His Britannic Majesty's Government must reserve the right to claim compensation for the Northern Exploration Company in the event of the award of the commission in favour of the company.'[282]

The Northern Exploration Co. sent a small expedition in summer 1922. The board had decided on a limited programme.[283] Sæther replaced the winterers with new ones to guard the properties in winter 1922/23. He then warned off trespassers, renewed the claims, undertook the necessary customary works, took care of stocks, equipment, and houses, visited other Spitsbergen properties, kept London in touch with developments, and prospected when time permitted. This amounted to almost £4,700.[284] Despite the reduced scale of the activities, depreciation and interest swallowed £6,000. By the end of the year, the offices had moved, albeit to another floor in the same building.[285] This undoubtedly happened in an effort to save costs. Barr *et al.* relate that Frantzen made a first offer to sell the company's properties to the Norwegian Government on December 14, but the Norwegians impractically desired confirmation that all claims were legitimate.[286]

In 1923, the situation was gloomy. The company's annual report stated that although the composition of the board had been steady, interest alone had absorbed £10,670, and a further sum of ten per cent for the depreciation of assets on Spitsbergen had been written off.[287] Facts which *The Times* did not fail to

---

[280] *The Times* (1922) 'Norway's claim over Spitsbergen', 9 August.
[281] *Colliery Guardian* (1922) 'The coal deposits of Spitsbergen and Bjornøya and their importance to Norway', p. 783.
[282] Foreign Office (1922) *Letter to the Norwegian Foreign Department, 6 December*, Norwegian Foreign Department Box 5173, National Archives of Norway, Oslo.
[283] Sæther, C. S. (1922) *Letter to the Northern Exploration Co., 22 October*, Privatarkivet 112 (Carl Sæther) Box 7: 1927 Correspondence with the Northern Exploration Company, Ltd, Regional State Archives, Tromsø.
[284] *Balance sheet made up to the 31st day of October 1922* (1922), Northern Exploration Company, Limited BT 31/32080/112730, National Archives, Kew.
[285] *Offices* (1922) Northern Exploration Company, Limited BT 31/32080/112730, National Archives, Kew.
[286] Barr, Newman, and Nesteroff (2012) p. 140.
[287] *Report of the directors* (1923) Privatarkivet 112 (Carl Sæther) Box 7: 1927 Correspondence with the Northern Exploration Company, Ltd, Regional State Archives, Tromsø.

publish.[288] In view of the Danish Commission being set up, the company would have to prove its titles, which involved a registration fee for each claim. The directors regretted that the depression had prevented the sale or lease of properties. They anticipated the return of confidence and the improvement in trade.

In July 1924, the Northern Exploration Co. was forced to take steps to reorganise its nominal capital.[289] The scheme of reorganisation was officially sanctioned on September 10.[290] The authorised sum of £1,000,000 was reduced to £530,765 10s. The paid-up capital shrank from £521,092 to £65,136 10s. This was achieved by writing 17s 6d off each £1 share and meant that the nominal amount of such a share sank to 2s 6d. Each unissued £1 share was concurrently sub-divided into eight shares of 2s 6d. About half a million shares were then offered to the shareholders pro rata at par. The issue was underwritten.[291] However, the response was extremely meagre, and the underwriters had to take up practically the whole issue.[292] Legal action was taken against some of the underwriters, and although the company won, it was difficult to collect the money it was owed. This seriously inconvenienced and damaged the company.

While the financial reorganisation took place, the 1924 expedition had nevertheless sailed for Spitsbergen.[293] Salaries, food for the winterers, freight, and offices charges in Tromsø were down to £509.[294] Still, Sæther was able to send a trial shipment of 20 tons of zinc blende from Bell Sound. Encouraged by the assays, the company permitted the agent to go ahead with a small development programme, the first since 1921, in summer 1925.[295] On the east coast, a party opened up the coal and built housing and a coal chute down the mountain for loading and shipping. The miners got 270 tons of coal. Although Hoel had demonstrated that the east coast coal had the highest calorific value of all Spitsbergen coal at the recent World Power Congress in England, the British Government subsidy of English coal had caused a general depression of the coal trade, English coal being sold at such low prices that it was impossible for Spitsbergen coal to get a foothold on the market. Under the circumstances, Sæther

---

288 *The Times* (1923) 'Company results. Northern Exploration', 13 March.
289 *The Times* (1924) 'City news in brief', 22 July, p. 20.
290 *Extract of the Stock Exchange official intelligence* (1925) Norwegian Foreign Department Box 5374, National Archives of Norway, Oslo.
291 The Spitsbergen Syndicate, for instance, was incorporated on February 13, 1924 for the sole purpose of underwriting the company's shares. It was no longer in business in February 1927, yet it was only gazetted on April 4, 1930. (Spitsbergen Syndicate BT31/28351/195712, The National Archives, Kew).
292 *Directors' report and accounts to 31st December, 1925* (1925) MC 655/23 791X1, Norfolk Record Office, Norwich.
293 There may have been a 1923 expedition, but information has not yet been found.
294 *Directors' report and accounts to 31st December, 1925* (1925) MC 655/23 791X1, Norfolk Record Office, Norwich.
295 *Directors' report and accounts to 31st December, 1925* (1925).

did not recommend working the coal properties. Similarly, Bevan saw no choice but to adopt a waiting policy until either demand increased or the cost of production dropped.

Unlike coal, base metals enjoyed reasonable profit margins for high-grade ores at the time.[296] So the company worked an outcrop of zinc blende in Bell Sound in 1925. On Zinc Island, formerly deemed unworkable, 240 tons of high-grade ore were mined at a cost of £5 per ton and sold at £11s per ton in Antwerp. Bevan believed that the mining costs could be brought down to £4 per ton if the ore was shipped directly to Europe instead of via Norway. Plans were made to sink a shaft in the hope of discovering other ore bodies. The outcome would guide future development. Bevan also found encouraging traces of zinc blende on the neighbouring Duck Island, which was accessible at all times and had sufficient space for surface plant.[297] In addition, an effort was made to trace zinc blende on the mainland, but this was hindered by lakes and swamps in the region.

Sæther arranged a provisional option on another mining property.[298] The Icefiord Lead and Zinc Mine was situated 25 miles north of Zinc Island. A shaft was being sunk to intersect an occurrence that reached the surface on the southern shore of Icefiord. Sinking would continue in winter 1925/26, and the ore body was expected no deeper than 30 feet. Bevan and the mining engineer Arthur Lewin also prospected extensively in Bell Sound, Green Harbour, and Safe Harbour. They noted galena and zinc on the company's property in Recherche Bay. The specimens gathered were not particularly rich, but Sæther nonetheless recommended that two or three miners should inspect the area. In Safe Harbour, samples of copper ore were obtained from a formation, which correlated with that of Copper Camp in St John's Bay. Bevan hoped to locate a workable deposit soon. Although the 1925 expedition consumed almost £4,000, it was the first and only one to pay for itself.[299] The proceeds from the sale of zinc blende were £3,687, from coal £243, and from furs £438, generating a total of £4,368. Yet, the price of coal at 18s was so low that the directors decided to temporarily stop mining.

The focus had formerly been on exploration rather than mining. Now the directors planned to open up zinc blende on Zinc Island and in Icefiord.[300] They also intended to explore for other base metals if the limited capital permitted. The

---

[296] *Directors' report and accounts to 31st December, 1925* (1925).

[297] Bevan, J. A. (1926) *Annual report on the mineral properties of Northern Exploration Co., Ltd., 1925,* Northern Exploration Company (N.E.C.), Norwegian Polar Research Institute, Tromsø.

[298] Bevan (1926).

[299] *Summary of proceedings at annual general meeting* (1925) MC 655/23 791X1, Norfolk Record Office, Norwich.

[300] *Summary of proceedings at annual general meeting* (1925).

new policy of active mining was pursued in summer 1926.[301] Due to adverse weather, it was impossible to begin sinking the shaft before May 29 (Fig. 7.29).

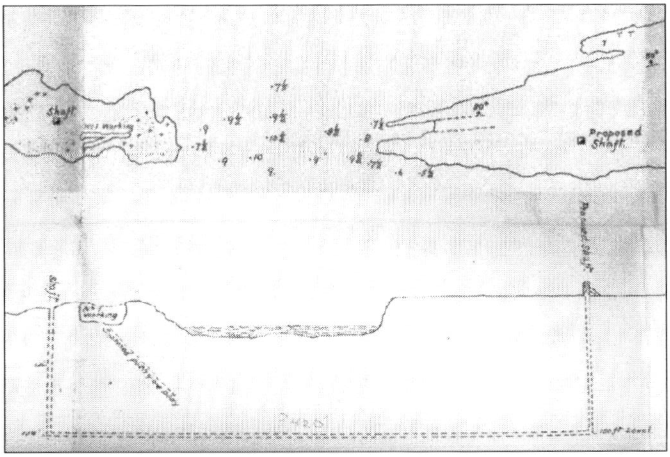

7.29 Details of a drawing showing proposals for Zinc Island (left) and Duck Island (right). (Source: Bevan, J. A. (1926) Annual report on the mineral properties of the Northern Exploration Co. Ltd., 1926, Northern Exploration Company (N.E.C.), Norwegian Polar Research Institute, Tromsø.)

Meanwhile, a suspension bridge was constructed between Zinc Island and Duck Island. It enabled direct communication between the workings and the plant and housing in all weathers. It proved efficient and saved time throughout the summer. By mid-June, the shaft on Zinc Island reached 34 feet when water was struck. Bevan halted operations until a small pump could be installed. The water rose to 16 feet and prevented the production of ore from this location. The site chosen for the shaft had been the only one available. Thus, the summer programme could no longer be carried out as planned. The bottom of Working No. 1 now lay below sea level, and only a narrow ridge of rock separated it from the sea. It was not possible to continue open-cast mining without striking water here, too. No. 2 lens was estimated to contain 30 tons of high-grade ore, but it was also unfavourably situated. Unable to progress on Zinc Island, Bevan decided to investigate Duck Island instead. Some work was done on a vein near the end of the season, but by the time the adit reached a length of 19 feet, the ore had all but disappeared. A prospecting pit was also sunk through five feet of overburden and into nine feet of rock. The workers encountered an appreciable dissemination of zinc blende, but the close of the season prevented further investigation.

In winter 1925/26, base metals had also been worked at the Icefiord Lead and Zinc Mine.[302] The shaft had originally been positioned by Arthur Lewin (Fig. 7.30), after which the Northern Exploration Co. took out an option on the mine.

[301] Bevan, J. A. (1926) *Annual report on the mineral properties of the Northern Exploration Co. Ltd., 1926*, Northern Exploration Company (N.E.C.), Norwegian Polar Research Institute, Tromsø.
[302] Bevan (1926).

7.30 Shaft as well as workshop and house at the Icefiord Lead and Zinc Mine. (Source: Photo Library np001973, Norwegian Polar Institute, Tromsø.)

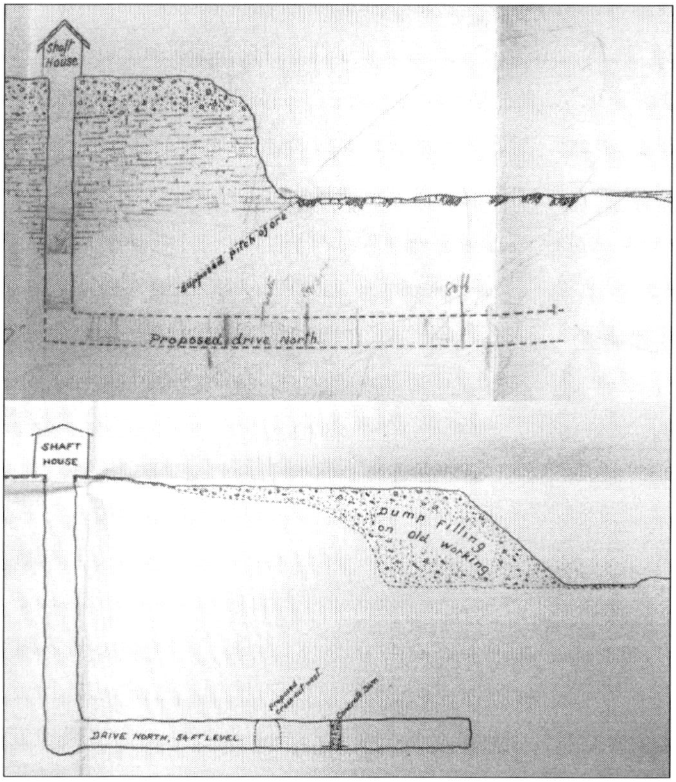

7.31 Composite drawing showing the proposed and the actual drive at the Icefiord Lead and Zinc Mine. (Source: Bevan, J. A. (1926) Annual report on the mineral properties of the Northern Exploration Co. Ltd., 1926, Northern Exploration Company (N.E.C.), Norwegian Polar Research Institute, Tromsø.)

The foreman in charge of the winter work had given instructions to drive straight for the mineral outcrop on the shore (Fig. 7.31). When Bevan arrived at the mine on June 30, 1926, the heading was too far advanced in wet ground to change the course of events. The work had to be discontinued due to water having been struck on the drive north. The drive was 80 feet long, and no ore had been encountered.

For the first 40 feet, the ground had been frozen solid, but it had thawed when it came into contact with sea water. In the first week of July, a concrete wall was erected in the tunnel at 54 feet from the shaft. The aim was to dam the water and enable a cross-cut. However, this failed due to an unhelpful rock temperature and the poor quality of the cement. Quick-setting cement would have been better suited. On July 9, therefore, all work was suspended in the shaft.

The set-backs in summer 1926 had presumably discouraged any winter work in 1926/27. Instead, the Northern Exploration Co. reflected on the previous season and formulated its priorities anew. In all, the expedition had cost £ 2,978. 140 tons of zinc blende were mined, 125.7 tons were shipped and sold to smelters at £13 5s 6d, generating £1,685, furs adding another £211.[303] In his report, Bevan offered some recommendations.[304] On Zinc Island, he did not want to propose sinking a shaft for anything other than prospecting, which could be done using hand pumps. Until the deposit could be proven, large schemes were premature. Yet, the mineral occurrences between Bell Sound and Icefiord warranted further expenditure on investigation. On Duck Island, diamond drilling at a small number of sites to no more than 300 feet would be most effective. On the mainland, electrical prospecting, which had been successful in locating ore in moraine-covered parts in northern Sweden, could be employed, followed by diamond drilling. Somebody pencilled the words 'very little encouragement' beneath these recommendations.

Britain's General Strike in May 1926 brought no meaningful change to coal mining on Spitsbergen, although Hoel temporarily mustered some optimism that the Bergin Process, which improved winning oil from coal, would revolutionise the coalfields.[305] Sæther's immediate tasks included the organisation of a programme of electrical prospecting and drilling for summer 1927. Unless money could be raised, there would be no such programme.[306] Even if money could be found, the investors decided what to spend it on. Broadly speaking, Sæther would be in charge of a team of contracted specialists. The parts for a drill were already in Tromsø, but the lack of funds meant that it could not yet be tested. Meanwhile, Sæther obtained quotes from specialists. A Swedish electrical prospecting firm

---

[303] *Balance sheet for the year to 31st October, 1926* (1926) Northern Exploration Company, Limited BT 31/32080/112730, National Archives, Kew; Bevan, J. A. (1927) *Letter to Northern Exploration Company, 21 March*, Northern Exploration Company (N.E.C.), Norwegian Polar Institute, Tromsø.

[304] Bevan, J. A. (1926) *Annual report on the mineral properties of the Northern Exploration Co. Ltd., 1926*, Northern Exploration Company (N.E.C.), Norwegian Polar Institute, Tromsø.

[305] Sæther, C. (1927) *Letter to the Northern Exploration Company, 3 January,* Privatarkivet 112 (Carl Sæther) Box 7: 1927 Correspondence with the Northern Exploration Company, Ltd, Regional State Archives, Tromsø.

[306] Sæther, C. (1927) *Letter to the Northern Exploration Company, 17 January*, Privatarkiv 112 (Carl Sæther) Box 7: 1927 Correspondence with Northern Exploration Company, Regional State Archives, Tromsø.

charged £400 for two engineers per month.[307] A German company could do the work for 6,000 marks, a British contractor for £160, the latter having a good reputation and the cheapest offer.[308] Sæther wondered if Bevan could not study electrical prospecting and carry it out by himself.[309]

In March 1927, the Northern Exploration Co. paid £110 to extend its option on the Icefiord Lead and Zinc Mine.[310] It had started negotiations with the British (Non-Ferrous) Mining Corporation, also called British Metal Corporation, for that firm to assess the zinc in Bell Sound and on Zinc Island. Since the arrangements took time, it was difficult to plan for the summer.[311] As things were, the company needed to get a Corporation engineer onto its property by June 10, but it was trying to delay until July 1. The Corporation had no intension to prospect, so it was up to the company to show the engineer the mineral occurrences and convince him of their potential. Although extremely short of funds, it needed to send a small crew northwards in advance of the engineer's arrival to clear the workings of snow and ice. The first boat of the season went to Spitsbergen on April 7.[312] Sæther therefore sent *Mina I* on April 20, mainly to keep up appearances and to relieve a watchman in Bell Sound.[313] The ship would be back in Tromsø in time for the company's use.

The secretary raised the funds to send the engineer at the last moment.[314] The company hoped that an option by the British Metal Corporation would make matters easier in the near future. The idea was that Arthur Lewin would accompany the man to Spitsbergen on a coal boat. Lewin had to make him comfortable and feed him well, a coal boat saving him the trials of going northwards on a smack. The workings would have been cleaned out on the men's arrival. If the engineer found things to be in order, then workers would go up with boring equipment in September, erect some shelter in Bell Sound, and drill throughout winter 1927/28.

---

[307] Sæther, C. (1927) *Letter to the Northern Exploration Company, 21 February*, Privatarkiv 112 (Carl Sæther) Box 7: 1927 Correspondence with Northern Exploration Company, Regional State Archives, Tromsø.

[308] Sæther, C. (1927) *Letter to the Northern Exploration Company, 7 April*, Privatarkiv 112 (Carl Sæther) Box 7: 1927 Correspondence with Northern Exploration Company, Regional State Archives, Tromsø.

[309] Sæther, C. (1927) *Letter to the Northern Exploration Company, 25 March*, Privatarkiv 112 (Carl Sæther) Box 7: 1927 Correspondence with Northern Exploration Company, Regional State Archives, Tromsø.

[310] Northern Exploration Co. (1927*) Letter to C. Sæther, 25 March*, Privatarkiv 112 (Carl Sæther) Box 7: 1927 Correspondence with Northern Exploration Company, Regional State Archives, Tromsø.

[311] Northern Exploration Co. (1927) Letter to C. Sæther, 28 March, Privatarkiv 112 (Carl Sæther) Box 7: 1927 Correspondence with Northern Exploration Company, Regional State Archives, Tromsø.

[312] Sæther, C. (1927) *Letter to the Northern Exploration Company, 7 April*, Privatarkiv 112 (Carl Sæther) Box 7: 1927 Correspondence with Northern Exploration Company, Regional State Archives, Tromsø.

[313] Sæther, C. (1927) *Letter to the Northern Exploration Company, 25 April,* Privatarkiv 112 (Carl Sæther) Box 7: 1927 Correspondence with Northern Exploration Company, Regional State Archives, Tromsø.

[314] Light, C. (1927) *Letter to Carl Sæther, 3 June*, Privatarkiv 112 (Carl Sæther) Box 7: 1927 Correspondence with Northern Exploration Company, Regional State Archives, Tromsø.

The current Corporation option was valid until September 30.[315] Upon receiving the engineer's report, the Corporation would notify the company before September 1, 1927 if it wanted to extend the option and open up the zinc blende.

7.9 Ratification of the Spitsbergen Treaty

Since the expedition in 1925, the administration of Spitsbergen had taken shape. The Norwegian Government had initially introduced a draft mining bill.[316] The citizens of all signatory powers were entitled to acquire and exploit coal and other mines. Provisions were being made for the expert management of mines and the protection of workers. Norwegian was going to be the official language, and all applications to the authorities had to be translated. *Bergmesteren* was free to ignore all applications in other languages. Taxes and duties in the form of a special fund, the Spitsbergen Fund, would be raised to cover the expenses of administration. The export duty would not exceed one per cent. In spring 1925, the Norwegian Cabinet proposed to combine Spitsbergen and Bear Island under the official name of Svalbard, which was to be considered an integral part of the Norwegian kingdom.[317] Locally, *sysselmannen* would carry the same function as any ordinary county governor. Norwegian civil and criminal rights would apply. On August 14, 1925, five years after the signing of the treaty, Norwegian sovereignty over Svalbard was officially proclaimed.[318]

For five years, the companies had been anxiously waiting for a decision regarding their claims. In conjunction with the ratification of the Spitsbergen Treaty, the Norwegian Government now issued a notice to all claimants.[319] Within the next three months, notifications of all claims had to be sent by the Government of the claimant to the Danish Commissioner charged with their examination. The notifications had to include the precise limits of the territory and a map on which the area was clearly marked. In addition, claimants had to pay one penny per acre to cover the expenses of the commission. The Northern Exploration Co., therefore, sent the particulars of their extensive claims to the Foreign Office. The company was hopeful that the expensive yearly tasks such as renewing claims and guarding properties, which involved lengthy sea journeys, would soon be redundant.

---

[315] *Draft of circular to debenture holders* (1927) Privatarkiv 112 (Carl Sæther) Box 7: 1927 Correspondence with Northern Exploration Company, Regional State Archives, Tromsø.

[316] *Mining Journal* (1924) 'Spitzbergen mining legislation', p. 198.

[317] *The Times* (1925) 'The government of Spitsbergen', 24 April, p. 11.

[318] *The Times* (1925) 'The new status of Spitsbergen', 14 August, p. 9.

[319] *London Gazette* (1925) 'Spitsbergen. Notice to all claimants', 14 August, p. 5427.

If the company intended to hold on to all titles, it would need to raise £15,000 for the examination and possibly further large sums afterwards.[320] Since the company and the Norwegian Government had privately settled the dispute with the Kings Bay Coal Co. over Kings Bay, however, the company chose to be compensated for the loss of earnings rather than pay for yet another territorial claim and work the currently worthless coal itself.[321] It was agreed that the Kings Bay Coal Co. should pay £37,000. The sum was received in November 1925 and was immediately handed over to the trustees of the noteholders. For the first time in the company's history, areas were being relinquished, but the chairman carefully pointed out that this did not mean that they were sold. The company retained the right to work such relinquished lands for a period of five years. If they were found to be commercially valuable, it would have to pay the necessary dues. The company also retained all houses and huts on such lands, which would be the base points for claiming such additional mining rights. Sæther and Bevan were working on notifications to the Norwegian Government of those areas, where the company wished to retain such rights. The approximate boundaries would be two and a half kilometres fronting the sea and four kilometres inland parallel to longitude and latitude.[322] Each hut would be situated in the middle of the water-front. The board believed the arrangement of withdrawing the company's extensive claims and substituting them for smaller ones served the Norwegians a political purpose rather than a practical one.

In summer 1926, Bevan accompanied Sæther around Spitsbergen to mark out so-called discovery points.[323] The Norwegian regulations from October 1925 suited the company because it did not have to file the requests for its claims instantly. Instead, it had six months to file discovery points in all areas. Filing granted the priority right to the claims in these areas for the next five years. In February 1926, Light sent the particulars of the discovery points to be registered to *bergmesteren*.[324] Although the majority was indeed marked by huts, the exception included Magdalena Bay, Möller Bay, Farm Harbour, and Lizet Rocks. Each discovery point covered ten square kilometres. *Bergmesteren* allowed Light's statement to be supplemented at a later date and granted a period of grace for the delivery of final details until the end of September 1926. In July, therefore, Sæther

[320] At a cost of one penny per acre, £15,000 was the equivalent of 5,625 square miles. This total appears to have included the 1920 acquisition of the majority of Edge Island.

[321] *Summary of proceedings at annual general meeting* (1925).

[322] Sæther, C. S. (1925) *Letter to Arnold Ræstad, State Advocate for Svalbard Matters, 2 December,* Privatarkivet 112 (Carl Sæther) Box 7: 1925 Correspondence with State Advocate A. Ræstad, State Regional Archive, Tromsø.

[323] Bevan (1926).

[324] Light, C. (1926) *Letter to the Mining Commissioner for Svalbard, 24 February,* Bergmesteren for Svalbard, 1926 Northern Exploration Company, State Regional Archive, Tromsø.

and Bevan used the ship to visit 30 places in order to undertake superficial examinations in the limited time available and to drive iron stakes into more or less promising outcrops.[325] Bevan subsequently reported that in most locations to the south of Bell Sound, there were no mineral occurrences of note and that the region was not comparable with the northern section between Bell Sound and Cross Bay. Nonetheless, by August 1926, the men had filed details of the discovery points depicted in Fig. 7.32.

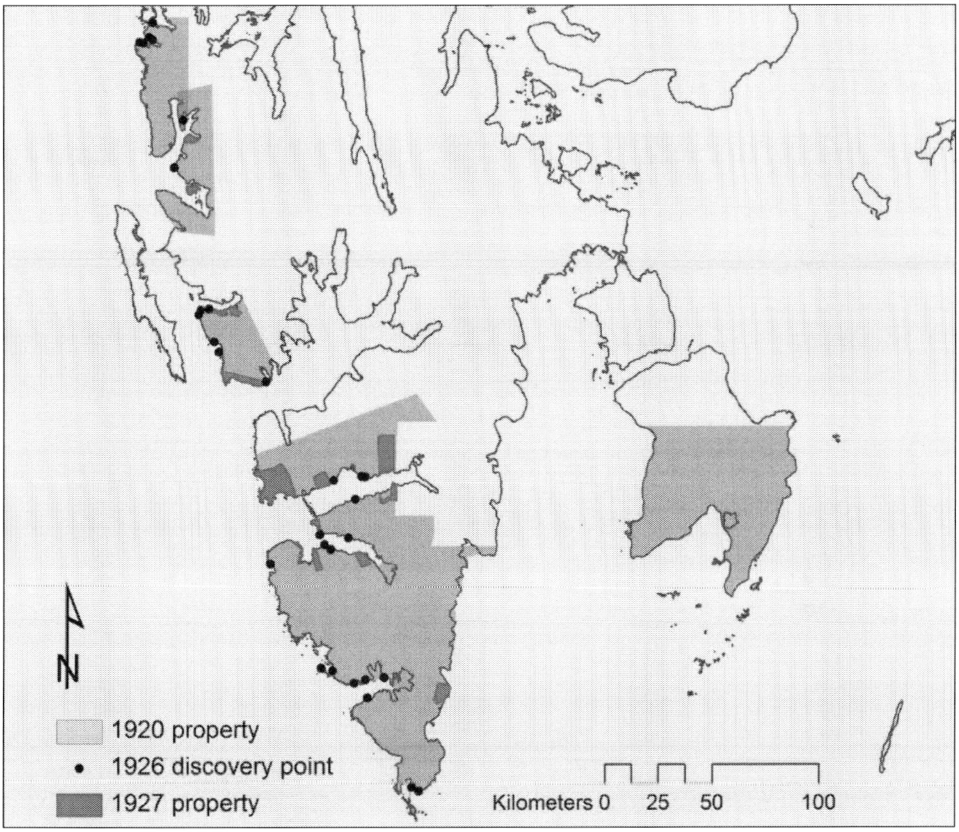

7.32 Map comparing the company's greatest territorial extent in 1920 with the discovery points registered in 1926 and the claims awarded in 1927. (Map: F. Kruse.)

Meanwhile, the Danish Commissioner settled the last remaining claim disputes. On May 7, Kristian Sindballe finally informed the company that their 16 areas were

[325] Sæther, C. S. (1926) *Letter to the Northern Exploration Co., 29 October*, Northern Exploration Company (N.E.C.), Norwegian Polar Institute, Tromsø.

now undisputed and would be recognised in his report.[326] Sindballe awarded 169,647 acres of the 175,232 aces claimed (Fig. 7.32).

7.10 Final efforts, sale, and dissolution

At the general meeting on July 11, 1927, Smith being ill and Frantzen abroad, Major Pery resided.[327] The negotiations with the British (Non-Ferrous) Mining Corporation had delayed the publication of the accounts. Although the financial situation had not yet recovered, the liabilities had been reduced by almost £32,500. The last two years had produced revenue of £5,372 8s 2d. During the General Strike of 1926, coal prices had risen abnormally, but the board had felt this to be transitory and had considered the time to be unsuitable for sale or floatation. Coal mines were still out of favour in Europe, and any attempt to use the fleeting opportunity had not been successful. Similarly, European and American stakeholders in marble were not prepared to finance a quarrying company.[328] On a good note, however, the preliminary investigations of the zinc and lead occurrences had been promising, and the properties were currently under option by the Corporation. The shareholders had granted a debenture of £1,100 to be able to send the engineer, in whom the company placed a lot of hope. 'If the report of the Corporation's Engineer is not sufficiently favourable, it is difficult at this juncture to foresee the future.'[329]

In August, Sæther reported back from the journey to Spitsbergen.[330] There had been some ice in the first week of work, but that had gone in the second week. Two miners and a cook were left on Zinc Island to work the zinc. They had worked there before and had the necessary experience. Six men and a cook were also left on Duck Island. In two shifts, they were to sink a shaft to a depth judged by the foreman and then drive towards Zinc Island. Sæther was anxious to start this work. Watchmen would be left at Calypso Bay and Marble Island with food and supplies for the winter 1927/28. Both the Kings Bay Coal Co. and Store Norske were cooperative. Sæther asked the Northern Exploration Co. to contact him if he were to go ahead with the winter programme.

---

[326] *Summary of proceedings* (1927) Privatarkiv 112 (Carl Sæther) Box 7: 1927 Correspondence with Northern Exploration Company, Regional State Archives, Tromsø.

[327] *Summary of proceedings* (1927).

[328] The question as to why the company failed to work Marble Island commercially or interest a subsidiary firm has not yet been answered adequately. Future research could, for instance, investigate the fault pattern on location or the downturn in British luxury building after the First World War.

[329] *Summary of proceedings* (1927).

[330] Sæther, C. (1927) *Letter to the Northern Exploration Company, 2 August*, Privatarkiv 112 (Carl Sæther) Box 7: 1927 Correspondence with Northern Exploration Company, Regional State Archives, Tromsø.

The Northern Exploration Co.'s hopes were shattered, however, by the Corporation engineer's final report.[331] He stated that 20 tons of ore had been taken from Zinc and Duck Islands in 1924, 240 in 1925, and 126 in 1926. A few tons had also been obtained from Icefiord. On Duck Island, the adit referred to by Bevan was completely filled with shingle. The shallow shaft had been examined, but no traces of ore had been found. There were minute traces of zinc in the dolomite but nothing else to report. On Zinc Island, 'no further work of importance can be done, on account of the Island's exposed position, small area, and the height above sea level.'[332] At Icefiord, the outcrop lay below sea level at high tide, was covered by shingle, and only partially exposed at low tide. Work here could only be done for a couple of hours at low tide. The conclusion was correspondingly brief, 'A few tons of pure zinc blende can be obtained from Zinc Island by fossicking, and from a mining point of view, a few hundred tons of high grade ore could be obtained, if preliminary treatment were possible, but this is out of the question. From inspection and from work done, there does not appear to be any evidence to show that these small deposits are anything but local, and further work is not recommended.'[333] Someone's pencil note on the margin lamented that this completely damned the property. Consequently, the Corporation regretted that it would not make use of the option and was not disposed to take matters further.[334]

The company had run out of options. The board notified the trustees of the debenture holders and the secured loan holders to ascertain their attitude regarding the future.[335] Requesting some more money to pay the outstanding debts for the summer programme, Sæther wrote, 'I hope you won't quit. The English flag ought to be shown up in the arctic, and your company is the only one still doing so.'[336] This rhetoric, however, was extremely outdated.

Following the withdrawal of the British Metal Corporation, the Northern Exploration Co. drafted a memorandum to its shareholders to outline their options.[337] After the Corporation had opted out, the trustees met with the board in September. The directors were given a three-month notice to comply with the

[331] Kerr, M. R. (1927) *Report on Spitzbergen, with special reference to the Northern Exploration Company's lead and zinc workings*, Northern Exploration Company (N.E.C.), Norwegian Polar Institute, Tromsø.

[332] Kerr (1927).

[333] Kerr (1927).

[334] British (Non-ferrous) Mining Corporation (1927) *Letter to the Northern Exploration Company, 12 August,* Northern Exploration Company (N.E.C.), Norwegian Polar Institute, Tromsø.

[335] Light, C. (1927) *Letter to the stockholders, 23 August*, Northern Exploration Company (N.E.C.), Norwegian Polar Institute, Tromsø.

[336] Sæther, C. (1927) *Letter to the Northern Exploration Company, September 2,* Privatarkiv 112 (Carl Sæther) Box 7: 1927 Correspondence with Northern Exploration Company, Regional State Archives, Tromsø.

[337] *Draft circular to shareholders* (1927) Privatarkiv 112 (Carl Sæther) Box 7: 1927 Correspondence with Northern Exploration Company, Regional State Archives, Tromsø.

conditions of the trust deed. Failure to comply by the beginning of December would bring in a receiver for the debenture holders. However, if the company could devise a scheme to work on Spitsbergen for another year, the notice might be extended. The directors thus informed the shareholders that a shaft could be sunk on Zinc Island to a depth of about 60 feet for no more than £3,000. The shaft would prove the ore body, and development work could commence in 1928. It was a possible solution for the current difficulties the company was in. A further £2,000 was required to pay off the debenture that had placed the Corporation engineer on Spitsbergen and the interest, provide a clerk and an office, and pay dues to Tromsø including the fees owed to the Norwegian Government. It was apparent that the new money had to come from the shareholders as the debenture holders would not offer any more funds. The shareholders were asked to indicate on a postcard if they were in favour of a reconstruction with an assessment of 6s per share on the present shares, if they would take up any 6% unissued debentures currently amounting to £6,000, or if they would take up any 8% preference shares of £1 each with the issue being limited to £10,000.

Carl Sæther had been an agent for the Northern Exploration Co. since 1919, and its demise meant an uncertain future. In mid-October 1927, the company was unable to send money to settle its accounts in Norway.[338] Sæther's most pressing bills were 4,200 kroner owed to Tinus Aune for food and 500 kroner for oil. He had been able to meet the one for oil, but Aune had started proceedings against the company and taken arrest of *Mina I*. This he did in friendly spirits though, owing to long and satisfactory business with the company since the Mansfield days.[339] Sæther was willing to reduce his salary for the sake of the shareholders, but if the company stopped prematurely, his sacrifice would be in vain. He suggested that things which did not sell in Tromsø might find a buyer in England, such as the medical supplies that came up in 1920 and had never been touched. The diamond drill could not be sold because it was missing the diamonds, and new ones cost almost as much as a new shot drill. He could send the copper wire down, but it might not be worth the freight and handling charges. Lloyd's agent in Tromsø might be able to value the company's two smacks and lighters for auction. The company replied that Sæther should sell anything he could at what to him was a fair price, but that he should not send anything to London.[340]

---

[338] Sæther, C. (1927) *Letter to the Northern Exploration Company, 12 October*. Privatarkiv 112 (Carl Sæther) Box 7: 1927 Correspondence with Northern Exploration Company, Regional State Archives, Tromsø.

[339] Sæther, C. (1927) *Letter to the Northern Exploration Company, 8 December*. Privatarkiv 112 (Carl Sæther) Box 7: 1927 Correspondence with Northern Exploration Company, Regional State Archives, Tromsø.

[340] Light, C. (1927) *Letter to C. Sæther, 9 November*. Privatarkiv 112 (Carl Sæther) Box 7: 1927 Correspondence with Northern Exploration Company, Regional State Archives, Tromsø.

The programme in summer 1927 had been the last. The withdrawal of the Corporation had been the final straw. The company was now primarily concerned with repaying its debts. On December 13, it moved offices to Old Broad Street in London in a quest to downsize further.[341] In March 1928, Pery resigned, Peirson, Smith, and Frantzen remaining on the board as of yet.[342] Very little activity was recorded in 1928. By the end of the year, the summary of share capital and shares showed a total debt of £47,789.[343] The appended balance sheet was only made up to the end of October 1926. Although the 1927 expedition had not been taken into account, Sæther had reported that it had cost £1481 12s 2d.[344] The proceeds from furs had been £144 4s 8d. The return from 30 tons of zinc blende had not been available, but he estimated the profit to not exceed £100. With little hope of any more revenue from Spitsbergen to relieve the financial burden, a receiver would recover the debts. On October 24, 1929, Crisp and L. E. Smith, trustees for a debenture stock of £60,000, appointed Gerald Dudley Smith.[345] Thereafter, the office moved to Surrey, and Peirson resigned, leaving only Smith and Frantzen.[346] Smith was also appointed receiver by Frantzen in December 1929 for a series of debentures issued by the Northern Exploration Co.[347] On February 28, 1930, Smith and Frantzen, too, resigned.[348] Merely Rendle, based at the Surrey address, remained for clerical assistance.

Rendle requested additional information from Sæther. The Norwegian indicated that the company's properties were treaty-protected with the British Government being a signatory to the Spitsbergen Treaty.[349] All titles had been granted by the Danish Commissioner and issued by the Norwegian Government. In addition, the company had entered into private contracts. The first contract was with the Norwegian Government. State servicemen would be given room and board

---

341 *Offices* (1927) Northern Exploration Company, Limited BT 31/32080/112730, National Archives, Kew.

342 *Directors* (1928) Northern Exploration Company, Limited BT 31/32080/112730, National Archives, Kew.

343 *Summary of share capital and shares* (1928) Northern Exploration Company, Limited BT 31/32080/112730, National Archives, Kew.

344 Sæther, C. (1927) *Letter to the Northern Exploration Company, 9 November.* Privatarkiv 112 (Carl Sæther) Box 7: 1927 Correspondence with Northern Exploration Company, Regional State Archives, Tromsø.

345 *Notice of appointment of receiver, 30 October* (1929) Northern Exploration Company, Limited BT 31/32080/112730, National Archives, Kew.

346 *Offices* (1929) Northern Exploration Company, Limited BT 31/32080/112730, National Archives, Kew; *Directors* (1929) Northern Exploration Company, Limited BT 31/32080/112730, National Archives, Kew.

347 *Notice of appointment of receiver, 10* December (1929) Northern Exploration Company, Limited BT 31/32080/112730, National Archives, Kew.

348 *Directors* (1930) Northern Exploration Company, Limited BT 31/32080/112730, National Archives, Kew.

349 Sæther, C. (1930) *Letter to Rendle, 9 October*, Northern Exploration Company (N.E.C.), Norwegian Polar Institute, Tromsø.

at a reasonable cost; the state was permitted to use the piers and other appliances for tying ships; and the company's radio stations were required to send messages through the radio station in Green Harbour. A second contract was with the Deutscher Seefischerei Verein to share the docking facilities at Safe Harbour. A third contract allowed the Spitsbergen Exkret Co. to take guano from the company's property at Middle Hook. Sæther attached a list of property names as they had been used by the company. The subsequent renaming by Norwegians had been 'mistaken national sentiment.'[350] In addition, Sæther had approached John D. Marsden, manager of a fishing company from Grimsby, 'on the proposition as of interest to Great Britain politically, and to his interest, trawling, specifically. The most cursory glance at the enclosed map will show him that north of Great Britain there are no British-owned spots outside of Canada.'[351] Spitsbergen was a navigable place, and petrol could be landed here for inter-continental flights. The arguments in favour of trawling were best known to Marsden himself. The company's properties were still part of Britain unless they passed into Norwegian hands. The interviews with Marsden seem to have lasted until the beginning of 1931, but they never amounted to anything.

There are few documents in the English language regarding the negotiations surrounding the sale of the company's properties to the Norwegian Government. There are considerably more Norwegian papers agreeing to the acquisition of sixteen claim-patents for the purchase price of 100,000 kroner. This sum, consented to by the debenture holders, converted to £5,080 13s 1d.[352] For his services in the proceedings, Sæther received a commission of ten per cent. Following the final transaction of the company, Smith ceased to act as receiver in June 1933.[353] The Northern Exploration Co. was dissolved by notice in the *London Gazette* on June 26, 1934.[354]

7.11 Summary and conclusion

The fieldwork results presented in Chapter 4 highlighted the local conditions under which the archaeological landscape of the Northern Exploration Co. took shape. The widespread physical remains populated an extensive yet one-sided actor-network (Fig. 4.63). The benefits of the archaeology fieldwork, however, became

---

350 Sæther, C. (1930) *Letter to Rendle, 9 October.*

351 Sæther, C. (1930) *Letter to Rendle, 9 October.*

352 *Receiver or manager's abstract of receipts and payments* (1932) Northern Exploration Company, Limited BT 31/32080/112730, National Archives, Kew.

353 *Notice of ceasing to act as receiver or manager* (1933) Northern Exploration Company, Limited BT 31/32080/112730, National Archives, Kew.

354 *London Gazette* (1934) 'Companies Act, 1929', 26 June.

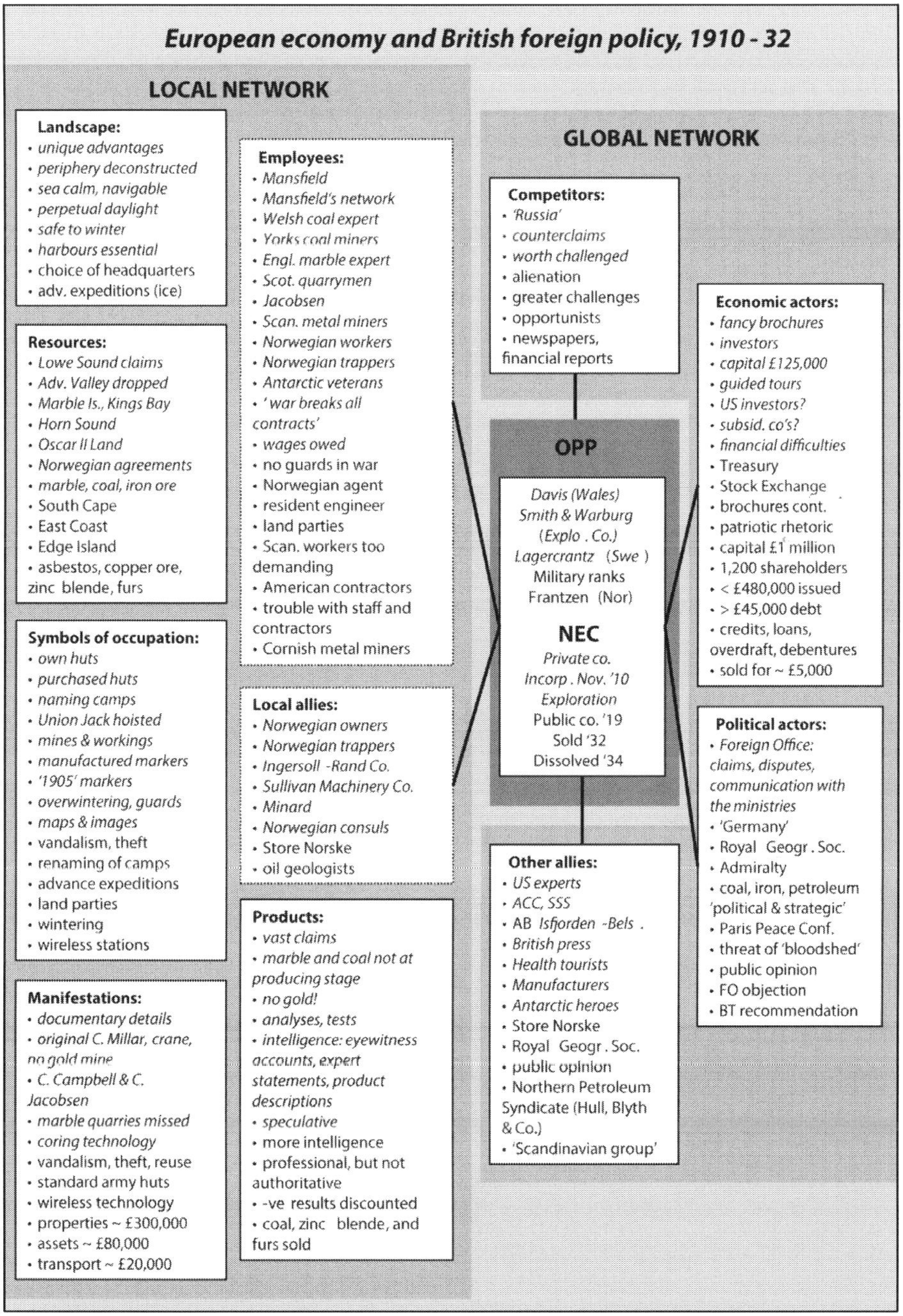

7.33 The company's two-part history is characterised by an actor-network before the First World War (in italics) and one after it. (Chart: F. Kruse.)

obvious during the subsequent comparison with the historical sources in this chapter, sometimes necessitating the re-interpretation of the documents. The complementary information from primary and secondary sources is summarised in an updated version of the actor-network (Fig. 7.33). The company's two-part history brought about by the First World War was visible in the field as distinct phases of building and prospecting. The severe change in the global context is even more evident in contemporary documents. It was characterised by the British war effort, the increased need for coal and iron, and the deterrence of German advances. The unpopular Treaty of Brest-Litovsk allowed the company to demand Britain's annexation of Spitsbergen in order to regain economic supremacy. War had supposedly hindered the development of its claims, which it now deemed pivotal to post-war reconstruction. This summary highlights selected aspects of the actor-networks before and after the conflict. The post-war one expanded rapidly, but it was overpowered by further changes at global level. Ultimately, the company's policies and strategies amounted to very little.

It must be stressed that the Northern Exploration Co. was founded as a private company for the employment of capital in exploration and prospecting for economic minerals. Subsidiary mining companies would extract them. The coal proprietor Davis undoubtedly knew the Welsh coal industry, but it is questionable if any of the directors had experience of other natural resources. More importantly, however, Smith and Warburg also sat on the board of *the* Exploration Co. In 1886, two Californian mining engineers enlisted the financial support of the influential Rothschild family of London and formed the Exploration Co.[355] It centralised the expertise of Californian engineers and engaged them in British ventures across world, thereby creating a tightly knit web of capital and expertise that linked Californian engineering and British enterprise in the late nineteenth century. The 1890s were a decade of gold discoveries and gold share booms on the London Stock Exchange. While in 1886, only 3.4 per cent of 237 new mining companies in Britain were formed specifically for the purpose of exploration, this figure reached 40 per cent of 961 new mining companies in 1895, a record year with gold booms in South Africa, Rhodesia, and Western Australia.[356] In the 1890s and the 1900s, the risky and speculative business was not ignored by the large merchant banks, and the Exploration Co. was a pioneer in the field. 11 of its 20 original subscribers were bankers. This included Smith. Two other directors of the Exploration Co. also sat on the board of the British South Africa Co., which exemplifies the imperial spirit in Chapter 2. The Northern Exploration Co. certainly wished to emulate the

[355] Teisch, J. B. (2011) *Engineering nature: water, development, and the global spread of American environmental expertise*, Chapel Hill: University of North Carolina Press.
[356] Turrell, R. V. and Van Helten, J.-J. (1986) 'The Rothschilds, the Exploration Company and mining finance', *Business History*, 28 (2), pp. 181-205.

Exploration Co. Yet, Warburg's untimely departure indicated that he did not think it would succeed.

The Northern Exploration Co. additionally offered directorships to foreigners who could consolidate its influence on Spitsbergen. As such, it exchanged a seat on the board with the Swedish mining company AB Isfjorden-Belsund, as part of the agreement to settle their dispute over Braganza Bay, and the Norwegian Frantzen was instrumental in the company reaching its maximum extent in 1920. After the war, men with a bearing on British politics became important, but the decorated newcomers to the board only showed a fleeting interest that waned with the signing of the Spitsbergen Treaty.

To maintain and expand its global network, the company's principal economic actant were its elaborate brochures, not to be confused with actual prospectuses. How they were presented to potential backers could no longer be discerned, but already before the war, shareholders agreed to increase the nominal capital to a substantial £125,000. Unfortunately, little is known about the ensuing 'employment of capital', an integral part of the company's approach. Its financial dealings are near enough invisible. The guided tours of the directors, shareholders, and other agents across the properties were designed to convince them of the value of the marble, coal, especially anthracite, and iron ore. The tours only ever took place late in the season, when the Arctic summer was at its most pleasant. Gold and oil find frequent mention but were never witnessed. The discussions with potential American investors and subsidiary mining companies must have been extensive, but no records have yet been found. The wish to display the company's possessions in the most favourable light without being able to attract a third party drew on the company's resources and led to financial difficulties before the war.

A buoyant post-war market provided the company with a second chance. The company went public, but the treasury was reluctant to grant fresh capital. Like many others, it succumbed to the wide-spread patriotic rhetoric. An immense £1,000,000 attracted no less than 1,200 subscribers; nonetheless, only half the amount had been issued when the economy stagnated. Unable to implement a more conservative financial strategy in time, the company's increasing debt went hand in hand with the stakeholders' decreasing trust. Over a decade, the lack of funds culminated in the sale of the properties to the Norwegian Government for only a fraction of the original costs.

Prior to the war, the company relied on the Foreign Office to register its claims, to mediate the claim disputes with Norway, and to communicate with other British ministries. The company needed the Foreign Office, but the same was not true the other way around. No one could have foreseen that the position of the ministry would be greatly altered in wartime, sacrificing any long-term political objectives in view of immediate military necessities.[357] The Foreign Office lost much of its influence and surrendered its power to the War Office and the Admiralty. Emerging from the war, the old diplomacy had largely been replaced by a new kind. Fundamental reforms followed between 1919 and 1921, in which the wartime structure was dismantled.[358] The Northern Department was created. Although it would deal with Spitsbergen matters, the concerns of the exploration companies were never at the top of its priorities.

At the close of the conflict, 'Germany' became the company's foremost political actant, the prevailing geopolitical construct of Britain's archenemy ruling its rhetoric. Resources such as coal, iron ore, and oil were now of political and strategic importance. Although the company may have added to the interest in Spitsbergen of bodies such as the Royal Geographical Society and the Admiralty, it never enjoyed their unconditional support. Even the grossly exaggerated threat of violence and bloodshed on the archipelago did not make an impression on the British legation in Paris. If the company hoped that the public could sway the British Government to annex the islands, it was disappointed. Its political connections were generally weak. With the advent of Norwegian administration, the Foreign Office communicated all British claims to the Danish Commissioner, after which political actors restricted themselves to the observation of – and occasional objection to – the company's financial business.

Where other allies were concerned, this chapter has shown that the company tried to interest a whole host of potential actors, while deliberately including others as actants in its rhetoric. By and large, these actors paid only limited attention to Spitsbergen. Where the company succeeded to enrol them into its network, it failed to make itself indispensible to the proceedings. It invested in the AB Isfjorden-Belsund, but the Swedes never paid any dividends nor did the company benefitted from the arrangement in other ways.

Rivalry in the form of claim disputes and others challenging the worth of the company's natural resources already existed before the war. In the British absence from Spitsbergen during the conflict, these challenges intensified. Thereafter, the company's approaches to attract new shareholders frequently backfired. Overly aggressive canvassing, for instance, may have appealed to the poorly informed masses, but it only served to alienate reputable investors. The conversion to a public company included the regular publication of business information, which was beneficial when the market was buoyant. Yet, the unrelenting reproduction of

---

357 Warman, R. M. (2003) 'The erosion of Foreign Office influence in the making of foreign policy, 1916-1918', *Historical Journal,* 15 (1), pp. 133-59.

358 Steiner, Z. and Dockrill, M. L. (2003) 'The Foreign Office reforms, 1919-21, *Historical Journal,* 17 (1), pp. 131-56.

lacking incomes and increasing debt was detrimental to the company's image and enhanced the damaging effect of the post-war depression.

Contemporary documents revealed a wealth of information regarding the company's employees and its local allies, but this was usually anecdotal and therefore difficult to interpret. Above all, the company will have wanted to replicate the tightly knit web of capital and mining expertise that characterised the Exploration Co., but its approach to staffing is unclear. Mansfield and Jacobsen were prospectors apt at locating natural resources. Although they could not be trusted to quantify a deposit reliably, they nonetheless led the pre-war expeditions. Mangham, a survivor of Mansfield's former network, proved to be a good foreman of a coal mine, but he knew nothing about gold, extracting iron pyrite instead, which calls his statements concerning any other resources into question. Coal and marble experts were presumably engaged for that reason. Oddly, only the marble expert remained on site for whole seasons at a time, while the coal expert was only taken on guided tours and did not oversee the works. An iron ore expert was not even hired. It was as if the company tried to avoid the devaluation of its properties in this way, which was, of course, counterproductive. It did, however, use experienced Yorkshire coal miners, Scottish quarrymen, and Norwegian and Cornish metal miners. Although they would not have been able to produce an output where there was nothing to produce, they ranked so low in the hierarchy that their opinion regarding the claims would not have mattered much.

After the war, cheap Scandinavian labour was no longer available, which changed the structure of the workforce in favour of British men. There was also an increasing demand in the global network for accountable information. The company therefore employed a resident engineer, who was assisted by a Norwegian agent. They repeatedly coordinated land parties headed by qualified engineers and geologists. Their reports were, by and large, negative.

The archaeological record provided a very accurate picture of the installations. Archival research added some interesting details. The original Camp Millar was not one of today's buildings. A second crane relocated from Marble Island had stood at the supposed gold mine, from which only iron pyrite was extracted. Hut foundations at Lægerneset and Iron Mountain Camp could be attributed to Camp Campbell and Camp Jacobsen. Some marble quarries in Kings Bay have in all likelihood been mistaken for natural features and remain unrecorded. While a number of boreholes have been mapped, primary sources recount problems with the technology and sub-surface faults. The standard army huts, which were inappropriate for the Arctic, were probably a contribution of the Admiralty to the post-war expeditions. As with inappropriate machinery elsewhere, the barracks were not reused, one exception being the Polish research station. In light of subsequent events, the wireless technology, let alone the seaplane, had been premature. They were among the many items that added to the unnecessarily high costs of the properties, assets, and transport.

Archaeologically and historically, the full range of the company's symbols to protect its claims becomes apparent. As such, it constructed its own houses in strategic locations on yearly expeditions while it also hired trappers to erect huts on its behalf. The cabins purchased from Norwegians such as Giæver, Svendsen, and Hagerup completed the list of buildings in its possession. They formed a string of claim markers along the west coast with additional outliers on the east coast and on Edge Island. The buildings were usually named after influential stakeholders, personal acquaintances, or company employees. There is no indication that any flag other than the Union Jack was hoisted on the company's properties. The mines and workings were a sign that the claims were being exploited. Whereas early claim signs tended to be haphazard, the company now made use of manufactured wooden posts and embossed metal plaques. The plaques stated that the territory in question had been in the company's possession since 1905. The historical truth was that Mansfield had first prospected in Bell Sound in 1905, that other claims were made in subsequent years, that the company was only founded in 1910, when it obtained these claims, and that the plaques were first erected in 1912, after which time they understandably attracted fierce criticism by the company's rivals. It would have been more appropriate to name the discoverer and the date of discovery followed by the later acquisition by the company. Effective occupation further found expression in the regular winterings by Norwegian trappers, although this barely happened during the war. Occasional maps and frequent photographs made the properties a reality for global actors. After the war, some camps were renamed in honour of current stakeholders. Besides the practical purposes of exploration, advance expeditions and land parties were able to extend the season and cover more ground. Combined with the wireless technology onboard the company's ships, a web of effective occupation therefore engulfed the far-reaching properties.

Effective occupation of vast properties that held mineral resources must also be seen as the company's principal product for the market. Before the war, it had proved marble and coal. Although it had not reached the producing stage itself, the developments were ready for subsidiary companies. The company had undertaken analyses and tests, and it had amassed intelligence in the form of eyewitness testimonies, expert statements, and product descriptions. Nonetheless, it remained speculative, so after the war, the demand for more intelligence was met by increasingly professional reports. Yet, the company failed to hire the most authoritative engineers and geologists in the field, and it did not sink the necessary boreholes to dispel all rumours and doubts. Whether it ever intended to do so can no longer be discerned.

The next occurrence on the global stage was the onslaught of the post-war depression. This time has not been considered in the actor-network graphic above because hardly any new connections were sought out, let alone formed. In view of its detrimental financial situation, the company was hard-pressed to maintain its existing network and keep the links with especially its economic actors from failing. Its shares were unpopular, the global market was dismal. The company's limited local programme of replacing the winterers, warning of trespassers, renewing the claims, and carrying out customary works and occasional prospecting was nonetheless costly. A first offer to sell everything to the Norwegian Government was already made in 1922, but it did not come into effect. No other sale or lease came about either, and the company was forced into financial reconstruction in 1924. It had to adopt a waiting policy regarding the coal, but the profit margin for base metals being better, it for the first time adopted an active mining strategy for zinc blende. The expedition in 1925 thereby paid for itself. This market temporarily attracted the attention of the British Metal Corporation.

The administration fee connected with the Danish Commission to allocate the claims was dear. For the first time, the company needed to consider the question of how much its vast territory was actually worth. The opportunist expansion had now reached a very real limit. There were no funds to hold the frontier, and areas needed to be relinquished. The recognition of the company's areas came a decade too late. The British Metal Corporation decided not to exercise its option, and the shareholders were no longer supportive. The last links in the company's network had been broken. It lastly needed to sell its properties and realise its assets, but it was no longer in a position to bargain. The sale to the Norwegian Government only raised a fraction of the original investments.

Conclusively, the Northern Exploration Co. had primary economic goals, which it failed to achieve before the war, when it was for all intents and purposes bankrupt. The new political climate after the war provided it with a second chance. The annexation of Spitsbergen was never an explicit aim, but the accompanying rhetoric suited it. It was granted a large capital, raising about half of it, but it failed to copy the successful formula of the Exploration Co., which also hinged on unquestionable mining expertise. The Northern Exploration Co. spent its funds on properties, assets, and transport prior to any economic minerals having been proven. It did not employ the necessary specialists, and it did not undertake trial boring in the coal and iron ore deposits. The conversion to a public company did not work out in its favour, and the chosen course of bringing a vast territory under British control, if not British sovereignty, was ill-advised. Not only did the company reach its explorative limits, it was unwilling to accept the negative reports that hardly any economic minerals could be found on its properties and to relinquish the worthless areas until the very last. A change to active mining came too late. While the financial dealings of the company are poorly understood, the root of the problem probably not only lay with an overconfidence in British capital but also with the directors' inability to correctly judge the intelligence provided by the local network and act accordingly.

8 The Scottish Spitsbergen Syndicate, Ltd. (1909-53)

8.1 Introduction

The last of the empirical chapters benefits from a large body of contemporary documents bequeathed to British archives. The Scottish Spitsbergen Syndicate was founded in 1909. After 44 years, it ceased to exist in 1953. It, too, has a history in two parts that is characterised by a private company before the First World War. Unlike the Northern Exploration Co., however, the first syndicate voluntary wound up in 1919, only to be reconstructed under the same name. A little later, this second syndicate converted to a public company. It not only survived the post-war depression and the ratification of the Spitsbergen Treaty; it also needed to respond to the expiry of its treaty properties in 1937 and the onslaught of the Second World War. This chapter reveals its motivations and approaches throughout the turbulent first half of the twentieth century. It highlights the most likely reasons for it not meeting its intended goals and its eventual sale and liquidation.

8.2 Formation of the first syndicate and chronological overview

William Speirs Bruce had conducted three scientific expeditions to Spitsbergen before he began to promote a business idea in 1907. For this purpose, he brought coal back to Scotland.[1] An Edinburgh laboratory observed that 'from the appearance of the coke and also from the general properties of the coal, it would evidently turn out to be a high-class gas coal. Its high calorific value also indicated that it should be a very good steam coal.'[2] Examiners in Glasgow concluded that 'the analysis shows that this is a steam coal of very good quality. It is very free from sulphur and ash, and is consequently a very clean coal. Its heating power is great, and it would be exceedingly well adapted for steam and navigation purposes.'[3] A comparison of Spitsbergen and Glamorgan coal also indicated that the Arctic sample measured up to the Welsh one.[4] These results undoubtedly encouraged Bruce and his co-promoter James Victor Burn Murdoch.

---

[1] The sample's exact origin is unknown.

[2] King, A. (1907) 'Letter to W. S. Bruce, 18 November' in: *Scottish Spitsbergen Syndicate* (1909), Scottish Spitsbergen Syndicate, Norwegian Polar Institute, Tromsø, p. 3.

[3] Tatlock, R. R. and Thomson (1907) 'Letter to W. S. Bruce, 11 November, in: *Scottish Spitsbergen Syndicate* (1909) p. 4.

[4] Tatlock, R. R. and Thomson (1907) 'Letter to W. S. Bruce, 25 November, in: *Scottish Spitsbergen Syndicate* (1909) p. 4.

From September 1908, Bruce and Burn Murdoch bundled essential information in support of their Arctic proposal into a circular titled *Scottish Spitsbergen Syndicate*.[5] Eleven pages concisely related that coal had for a time been known on the archipelago and was commercially exploited. The main actors were the Arctic Coal Co. and the Spitzbergen Coal & Trading Co., who chose not to advertise their enterprises. Although the seams were of considerable thickness and could be worked at adit level, the promoters elucidated three main problems for the benefit of their readers. These problems concerned title, climate, and market.

Potential investors were unambiguously informed that the only title to a claim in the no man's land was effective occupation and the unwritten law of mining camps throughout the world. It was likely, however, that all existing rights would be maintained in an international settlement. The archipelago lay well within the Arctic Circle, but the Gulf Stream rendered it freer from ice than other northern lands. Navigation was possible for four months, and mining could be carried out throughout the year. With regards to a market, 'there is in Spitsbergen [...] almost every possible geological formation from Archaean to Pleistocene, and therefore, the probability of the existence of other valuable minerals besides coal. Coal on the spot is surely the key to using the rest of the mineral products which actually exist in Spitsbergen.'[6] The local demand for coal was met by annual whaling fleets as well as tourist and other cruises. Export of coal also looked promising.[7] All labour, provisions, and building material needed to be imported, but there were no duties and taxes. Norwegian and Russian miners were employable at 4s 6d a day, although foremen were better sought in Scotland.

According to the promoters, 'Spitsbergen is worth having a grip of.'[8] They knew of two large areas, which by 1908 had not been claimed by anyone. They proposed to equip and send an expedition to occupy this territory in early 1909. The costs of this expedition would approximate £3,000 and would include ship, staff, and wintering party. In order to raise these funds, the men suggested the formation of a syndicate with a total capital of £6,000, of which shares representing the cost of the expedition would be allotted to subscribers in the usual way. Bruce was already approaching potential subscribers. His letter to the Scottish geologist and geographer Henry Moubray Cadell, for instance, read as follows,

---

[5] *Scottish Spitsbergen Syndicate* (1909), Scottish Spitsbergen Syndicate, Norwegian Polar Institute, Tromsø.
[6] *Scottish Spitsbergen Syndicate* (1909) p. 2.
[7] Export was envisaged to Archangel, where there was a direct railway to Moscow; to Norway, where the state railways were already using Spitsbergen coal while the growing iron industry would require more; to the Baltic, in light of respective distances from Cardiff and the archipelago; and to Hudson's Bay, where a new Canadian railway was under construction.
[8] *Scottish Spitsbergen Syndicate* (1909) p. 2.

> Herewith I take the liberty of sending you a circular that may be of considerable interest to you. As you are probably aware I have now made four voyages to Spitsbergen, as well as other voyages to polar regions. I am probably as well acquainted with Spitsbergen including its geology as any other British subject. I therefore trust you may peruse the circular and express the hope that you may see your way to take part in the formulation of a small syndicate.[9]

Bruce and Burn Murdoch would receive salaries as expedition leaders and an amount of fully-paid shares to recoup past outlay.

The reactions to the circular were varied. Some addressees could appreciate the value of the analysis but did not see their way to participating.[10] Paul Rottenburg[11] thanked Bruce for the attention the latter had shown to the German Society and enclosed £300 conditional on £2,000 being raised and 'the venture not being another Christmas Island as with the phosphates.'[12] For Robert Morton Paterson, a chemical manufacturer, the proposal had 'come at a time when I am "hardup", likely when the prospecting is done and a larger venture is arranged satisfactorily, I may be in a better fettle. I enclose form filled up for £50 towards preliminary experiments.'[13] Similarly, Thomas Leslie Usher at Park Brewery had 'no funds to play with.'[14] Yet, he doubted that Bruce would make any play of it. Before long, Rottenburg, Paterson, and Usher sat on the board of directors.

The first Scottish Spitsbergen Syndicate was incorporated on July 15, 1909.[15] It aimed to purchase existent rights and claim additional territory on the islands to carry out the actual business of mining as well as supplementary dealings in game, skins, and hides or as merchants, ship owners, builders, or hotel

[9] Bruce, W. S. (1908) *Letter to H. M. Cadell, 15 July.* Cadell of Grange papers, Acc. 5318, National Library of Scotland, Edinburgh.

[10] W. S. Bruce papers, box 10/122, National Museums of Scotland Library, Edinburgh.

[11] 'A son of Franz Rottenburg and born in 1846 at Danzig, where his forefathers have lived for three centuries, Dr. Rottenburg was educated at the Royal College in his native city, and, though now a naturalised Briton, still takes proper pride in the place of his birth. Following the occupation of a chemical merchant, he has filled the office of President of the Chamber of Commerce, and he also takes considerable part in the intellectual life of Glasgow. He has been President of the Civic Society, Vice-President of the Philosophical Society and of the Marine Biological Association, Chairman of the Glasgow Branch of the Royal Scottish Geographical Society, and an Associate of the Society of musicians. [...] He is Conservative in politics [...].' (Eyre-Todd, G. (1909) *Who's who in Glasgow in 1909*. Glasgow: Gowans & Gray Ltd.)

[12] W. S. Bruce papers, box 11/138, National Museums of Scotland Library, Edinburgh. Rottenburg had presumably been a stakeholder in the Christmas Islands, but 'the rush to take advantage of high world phosphate prices led the [Christmas Island Phosphate Co.] to cut corners, and a serious outbreak of beri-beri, caused by vitamin deficiency, led to almost 550 deaths from the disease, almost all [Chinese] coolies, in the period 1900-04.' (Available at: http://www.christmas.net.au/history.html (Accessed: 29 June 2011).)

[13] W. S. Bruce papers, box 11/134, National Museums of Scotland Library, Edinburgh.

[14] W. S. Bruce papers, box 11/140, National Museums of Scotland Library, Edinburgh.

[15] *Certificate of Incorporation* (1909) The Scottish Spitsbergen Syndicate Limited, BT2/7201/1, National Archives of Scotland, Edinburgh.

keepers.[16] Extensive arrangements with subsidiary companies were envisaged. The private company would have no more than 50 members; the public would not be able to subscribe; and the directors needed to hold at least two shares, although the first directors could act prior to acquiring these.[17] The original share capital of the syndicate was not £6,000 but £4,000, divided into 400 shares of £10 each.[18] The solicitor Charles Hanson Urmston was one of two original subscribers, and by August 1909, he was a director together with Rottenburg, Paterson, John Henry Davidson, and Thomas Barnby Whitson, the latter in a double role as secretary.[19] Davidson was a gentleman in the timber business. Whitson had previously backed Bruce's work in the Antarctic, where a cape is named after him. A chartered accountant for Whitson & Methuen, his firm's address in Edinburgh was the syndicate's first offices.

Bruce and Burn Murdoch transferred their rights and interests to the Scottish Spitsbergen Syndicate on July 16, 1909.[20] Although they had referred to only two unclaimed areas, five claims changed hands (Fig. 8.1). They included (1) an area south of Bell Sound; (2) Prince Charles Foreland; (3) an area between Kings Bay and Icefiord; (4) an area between Wiche Bay and Inglefield Glacier; and (5) the whole of Barents Island. Bruce and Burn Murdoch received £2,000. It was paid in 200 shares worth £10 each, a fraction of which the men allotted to other nominees. In addition, the syndicate would fit out an expedition to leave no later than July 22, 1909, over which Bruce, receiving a salary of £200, would have the fullest power and control. Among the witnesses to the agreement was the botanist and marine biologist Robert Neal Rudmose Brown, who had attended Bruce's *Scotia* expedition, after which the men remained close friends. The syndicate's seal was a large, bold X, reminiscent of St Andrew's Cross – St Andrew being the patron saint of Scotland.

The syndicate's 400 shares were taken up completely by only 30 subscribers.[21] Half of the shares had been allotted to Bruce and Burn Murdoch during the above transfer of the claims; the other half was purchased by others and would finance the expedition. From the list of subscribers, it is clear that Bruce had mainly recruited among his earlier Antarctic connections, his extensive scientific

---

[16] *Memorandum of Association* (1909) The Scottish Spitsbergen Syndicate Limited, BT2/7201/1, National Archives of Scotland, Edinburgh.

[17] *Articles of Association* (1909) The Scottish Spitsbergen Syndicate Limited, BT2/7201/1, National Archives of Scotland, Edinburgh.

[18] The high share price may have been another limiting factor to membership.

[19] *Register of directors or managers* (1909) The Scottish Spitsbergen Syndicate Limited, BT2/7201/1, National Archives of Scotland, Edinburgh.

[20] *Agreement between Bruce, Burn Murdoch & the Scottish Spitsbergen Syndicate* (1909) The Scottish Spitsbergen Syndicate Limited, BT2/7201/1, National Archives of Scotland, Edinburgh.

[21] *Return of allotments on the 16th of July 1909* (1909) The Scottish Spitsbergen Syndicate Limited, BT2/7201/1, National Archives of Scotland, Edinburgh.

network, and the Scottish elite. Those who had been of practical assistance during previous expeditions and during the founding of the syndicate also seem to have been rewarded. Others, such as Bruce's wife, were family and friends.

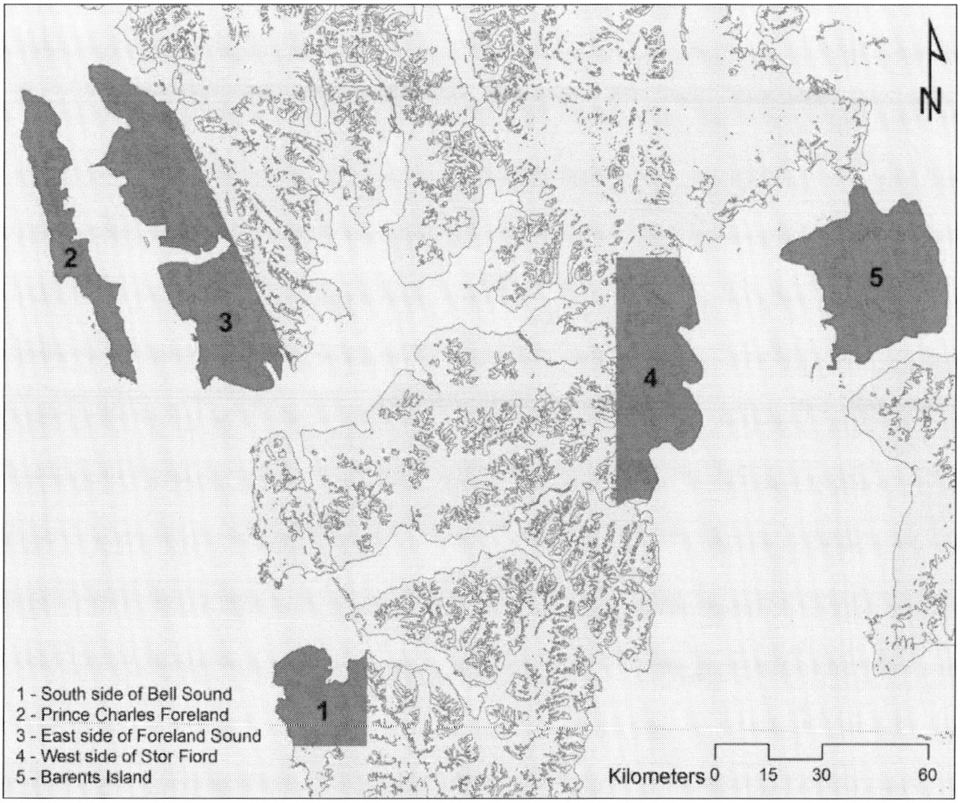

8.1 Map showing the claims transferred to the Scottish Spitsbergen Syndicate. (Source: Agreement between Bruce, Burn Murdoch & the Scottish Spitsbergen Syndicate (1909) The Scottish Spitsbergen Syndicate Limited, BT2/7201/1, National Archives of Scotland, Edinburgh; Map: F. Kruse.)

The timeline of this first Scottish Spitsbergen Syndicate (Fig. 8.2) is subject to the same political context of the First World War and its effects on British coal prices that has already found mention in previous chapters. The sinking of the *Titanic* had greater impact on the global scale than may initially be thought. On the regional scale, the timeliness of the settlement of the Spitsbergen Question, which was not achieved in three international conferences, was underlined by companies forming and expeditions occurring in quick succession. Against this backdrop, the newly founded syndicate undertook its maiden voyage and staked out further claims. A repeat journey at a much smaller scale followed in 1912, and an increase in capital in 1914 allowed for a third trip to take place. After being absent from the islands during the conflict, the Scots met the changed post-war context with the reconstruction into a second syndicate, deserving of its own timeline (Fig. 8.7).

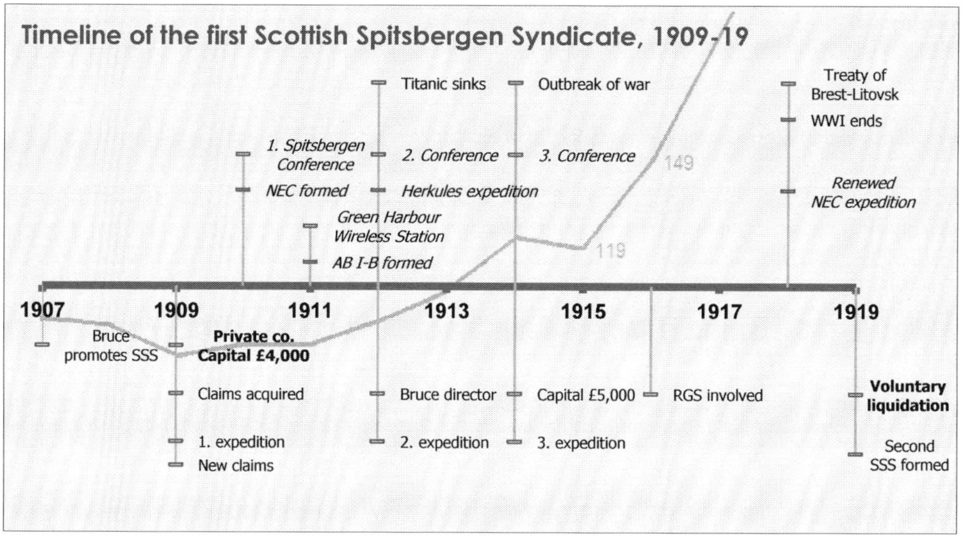

8.2 Timeline of the first Scottish Spitsbergen Syndicate, 1909-19. Events below the bar are company-specific. Above the bar, events in italics are relevant to Spitsbergen, while others are thought to have defined the historical context. The grey line indicates the indexed British coal price movements, whereby the bar denotes 1913 = 100, and the bottom edge is 0. Actual percentages have intermittently been added for clarity. (Sources for the coal price development, see Fig. 2.2 and Fig. 7.8; Chart: F. Kruse.)

8.3 The global network of the first syndicate

8.3.1 Economic actors

After the syndicate spent its funds on the 1909 expedition, making a new claim in Sassen Bay, the initial excitement of the shareholders quickly died down. There are few sources for 1910 and 1911, suggesting equally few activities. When the possibility of tourism on Spitsbergen presented itself as an alternative income to mining at the beginning of 1912, the correspondence between Bruce and Brown picked up again. During the preparations for a second voyage, Brown tried to gather information about the Spitzbergen Coal & Trading Co. This was made difficult by the fact that any enquiry would send the share prices up and entice shareholders to hang on to them, although the mine was practically worthless compared to the American one across the bay, 'and they know that.'[22] The Scots merely wanted to have some of the houses. The expedition could then concentrate on Sassen Bay, unless they could afford more staff to send into Stor Fjord.

Despite these plans, Brown was aware that all was not well with the syndicate. In February 1912, he wrote,

> Certainly if the SSS has lived so long it would be a pity to give up at this moment, when something may be made of it, and when its existence may be important to [the Arctic] Exploration Co. [...] Should Norway get Spitsbergen it is only by strong British claims that we can hope to escape heavy taxes and all sorts of expenses in our new venture. We must keep SSS alive if only on paper. We should get another secretary, for Whitson never pretended to believe in Spitsbergen.[23]

By May 1912, there had been an unexpected rupture in the North Atlantic economic context, due to which the syndicate's plans for an expedition faced new obstacles. The RMS *Titanic* had sunk on April 14, and its collision with an iceberg had literally sent ripples through the shipping world. Insurance, especially for journeys into the Arctic, was hard to come by.[24] Brown was exasperated, 'I can't help feeling that if the underwriters are panic stricken – as they certainly are – there must be other brokers, Jews perhaps, who have kept their heads on despite the "Titanic" and will insure.'[25] Personally, he could not see the usefulness of another journey to Spitsbergen and hoped sincerely that it would entail some material gain for the syndicate.

To better their chances, Whitson's secretarial duties were bestowed on the law agent James Alfred Philp, which incorporated a change of address to the offices of the syndicate's solicitors Aitken & Methuen.[26] Philp promptly circulated a letter, outlining the aim of the ensuing expedition to substantiate the syndicate's claims against possible Norwegian counterclaims and asking, if any receivers would like to subscribe.[27] Burn Murdoch and his mother Elizabeth Susan Rattray, for example, contributed £5 each.[28] Despite his replacement, Whitson continued on

---

[22] Brown, R. N. R. (1912) *Letter to W. S. Bruce, 9 January,* Robert Neal Rudmose Brown Collection, MS 101/19/28, Scott Polar Research Institute, Cambridge.
[23] Brown, R. N. R. (1912) *Letter to W. S. Bruce, 3 February,* Robert Neal Rudmose Brown Collection, MS 101/19/28, Scott Polar Research Institute, Cambridge.
[24] Brown, R. N. R. (1912) *Letter to W. S. Bruce, 19 May,* Robert Neal Rudmose Brown Collection, MS 101/19/43, Scott Polar Research Institute, Cambridge.
[25] Brown, R. N. R. (1912) *Letter to W. S. Bruce, 25 May,* Robert Neal Rudmose Brown Collection, MS 101/19/28, Scott Polar Research Institute, Cambridge.
[26] *Notice of change in the situation of the registered office* (1912) The Scottish Spitsbergen Syndicate Limited, BT2/7201/13, National Archives of Scotland, Edinburgh.
[27] Philp, J. A. (1912) *General letter, 12 July*, R. N. R. Brown S.S.S. Correspondence. MS 356/95, Scott Polar Research Institute, Cambridge.
[28] Burn Murdoch, J. V. (1912) *Letter to W. S. Bruce, 12 July,* J V Burn Murdoch Collection, MS 101/64/8, Scott Polar Research Institute, Cambridge.

the board, but the timber merchant Davidson had resigned, and the chemical manufacturer Paterson had died. Paterson's place was now taken up by Bruce.[29] Although Bruce preferred a consulting role and never broadcasted his directorship, he nevertheless retained it until the syndicate's reorganisation in 1919.

In 1914, the Scots made preparations for a third expedition to the archipelago. Bruce sent a letter to the Royal Geographical Society in London, stating that 'it is likely that I may be employed by a Syndicate, to do some prospecting work, and I understand that the Syndicate wishes to get shareholders, some of whom might take an active part in the expedition, under my leadership.'[30] Secretary Keltie replied that in order to find investors who might like to go, Bruce should contact the travel editor of *The Field*, a magazine about country and field sports and the country lifestyle.[31] A month later, Bruce informed Keltie that there would also be room for scientific investigations in Stor Fjord, trusting that the Royal Geographical Society would assist him in doing this work 'under the auspices of the Union Jack.'[32]

Prior to the expedition's departure, there was some movement on the syndicate's board and in its finances. Rottenburg, Urmston, Whitson, and Bruce were joined firstly by the solicitor Alfred Niven Gillies Aitken, to whom Bruce had already nominated three shares in 1909, and secondly by the artist William Gordon Burn Murdoch, cousin of J. V. Burn Murdoch and Bruce's lifelong friend, who had originally purchased 15 shares. An extraordinary resolution was passed in June 1914 to increase its capital from £4,000 to £5,000. The additional capital was to be issued as 100 preference shares of £10 each.[33] The list of ordinary shareholders did not change significantly; a list of preference shareholders was not recovered.[34]

8.3.2 *Political actors*

The promoters originally contacted the Foreign Office prior to forming the Scottish Spitsbergen Syndicate. On February 6, 1909, Bruce and Burn Murdoch wrote to the ministry with regards to their claims.[35] Like other British companies before them, they presumably sent a notification of the areas as well as a request for

---

[29] *Register of directors or managers* (1912) The Scottish Spitsbergen Syndicate Limited, BT2/7201/14, National Archives of Scotland, Edinburgh.
[30] Bruce, W. S. (1914) *Letter to S. Keltie, 14 April,* RGS/CB8/Bruce, RGS/IBG Archives, London.
[31] Keltie, S. (1914) *Letter to W. S. Bruce, 16 April,* RGS/CB8/Bruce, RGS/IBG Archives, London.
[32] Bruce, W. S. (1914) *Letter to S. Keltie, 15 May,* RGS/CB8/Bruce, RGS/IBG Archives, London.
[33] *Extraordinary resolution* (1914) The Scottish Spitsbergen Syndicate Limited, BT2/7201/19, National Archives of Scotland, Edinburgh.
[34] *Summary of share capital* (1914) The Scottish Spitsbergen Syndicate Limited, BT2/7201/22, National Archives of Scotland, Edinburgh.
[35] *Memorandum of Hertslet* (1910).

official recognition and protection. Within three days, they received the assurance that the British Government would only participate in an international settlement on the condition that existing British claims were safeguarded.[36] On August 12, 1909, the syndicate's solicitors renewed the communication, informing the Foreign Office that the promoters' rights had been transferred to their clients and attaching an Admiralty chart with the relevant areas marked in red.[37]

On January 27, 1910, the Scots forwarded a revised claim map based on the new acquisitions in the previous year to the Foreign Office.[38] They had found the claims to comprise valuable gypsum, coal, and oil shale. The syndicate undoubtedly wished for them to be legitimised, but on February 4, the Foreign Office saw no immediate prospects of a conference convening to discuss the issues. 'If such a conference did eventually assemble, the considerations set forth in [Whitson's] letter would receive careful attention when the instructions to the British delegate were drawn up.'[39] Meanwhile, Foreign Secretary Grey was interested to hear to what extent the Scottish claims could be said to have been occupied and worked, which could support the syndicate's title.

While the preparations for the expedition in 1912 were in progress, Bruce and Brown contemplated Britain's annexation of Spitsbergen. At the heart of it lay their desire to breathe new life into the business by ascertaining indisputable legal rights over their claims. They therefore tried to politically motivate their stakeholders and potential investors. Brown's public lectures drew attention to the historical British claim of 1614 on the authority of Conway's *No Man's Land.*[40] Although this claim was ultimately forgotten and allowed to drop, he did not believe it was ever withdrawn and should therefore still hold. When he noted that Germans had claimed Cross Bay, Magdalena Bay, and Hamburg Bay, he implied to Bruce that they might achieve British annexation yet if they could only create a German scare.[41] Similarly, he stressed,

> What we must do in order to overcome British apathy is to get the Norwegian pirates (if there are any) on our claims to insult the British flag. Then we will have the press with us. So get a Union Jack ready and we'll hoist it on our claims and let them haul it down if they dare. This I mean seriously so try to beg, borrow or steal

---

[36] At the time, the British Government was arguably the most powerful global political actor. Without British participation, any international agreement would practically be negligible, a fact that the Foreign Offices repeatedly used as diplomatic leverage.
[37] *Memorandum by Hertslet* (1910).
[38] *Memorandum by Hertslet* (1910).
[39] *Memorandum by Hertslet* (1910).
[40] Brown, R. N. R. (1912) *Letter to W. S. Bruce, 1 June.* Robert Neal Rudmose Brown Collection, MS 101/19/47, Scott Polar Research Institute, Cambridge.
[41] Brown, R. N. R. (1912) *Letter to W. S. Bruce, 2 July.* Robert Neal Rudmose Brown Collection, MS 101/19/52, Scott Polar Research Institute, Cambridge.

> one. A. M. might be able to help you to get one. For this purpose it must be the Union Jack and not the St Andrews Cross.[42]

Evidently, Brown was acutely aware of the geopolitical connotations of the British flag and how to utilise them in the syndicate's favour.

Bruce began to campaign among influential actors outside the Foreign Office. Following the 1912 expedition, he wrote to the traditionally apolitical Royal Geographical Society in London and warned,

> If Britain does not look out Russia will grab the place, have an excellent supply of Welsh coal and be within two days steaming of a modern cruiser to British coasts. This in our opinion should be checked and we intend to do all we can to induce the Government to wake up out of its usual lethargic state. [...] There is no harm in letting any specially interested person know the contents of this letter but it may be wise for a little while not to make anything I have said in it public as we may have to make important diplomatic moves within the next few weeks.[43]

On this occasion, Bruce could not rouse the secretary, who refrained from committing the Royal Geographical Society to a political course. Instead, Keltie suggested approaching Winston Churchill. So by September 16, Bruce had requested to be put in touch with the then First Lord of the Admiralty.[44] Yet, the statesman's reply was not what Brown had hoped for. He complained, 'I fail to understand either his statement that a claim to Spitsbergen would be "useless" or that it would not be "possible". I don't think much of his "expert advisors."[45] It has unfortunately not been possible to identify Churchill's advisors.

By October, Bruce had also directly addressed Alfred Harmsworth, the newspaper mogul behind the Jackson-Harmsworth expedition 1894-7 and an admirer of Bruce's polar work. In 1905, he had been raised to the peerage as Baron Northcliffe and by 1912, he owned the *Daily Mirror*, the *Observer*, *The Times*, and the *Sunday Times* among others. He was and would be an influential figure in society. Bruce again outlined the threat to Britain if Russia or Germany annexed Spitsbergen, stressing that there would be a coaling station within 53 hours steaming off the Royal Naval Dockyard at Rosyth.[46] Northcliff's reply is not known. None of his papers appear to have run the story.

---

[42] Brown, R. N. R. (1912) *Letter to W. S. Bruce, 23 July.* Robert Neal Rudmose Brown Collection, MS 101/19/54, Scott Polar Research Institute, Cambridge.
[43] Bruce, W. S. (1912) *Letter to S. Keltie, 13 September*, RGS/CB8/Bruce, RGS/IBG Archives, London.
[44] Bruce, W. S. (1912) *Letter to S. Keltie, 16 September*, RGS/CB8/Bruce, RGS/IBG Archives, London.
[45] Brown, R. N. R. (1912) *Letter to W. S. Bruce, 18 October*, Robert Neal Rudmose Brown Collection, MS 101/19/59, Scott Polar Research Institute, Cambridge.
[46] Bruce, W. S. (1912) *Letter to Baron Northcliffe, 24 October*, W. S. Bruce papers, box 10/133, National Museums of Scotland Library, Edinburgh.

It was plainly difficult to muster interest in the syndicate, let alone active support. The incoming secretary Philp sided with the despairing Brown. He offered an explanation as to why no one had come forward. He wrote, 'I fully endorse all that you say in regard to "Backward Britain" but I think the reason for its apathy lies in the fact that it has been accused of territorial grabbing, and is on that account somewhat reluctant to further aggrandise. At the same time, however, there exists a doubt as to whether Spitsbergen is not one of our oldest possessions.'[47]

In the meantime, the Governments of Britain and Norway strove to dissolve ongoing claim disputes. The Norwegians requested information on the Scottish claim between Agardh Bay and Sassen Bay in February 1913, which the Foreign Office would only provide after consultation with the syndicate.[48] In April, the firm sent a revised chart, but without an expedition having taken place, this chart was outdated within a month. The directors had become aware that the area south of Bell Sound (Fig. 8.1) had been counterclaimed. They felt that 'there might be difficult questions arising as to priority in this region, and therefore informed His Majesty's Government that they would advise the shareholders to depart from their claims to the area in question should any difficulty arise.'[49] In July, the shareholders formally abandoned and renounced their rights and title.[50]

The resolution surrendered the territory to the Northern Exploration Co. It was not entirely selfless since the syndicate entertained the idea of an amalgamation with the London-based company. It is questionable whether it would have relinquished it, had the counterclaim been Scandinavian. Subsequently, the Foreign Office requested a revised map, in which the claims of any other company should be omitted, which it could in turn use to settle the disputes with Norway.[51] It is possible that this revision entailed the areas depicted in Fig. 8.3. It is noteworthy that of the original claims, two had in fact been abandoned, giving rise to the notion that the two areas the promoters initially referred as unclaimed lay on Prince Charles Foreland and in Stor Fjord. Bell Sound had been renounced in favour of the English, but Oscar II Land may have been deserted less noisily because the region had first been surveyed by Norwegians, who were currently defending their rights vehemently against the encroachment of Northern Exploration Co. Despite

[47] Philp, J. A. (1913) *Letter to R. N. R. Brown, 7 May*. R. N. R. Brown S.S.S. Correspondence. MS 356/95, Scott Polar Research Institute, Cambridge.
[48] Grey, E. (1913) *Letter to Vogt, 8 May,* Norwegian Foreign Department Box 5373, National Archives of Norway, Oslo.
[49] *Memorandum by Hertslet* (1913).
[50] *Extraordinary resolution* (1913) The Scottish Spitsbergen Syndicate Limited, BT2 7201/16, National Archives of Scotland, Edinburgh.
[51] Cardonnel Findlay, M. de (1913) *Letter to Ihlen, 12 August*, Norwegian Foreign Department Box 5373, National Archives of Norway, Oslo.

the syndicate's compliance, the Foreign Office was unable to settle the disputes before the outbreak of war in 1914.

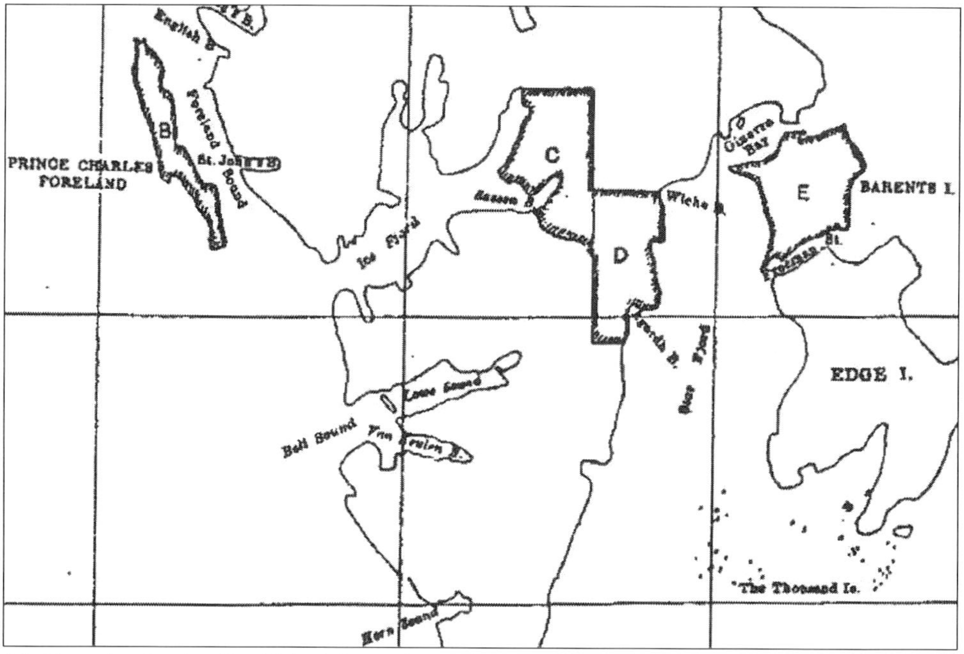

8.3 Map showing the Scottish claims prior to the First World War. (Source: Extracts from H. M. Cadell's chronicle, Vol. 4 1910-1925, Cadell of Grange papers, Acc. 5381, National Library of Scotland, Edinburgh.)

8.3.3 *Other allies*

The possibility of tourism on Spitsbergen presented itself at the beginning of 1912. A meeting between the syndicate's promoters, Joseph Foster Stackhouse, and an unidentified J. Reid resumed at the Conservative Club in Edinburgh. Afterwards Burn Murdoch reported to Brown.[52] Reid had been one of the few people he had been able to interest in Spitsbergen. Three avenues now presented themselves, which were firstly, the amalgamation of all British claims and an alliance with the Americans; secondly, a tourist company along the lines proven by Stackhouse; and thirdly, a hotel company. An amalgamation seemed out of the question as there was no money in it until the claims could be secured. Stackhouse had led the Jan

[52] Burn Murdoch, J. V. (1912) *Letter to R. N. R. Brown, 8 January*, J. V. Burn Murdoch Collection, MS 101/64/4, Scott Polar Research Institute, Cambridge.

Mayen Expedition in 1911 to make meteorological observations.[53] Besides scientists, it comprised two colour photographers, a mountain climber, a cinematograph operator, and an artist.[54] The implication was that they attended the expedition at their leisure and thus indicated the possibility of tourism not based on indiscriminately shooting animals.[55] Stackhouse charged £37 and made a small loss; he could have charged more at a profit.

Stackhouse pointed out that the business needed little capital as the fares would be paid up front. A hotel would not need a great outlay either. Reid offered to handle the incorporation at no extra cost if the men contributed towards stamp duty and the like. The plan emerged that the Arctic Explorations Ltd. would have a capital of £2,000, the majority of which would be issued as preference shares for cash. The rest would be ordinary shares issued for services rendered. Stackhouse would get the lion's share of the latter to distribute as he saw fit while he acted as general manager of the tourist business. He was anxious to publish a timetable and a flyer, so he proposed to start in early July 1912, dump a pre-fabricated building and some builders on Spitsbergen, continue to Jan Mayen with a first group of tourists, and return to England. A second group could then be taken to the newly erected hotel on Spitsbergen. An alternative to buildings from England would be to purchase a house from the Spitzbergen Coal & Trading Co., which was practically bankrupt, its shares having been offered to Brown at 1s each. The Scots, therefore, 'might buy their whole outfit at the mouth of Advent Bay. They have a huge store and numbers of houses calculated to stand the Arctic winter.'[56] Although Arctic Explorations was never founded, at least not in Britain, there was talk of it in October, when Stackhouse said he had been promised £2,900 and expected £5,000 before coming up to Edinburgh.[57] By March 1913, the stakeholders were still waiting for this to materialise.

---

[53] Commander J. F. Stackhouse, US Navy retired, F.R.G.S, F.R.S.S., would later plan an expedition to the Antarctic but found himself a saloon passenger on the RMS *Lusitania* when it was sunk by a German submarine in May 1915. He was among the 1,198 casualties. He is said to have explained why the ship would not be torpedoed when it happened.

[54] Russel, W. S. C. (1911) 'Jan Mayen Expedition of 1911', *Bulletin of the American Geographical Society,* 43 (12), pp. 881-90.

[55] The conservation of threatened wildlife is a recurring theme throughout this chapter but has not been looked into in great detail. Bruce was probably an influential member of the Royal Zoological Society of Edinburgh (est. 1909) since he was also a co-founder and the first vice-president of Edinburgh Zoo (est. 1913), which he occasionally supplied with specimens. (The Gazetteer for Scotland (2013) William Speirs Bruce 1867-1921. Available at: http://www.scottish-places.info/people/famousfirst109.html (Accessed: 18 July 2013).)

[56] Burn Murdoch, J. V. (1912) *Letter to R. N. R. Brown, 8 January*, J. V. Burn Murdoch Collection, MS 101/64/4, Scott Polar Research Institute, Cambridge.

[57] Burn Murdoch, J. V. (1912) *Letter to W. S. Bruce, 26 October,* J. V. Burn Murdoch Collection, MS 101/64/9, Scott Polar Research Institute, Cambridge.

Only in March 1913 did the syndicate become aware of the Northern Exploration Co. Seemingly through hear-say, the Scots had learned of 'a company having been registered in London with a capital of a million.'[58] Burn Murdoch found this difficult to believe, and Bruce asked an acquaintance to make enquiries.[59] The Spitzbergen Mining & Exploration Syndicate had been gazetted in February 1911, and Mansfield had controversially sold the claims to the Northern Exploration Co.[60] Bruce's acquaintance could not verify the company's capital. Meanwhile, the Scots contemplated dispatching a small expedition in 1913, if only to 'avoid shipwreck and extinction.'[61] Ideally, they would identify a capitalist to take over either the syndicate or at least the gypsum business. Alternatively, they should merge with a company that had money. In this respect, the Arctic Coal Co. had been named, and if the capital of the Northern Exploration Co. could be confirmed, it was also a potential candidate.

8.3.4 Competitors

While allies were not easily identified, let alone recruited into the network, the Scottish Spitsbergen Syndicate was not short of competitors. In this regard, the main achievement of the expedition in 1912 was the intelligence gathered. Bruce and Brown 'were able to collect a great deal of valuable information about the present state of affairs in Spitsbergen by personal contact with miners, prospectors, sailors and others. This information could not have been obtained except by a visit to Spitsbergen.'[62] Consequently, 12 of their 14-page report concerned this information. Several nationalities were active, and trespassing was rife. Although the majority of trespasses evoked no comment, the Scots felt that it was unforgivable where buildings had been erected and stores deposited. If uncurbed, they predicted, '[...] that this jumping of claims as well as the overlapping of ill defined territories is bound within a very few years to cause serious conflicts, if not actual warfare, in Spitsbergen.'[63] Such a statement undoubtedly arrested the attention of the board.

---

[58] Burn Murdoch, J. V. (1913) *Letter to R. N. R. Brown, 9 March,* J. V. Burn Murdoch Collection, MS 356/75, Scott Polar Research Institute, Cambridge.
[59] See Chapter 7.
[60] Ross, J. (1913) *Letter to W. S. Bruce, 4 June*, Scottish Spitsbergen Syndicate (S.S.S.) Norwegian Polar Institute, Tromsø.
[61] Burn Murdoch, J. V. (1913) *Letter to R. N. R. Brown, 9 March,* J. V. Burn Murdoch Collection, MS 356/75, Scott Polar Research Institute, Cambridge.
[62] *Report to the directors of the Scottish Spitsbergen Syndicate on the visit of the Syndicate claim in August, 1912* (1912) Scottish Spitsbergen Syndicate, Norwegian Polar Institute, Tromsø, p. 3.
[63] *Report to the directors* (1912) p. 4

'The Norwegians' seemed to be the Scots' nemesis. Their claims had only grown nominally in recent years and included the shore-based stations of two whaling companies. However, the wireless station and post office in Green Harbour, built by the Norwegian Government in 1911, increased their stakes in the archipelago significantly. It kept a staff of six, although business was slow. The initial cost and annual upkeep could not possibly be in proportion to its income. Norwegian claims commonly overlapped with others, and the Scots regarded a trespass at Advent City as 'a definite act of aggression and theft. [If] allowed to pass unchallenged, [it] will constitute a precedent which will lead to no property being safe in Spitsbergen.'[64] In addition, Norwegian hunters, through the use of poison, had depleted animal life on the islands. Prospectors could no longer rely on this resource during their expeditions, nor did those hunters shy away from caches of food and other property that did not belong to them.

The Swedish AB Isfjorden-Belsund had been engaged in coal mining in Braganza Bay for some time, but the Scots could not ascertain if that particular claim overlapped with other British interests.[65] In 1911, the company had erected a claim board on the north coast of Temple Bay and Sassen Bay, which was removed by the syndicate's expedition in 1912. In communications with the Swedish mining engineer, the Scots learned that the Swedes had taken and tested 100 tons of gypsum from the claim. Although of great quality, they did not intend to work the deposit, coal being more attractive. Hence, the Scots got the impression that the Swedes had abandoned the area by 1912.

With regards to Russia, Rusanov's *Herkules* expedition in 1912 'shows the eagerness of the Russian Government at least to participate in the division of Spitsbergen if not acquire a useful coaling station on the Atlantic [...]. The advantage to Russia of a naval coaling station in Spitsbergen is too obvious to require comment. [...] No less obvious is the danger to the balance of power in Europe in such an acquisition on the part of Russia.'[66] On the other hand, and despite the idea of creating a German scare, Bruce and Brown did not know if the German activity they observed was politically motivated.

When war erupted, the Foreign Office had been assessing the Scottish claims. In its memorandum to the Norwegian Foreign Department, it firstly dismissed the Stavanger Co.'s counterclaim on Prince Charles Foreland (14 in Fig. 7.2) by referring to Bruce's scientific expeditions to the island in 1906, 1907, and 1909, during which he flew the Union Jack and St Andrew's Cross.[67] Secondly, it

---

[64] *Report to the directors* (1912) p. 7.

[65] The Swedes had claimed four areas since 1910: Braganza Bay, Bünsow Land, Erdmannflya, and Pyramiden. The Bünsow Land claim, in fact, overlapped with a Scottish claim in the same area.

[66] *Report to the directors* (1912) p. 10.

[67] *Memorandum* (1914) Norwegian Foreign Department Box 5174, National Archives of Norway, Oslo.

rejected Hoel's claim on the Foreland from 1909 (3 on the map) on the same grounds as well as place-name evidence.[68] Hoel had erected a claim board, but the British were critical of its location and material. It was found behind a moraine and apparently consisted of the lid of an old packing case. Thirdly, the Norwegian Exploration Co. of Kristiania had taken possession of Sassen Bay in July 1911 and of Klaas Billen Bay and Temple Bay in August 1911 (7 and 8 on the map). The Sassen Bay claim infringed on the Scottish territory originally staked out in 1909 and revisited in 1912. That year, Bruce and Brown discovered a claim board by J. Berner Sverdrup that post-dated the original Scottish occupation. It was also only made of a packaging case.

8.4 The local network of the first syndicate

8.4.1 Claims and natural resources

The funds for the 1909 expedition were limited to £2,000. A cost estimate therefore considered the shipping requirements, a largely scientific staff led by Bruce, and provisions for three weeks, which amounted to £1,850.[69] This left some money to cover unforeseen expenses. Prior to its departure, the Scottish Arctic Expedition, as it was occasionally called, attracted a little media interest.[70] Some papers reported that it had been entirely fitted out in Edinburgh and Leith, placing emphasis on the fact that it was a Scottish undertaking, with material assistance from the Admiralty, the Royal Observatory of Scotland, the National Physical Laboratory, and the Scottish Oceanographical Laboratory. Although this sounded impressive, the assistance may have ranged from the permission to proceed via shipping supplies to scientific instruments. The Scottish Oceanographical Laboratory was Bruce's creation and vocation. Nevertheless, these institutions and their stakeholders endorsed the expedition.

On July 19, the expedition left Scotland to continue the topographical survey of Prince Charles Foreland and adjacent regions. There was no public mention of commercial interests. The ship was the SY *Conqueror*, a steam trawler reconstructed as a yacht within a week. She anchored in Recherche Bay on July

[68] 'As a striking incident in the propriety of the British claim, it may be mentioned that Mount Jessie which the Norwegians use as the southern land mark of this claim, was actually named by Dr [*sic*] Bruce after his wife [...].' *Memorandum* (1914).

[69] *Accounts for expedition* (1909) W. S. Bruce papers, box 11/141, National Museums of Scotland Library, Edinburgh

[70] *The Times* (1909) 'The Scottish Arctic Expedition, 21 July, p. 14; *British Medical Journal* (1909) 'Scottish Arctic Expedition', 24 July, p. 229; *Mining Journal* (1909) 'Spitzbergen', 31 July, p. 144.

31, where at least two parties went out in search of coal, but none was found.[71] On August 1, they arrived on the Foreland. Four days later, they observed one of Gunnar Isachsen's parties surveying the west of Spitsbergen.[72] A trip to Kings Bay on August 8 was presumably the beginning of an excursion in order to renew the syndicate's claims and occupy new territory. By August 19, Brown was back on the Foreland, where a claim board had been erected by Hoel and Isachsen. The Scots pulled it down and replaced it with a board of their own.[73]

In addition to the properties named in the above agreement, the syndicate now occupied a new claim between Agardh Bay and Sassen Bay. It was bound by 78° 44' north, to the south and west by the Sassendal River, Sassen Bay, and Klaas Billen Bay, and by 18° east. During the occupation, '[Bruce and Hannay] crossed over from the main land from Sassen Bay to Stor Fjord past the head waters of Sassendal River and over Rabot and Hayes Glaciers. They also made further excursions to the north of their main camp lying about one mile west of Rabot Glacier, and also to the south-east of the Ivory Gate valley, through the valley and on its northern hills.'[74] Grey later notified his Norwegian counterpart of the careful steps the Scots had taken to prove their occupation,

> [...] posts have been erected on the various areas to show the proprietary right of the Syndicate and that numerous expeditions have been made to the districts claimed – both of which facts serve to prove that the title is being kept alive. Small depots were left at the Rabot Glacier camp, and also about half way between the camp and the mouth of the Sassendal River. A large well-made post and board, painted and varnished, was erected about half a mile inland from the Temple Bay shore of Sassendal. A record was left also at the Rabot Glacier camp and near the shore of Stor Fjord.[75]

[71] Brown, R. N. R. (1909) *Journal kept by Robert Neal Rudmose Brown, Spitsbergen 28 July to 14 September 1909,* Robert Neal Rudmose Brown Collection, MS 356/2/1, Scott Polar Research Institute, Cambridge.

[72] Gunnerius Ingvald Isachsen (1868-1939) was a Norwegian military officer and polar scientist. His first Arctic voyage took place in 1898. From 1906 to 1910, he led topographic and bathymetric surveys on Spitsbergen. Some were paid for by the Prince of Monaco, who decreed that Bruce should map the Foreland and Isachsen the neighbouring mainland. Subsequently, a quarrel erupted, and Bruce accused Isachsen of introducing errors into his maps and systematically replacing British with Norwegian nomenclature (RGS/CB8/Isachsen, RGS/IBG Archives, London). From 1909, Isachsen's expeditions were government-funded, and he founded Norway's systematic research on Spitsbergen. Later, he was Norway's technical delegate in the Spitsbergen Question.

[73] Evidence for this ongoing dispute can be found in two sequential photographs. The first shows a well-built claim board erected by Bruce on the Foreland in August 1909. More snow having melted in the background of the second shows that the season had progressed by the time the original text had been poorly scraped off and replaced with a Norwegian rendition, signed A. Hoel, dated simply 1909. (Photo Library np003894 & 00958, Norwegian Polar Institute, Tromsø.)

[74] Grey, E. (1913) *Letter to Vogt, 8 May*, Norwegian Foreign Department Box 5373, National Archives of Norway, Oslo.

[75] Grey, E. (1913) *Letter to Vogt, 8 May*, Norwegian Foreign Department Box 5373, National Archives of Norway, Oslo.

At the end of summer, Norwegian newspapers reported that the expedition had returned from Spitsbergen.[76] The *Conqueror* arrived in Tromsø on September 7. The Scots were said to have formally annexed the Foreland, where large deposits of coal and also petroleum had been discovered, and they had been able to explore Icefiord, Sassen Bay, and Mohn Bay. A hint of petroleum had actually been found in Stor Fjord, but for the time being, its exact location was less important to potential investors than the repercussions of its presence.

Two summers elapsed before a follow-up expedition. It occurred late in the season and did not last as long. Leaving Edinburgh on July 27, 1912, it only consisted of Bruce and Brown.[77] They travelled on commercial ships to Norway. Their transfer from Tromsø was interrupted by a strike at the American mine upsetting the usual service of coal boats. Instead, they made use of the Norwegian mail boat and reached Green Harbour on August 11, where they visited the coal mines. On the following day, the mail boat made a stop in Advent Bay, where the Scots inspected the development of the Arctic Coal Co. Thereafter, they were ferried to Bjona Haven in Sassen Bay. A camp site was chosen, and two boats, equipment, and stores were brought ashore (Fig. 8.4).

8.4 Bruce (second from right) and Brown (right) arriving at Bjona Haven in Sassen Bay in 1912. (Source: Glasgow Digital Library. Available at http://sites.scran.ac.uk/voyage-of-the-scotia/scotia.vs043-010.htm. (Accessed: 18 July 2013).)

Within a week, Bruce and Brown examined their claims in Bünsow Land from Gips Valley to the Post Glacier. Having collected samples on their last visit in 1909, they did not repeat the exercise. Although they found their claim board to be intact,

[76] *Memorandum by Hertslet* (1910).
[77] *Report to the directors* (1912).

adding a note as to this year's call, they came across a Swedish and a Norwegian marker, which were later and therefore removed.[78] The men made a detailed topographic survey of Bjona Haven and charted its anchorage with a view to letting or selling the area to their hotel company. In Bruce's opinion, 'Bjona Haven affords the most admirable site for this purpose in the whole of the Fjord on account of the beauty and varied nature of the view, its admirable harbour, the nature of the ground and its accessibility to other interesting places.'[79] On the cape, they built a large flagpole, hoisted the Union Jack, and left it flying. These measures as well as leaving a cache of stores and two boats assured the syndicate's continued effective occupation. The encountered claim-jumping only strengthened their belief that their claim was worth having.

On August 20, they expected to be picked up by a Norwegian tourist boat. When it failed to arrive, the men launched one of their own boats in an attempt to reach Advent Bay. At Windy Point in southern Sassen Bay, the onward journey was thwarted by strong winds. Continuing on foot, they marched 60 miles over the terrain of the Spitzbergen Coal & Trading Co., observing the derelict Advent City and reaching Longyear Valley on August 26. They were able to charter a motor sloop to gather their equipment at Bjona Haven. Aboard the American *W. D. Munroe,* they returned to Tromsø on August 28 and were back in Newcastle on September 6.

Another year passed before the syndicate launched its third expedition. Although the Scottish Oceanographic Laboratory, the Prince of Monaco, the Royal Geographical Society, and the Admiralty lent their support[80], the expedition would in all probability not have taken place, had not the syndicate increased its capital by £1,000 in June 1914. As before, the estimated costs concentrated on transportation, scientific staff, and provisions, and totalled £950.[81] It once more succeeded to stay within its limited budget, the most expensive item being the return expenses of five men travelling between Edinburgh and Tromsø.

In July and August 1914, *The Times* briefly reported that a Scottish voyage was destined for Spitsbergen.[82] Bruce, J. V. Burn Murdoch, and the geologist R. M. Craig had travelled on a Norwegian steamer to Bergen, where they met with the young zoologist John H. Koeppern. Not only did Bruce not receive a salary as

[78] The Swedish claim board had been placed by the A/B Isfjorden-Belsund in 1911. The Swedes had investigated the gypsum deposits but no longer intended to mine these. The Norwegian claim board was placed by Sverdrup of Kristiania as recently as July 25, 1912.
[79] *Report to the directors* (1912).
[80] Bruce, W. S. (1914) *Report of an expedition to Spitsbergen, July to September, 1914*, Scottish Spitsbergen Syndicate, Norwegian Polar Institute, Tromsø.
[81] *Stor Fjord Expedition 1914* (1914) RGS/CB8/Bruce, RGS/IBG Archives, London.
[82] *The Times* (1914) 'Unknown islands. Dr. Bruce's expedition to Spitzbergen', 13 July, p. 6; *The Times* (1914) 'Scottish Spitsbergen expedition', 1 August, p. 5.

expedition leader; his expenses were covered by Koeppern, who paid £75 each. Bruce hoped to recover the cost later by the sale of specimens.[83] *The Times* related that the expedition would be fitted out in Tromsø. Bruce had taken the engines for two motor boats, which his party, assisted by four others, intended to use to make extensive soundings and map uncharted islands. Without the mention of any commercial reasons for going, the survey alone constituted a worthy cause in Britain, a maritime nation priding itself on its state-of-the-art Admiralty charts. Bruce chartered the sailing ship *Pelikane* in order to proceed to Stor Fjord, where he would drop a geological party led by Craig. Bruce himself planned to take the *Pelikane* to Green Harbour, Ebeltoft Haven, and Prince Charles Foreland. In addition to landing her cargo, he would presumably check on Norwegian and German activities as well as the syndicate's claims before undertaking further hydrographical work in Stor Fjord. The newspaper stated that ice conditions on the east coast had previously been bad, but a clear sea had recently been reported. This soon turned out to be incorrect.

The geological work in Stor Fjord was in fact intended to investigate the oil shale deposits, which had first been noted by Bruce on Barents Island in 1898 and again by Bruce and Hannay in Agardh Bay in 1909.[84] Stor Fjord, however, was completely blocked with heavy pack ice as far south as the South Cape. The Scots decided to discharge a load of paraffin in Green Harbour instead. On August 12, they reached the Norwegian wireless station and got news of the outbreak of war. This put Bruce in a precarious situation.[85] While contemplating whether to sail back to Norway at the risk of being captured in Norwegian waters or to sail straight for Scotland, Bruce and Burn Murdoch were nonetheless able to inspect the property of the Northern Exploration Co. in Kings Bay. Marble Island lay deserted. The Scots doubted, for reasons unknown, that the marble was of sufficient value to be economically useful. During a further delay caused by bad weather and bad navigation, the expedition was also able to further examine Prince Charles Foreland, which revealed indications of iron ore along the east coast.

[83] W. S. Bruce papers, box 10/130, National Museums of Scotland Library, Edinburgh.
[84] Bruce, W. S. (1914) *Report of an expedition to Spitsbergen, July to September, 1914*, Scottish Spitsbergen Syndicate, Norwegian Polar Institute, Tromsø.
[85] He later reminisced, 'One thing was certain, that to go back to [Stor Fjord], as I had intended, was out of the question, and what was I to do with twenty tons of coal which the owner of the ship had to deliver at the German meteorological station in Cross Bay. [My old friend Captain Staxrud of the Norwegian army] went through the ship's contract with me and was firmly of the opinion, and I think he was right, that according to the contract, in order to have the ship entirely at my disposal I was compelled to go to Cross Bay and deliver the cargo. I decided to do this because I wanted to be in absolute command of the ship and give my skipper orders which he was compelled, according to our contract to obey [...].' (Bruce (1914) pp. 3-4.)

8.4.2 Employees and local allies

The Scottish Spitsbergen Syndicate had few employees and fewer local allies. The accounts of the 1909 expedition initially listed seven staff, but additional sources raised this number (Appendix 2). Of these, the geologist Alastair C. B. Geddes was the son of Patrick Geddes, the professor who had a profound influence on Bruce's university days. The surveyor John Mathieson had been obtained from the Ordnance Survey, and Ernest Alexander Miller was a meteorologist, who had accompanied Bruce on Prince Charles Foreland in 1906. Angus MacEwen Peach was the son of the eminent geologist Benjamin Peach. Together, they gave the voyage a scientific rather than commercial character. A special role was reserved for Gilbert Kerr. The piper had already been on Bruce's *Scotia* expedition.[86] His skills were also employed on Spitsbergen (Fig. 8.5).

8.5 Gilbert Kerr playing the bagpipes on land claimed by the Scottish Spitsbergen Syndicate. (Source: Images for All, Royal Scottish Geographical Society, Perth.)

While Bruce and Brown conducted their fieldwork in 1912 largely alone, the expedition in 1914 foresaw four staff in addition to four assistants. Bruce had hoped to again bring Mathieson, but he must have been unavailable. Burn Murdoch therefore took his place. While the geologist Robert M. Craig was inconsequential after the voyage, the zoologist Koeppern, whose father was a friend of Rottenburg and whose uncle was the German minister in Kristiania[87],

[86] A popular photograph shows Kerr playing his bagpipes in snowy surroundings next to a penguin.
[87] W. S. Bruce papers, box 10/130, National Museums of Scotland Library, Edinburgh.

would change his name to Kenneth in 1918 and continue his association with the syndicate into the 1950s.

Captain Staxrud's involvement in translating the contract with the captain of the *Pelikane* hinted at some local assistance, but during the three pre-war expeditions, Bruce's parties barely reported on active alliances. It is unsurprising, therefore, that the Northern Exploration Co.'s attempt to bring several companies together to discuss the Spitsbergen Question was met with Store Norske's reply never to have heard of a Scottish syndicate on the islands.[88]

8.4.3 Manifestations and products

Before the war, the Scottish Spitsbergen Syndicate used few installations with which to express the effective occupation of their claims, thus protecting them against competitors. So far, the careful positioning of well-built claim boards, the construction of flag poles, and the hoisting of flags have found mention. Examples of the syndicate's claim boards survive in the collections of the Svalbard Museum in Longyearbyen (Fig. 8.6).

8.6 A well-made Scottish claim board dated to 1909. (Source: Førisdal, L. and Lien, H., No man's land: the Scottish Spitsbergen Syndicate Ltd – SSS. Available at: http://www.svalbardmuseum.no/skilt/index.php?skiltselskap=18&lang= (Accessed: 1 July 2011).)

The small-scale endeavour used only the most basic scientific methods, seemingly did not build any huts on Spitsbergen before the war, and did not produce an output measurable in tons. The principal outcome of the expedition in 1909 was the assertion of existing rights and the acquisition of new territory in Sassen Bay and Agardh Bay. This looked imposing on a map, but it did not amount to any financial gain for the shareholders. Besides the reported presence of mineral occurrences,

[88] See Chapter 7.

the expedition also appears to have brought back plenty of eider down.[89] The Scots requested a quote from Davis Feather Mills in London, presumably for cleaning the down, but the firm replied it was too busy to provide one.[90] As a result, the sample given to the Royal Blind Asylum was full of bits of nest, and the asylum was not interested in it. Besides that, there was little demand and too much stock.[91] After these set-backs, the syndicate did not appear to have tried the commercial exploitation of eider down again.

The staff in 1912 had been greatly reduced to only two men, who admitted that not the survey work at Bjona Haven but the intelligence gathered was the journey's greatest achievement as well as its justification. The majority of their report to the directors concerned this information. For the time being, Bruce and Brown based their geopolitical constructs and company propaganda on this. Similarly, Bruce reported that the 1914 expedition, during which small motor boats had first been used for speedier survey work, had been exciting from a political point of view but unsuccessful economically.[92] Despite these meagre results, he was convinced that the oil shale deposits warranted systematic investigation, particularly in view of the increased use of oil as fuel. This, however, would have to wait until after the war.

8.5 The First World War

Britain was engaged in the war effort, and the Scottish Spitsbergen Syndicate was no exception. It was quiet around the group until the Northern Exploration Co. requested a favour. The English had previously sent a letter to the syndicate, noting that their claims were marked incorrectly on one of its maps. They were eager to see this rectified because 'the more we can show as claimed by Britain, the stronger the case for British annexation.'[93] In October 1916, Maples again contacted Brown, thinking the latter shared the belief that shipping was possible all year round. To convince the sceptics, Maples asked for his professional opinion.[94]

---

[89] During breeding, female eider ducks pluck their down feathers to insulate their eggs. When the nests are abandoned, the down can be collected safely and sustainably. Because it is soft, light, and warm, it is used to stuff duvets and pillows. Featherbeds have a long history, for instance, in Germany, but eiderdowns, as the quilts were known, only began to substitute the heavy woollen blankets of Britain in Victorian times. This substitution was slow and may not have been sufficiently advanced to constitute a stable market in 1909. Today, eiderdown continues to be rare and costly.

[90] W. S. Bruce papers, box 10/122, National Museums of Scotland Library, Edinburgh.

[91] W. S. Bruce papers, box 11/138, National Museums of Scotland Library, Edinburgh.

[92] Bruce (1914).

[93] Maples, J. R. (1915) *Letter to R. N. R. Brown, 26 July*, Robert Neal Rudmose Brown Collection, MS 356/30/2, Scott Polar Research Institute, Cambridge.

[94] Maples, J. R. (1916) *Letter to R. N. R. Brown, 28 October*, Robert Neal Rudmose Brown Collection, MS 356/30/3, Scott Polar Research Institute, Cambridge.

Brown took a fortnight to respond, 'I do not feel justified in giving any information about the navigation of Spitsbergen waters when I don't know for what purpose it is wanted and into whose hands it may fall.'[95] The Northern Exploration Co. had in any case already quoted him in 1914.[96] Its current appeal may have been rooted in the desire to cooperate more closely with the Scots. The Scots, in turn, renewed their endeavour to summon political support. Perhaps they feared the Northern Exploration Co. could promote a flawed image of Spitsbergen and damage the syndicate's interests if they did not act.

On November 6, 1916, Bruce and J. V. Burn Murdoch visited Arthur Robert Hinks, the new secretary of the Royal Geographical Society. On the following day, they put their concerns regarding Spitsbergen in writing. The document listed Bruce's reasons for the British annexation of the archipelago:

1. British Companies, the chief of which are the Scottish Spitsbergen Syndicate Ltd., the Spitsbergen Mining and Exploration Syndicate Ltd., and the Spitsbergen Coal & Trading Coy. Ltd. and the Northern Exploration Coy. Ltd., claim about 4000 square miles of territory in Spitsbergen.
2. These claims along with the ground occupied by the American Coal Coy. at Advent Bay form a block of territory that could be administered by one body if Spitsbergen is to be divided up.
3. These claims are not immune from aggression, especially on the part of Norwegian and Swedish Companies and subjects. In several cases cachés and stores of provisions have been robbed, imperilling life, and valuable property wilfully destroyed.
4. These claims contain deposits of coal, oil shale, iron, gypsum, marble and possibly gold. On further prospecting other minerals will certainly be found.
5. These claims are part of a country which has been rich in game, viz: fur-bearing animals – Arctic and blue foxes, and reindeer on the land; seals, walrus, white whales, eider ducks, etc. on the coast.
6. These animals have been almost exterminated by hunters, (mostly Norwegians) who, besides using the gun, knife and trap, indulge in wholesale poisoning of animals.
7. British Companies claim adequate protection against Swedish, Norwegian and other aggressors both as regards mineral deposits and game in their territory claimed.
8. A close period for game throughout Spitsbergen for a term of years is very desirable; indeed, it is essential. Otherwise the game will be exterminated, chiefly by Norwegian so-called hunters and by careless tourists.
9. Next to Britons, American citizens have the most important claims on Spitsbergen, working as they do the priceless tertiary coal measures of Advent Bay. This summer the Arctic Coal Coy. of Boston, U.S.A. is said to have exported 45,000

[95] Brown, R. N. R. (1916) *Letter to J. R. Maples, 12 December*, Robert Neal Rudmose Brown Collection, MS 356/30/4, Scott Polar Research Institute, Cambridge.
[96] *Some particulars respecting the Co.'s properties in Spitsbergen with reports and extracts relating thereto* (1914). Northern Exploration Company (N.E.C.), Norwegian Polar Institute, Tromsø, p. 65.

tons of steam coal of high calorific value, almost, if not quite, equal to South Wales coal.

10. Agents on behalf of the Russian Government are said (on good authority) in the summer of 1912 to have attempted to purchase from the Americans part of their coal measures. Such purchases by Russians and Norwegians are now reported as carried through.
11. The American Coal mine is 1,432 nautical miles from Rosyth, a distance that would be covered by ships steaming 20 knots in less than three days, or by a cruiser doing 27 knots, like H.M.S. "New Zealand" in about fifty-three hours.
12. Germany has made a claim of land for a Zeppelin Airship Station and a harbour for their tourist steamers!
13. Germany has at the present time a high altitude Air Station and has also at the present time a Wireless Station there!
14. The west coast of Spitsbergen is influenced by the current of relatively warm Atlantic water that keeps the west coast of Norway free of ice all winter. It is therefore much less blocked with ice than the other coasts of Spitsbergen, and the coal mines there are worked all winter: the sea is open in summer, and even in the depth of winter ships could be brought in there with the help of an ice-breaker like the Russian "Ermack", and with the help of two or three leading lights.
15. Britain had already annexed Spitsbergen in 1615, and this claim has not been repudiated.[97]

Notably, several details of Bruce's reasons were either outdated or simply wrong.[98] In addition, most of the reasons given were actually only observations and did not answer the question *why*, in view of a certain circumstance, Britain should annex the islands. The Arctic being rich in blue foxes, for instance, did not justify territorial occupation. Nor was Germany supposedly having a wireless station in itself a rational cause. At the time, however, it was perceivably self-evident that Britain alone could ensure the protection of endangered Arctic wildlife and spoil German plans of world domination. Hinks now brought the matter before the Council of the Royal Geographical Society.

Meanwhile, the Scottish Spitsbergen Syndicate and the Northern Exploration Co. jointly addressed the Foreign Secretary. This was perhaps ill-timed because Grey would be succeeded by Arthur Balfour, formerly Prime Minister and most recently First Lord of the Admiralty, within a month. Even so, Bruce and Davis made a plea for British annexation, arguing that the interests of Norway, Russia, and now Germany might seriously jeopardise Britain.[99] Firstly, the Arctic Coal Co.

[97] Bruce, W. S. (1916) *Letter to Hinks, 7 November*, RGS/CB8/Spitsbergen, RGS/IBG Archives, London.
[98] The Spitzbergen Mining & Exploration Syndicate, for instance, had been dissolved in 1911, while the Spitzbergen Coal & Trading Co. had been inactive since 1908 and was about to be liquidated. Interestingly, Bruce corrected their spelling despite the fact that this was not how they had been registered. Furthermore, the Arctic Coal Co. had been sold to the Norwegian Store Norske; the German scientists had abandoned their weather station at the outbreak of war; and the first British annexation occurred in 1614 not 1615.
[99] Bruce, W. S. and Davis, F. L. (1916) *Letter to Grey, 18 November*, William Speirs Bruce Collection MS 101/103, Scott Polar Research Institute, Cambridge.

and the Spitzbergen Coal & Trading Co. had been bought by Norwegian firms (an adjustment to Bruce's assertions made a fortnight earlier). Secondly, Norway ran a wireless station on the islands. Thirdly, a purely German station in north-west Spitsbergen has carried on its communication since the outbreak of war. Fourthly, Norway was planning a wireless station on Bear Island. Fifthly, there had been Russian activity since 1912. Finally, without governmental protection, British development was seriously embarrassed. The men evidently hoped to reinforce their plea with the reminder that the Foreign Office had previously promised adequate safeguards to the English in 1910 and the Scots in 1912.

The attempt to influence foreign policy now received the genuine backing of the Royal Geographical Society. On December 1916, Hinks informed Bruce,

> You will be glad to know that the Council yesterday decided to send the following resolution to the Foreign Office: That the Council of the Royal Geographical Society beg leave to represent to His Majesty's Secretary of State for Foreign Affairs the urgent importance of taking immediate steps to safeguard British interests, political, strategic, and commercial, in Spitzbergen, and to urge that the matter be adjusted with [our] Allies before the termination of the war.[100]

Bruce wondered if Hinks had a chance to contact Maurice de Bunsen, Assistant Undersecretary at the Foreign Office between March 1915 and March 1918, or any other influential Foreign Office or Admiralty official. As far as Bruce could make out, '[...] the Admiralty agree with our view that it is a danger economically and strategically to let Spitsbergen slip through our fingers and come under control of any other country, neutral or otherwise, for lack of proper protection from our Foreign Office.'[101] A subsequent telephone conversation with de Bunsen established that 'he believes [the Foreign Office] came to a definite decision not to annex. [...] He thinks that the decision taken not to annex was based to a large extent on the results of the Conference in 1914.'[102]

President Freshfield of the Royal Geographical Society now composed his unusual political letter, which was sent to the Foreign Office on January 10, 1917. Balfour acknowledged its receipt, stating that it was receiving the careful consideration of his Majesty's Government.[103] On March 7, the Foreign Office informed Freshfield that '[...] the question of Spitzbergen, which cannot be decided

---

[100] Hinks, A. R. (1916) *Letter to W. S. Bruce, 19 December*, RGS/CB8/Spitsbergen, Annexation of, RGS/IBG Archives, London.
[101] Bruce, W. S. (1916) *Letter to A. R. Hinks, 19 December*, RGS/CB8/Spitsbergen, Annexation of, RGS/IBG Archives, London.
[102] *Letter to the President (RGS), 22 December* (1916) RGS/CB8/Spitsbergen, Annexation of, RGS/IBG Archives, London.
[103] Foreign Office (1917) *Letter to the President (RGS), 24 January*, RGS/CB8/Spitsbergen, Annexation of, RGS/IBG Archives, London.

without discussion with several foreign Governments, is not one in which any definite settlement can be come to now, as the present moment is not opportune for bringing forward matters of this nature.'[104] Although Freshfield's letter had no immediate effect on foreign policy, it was published in *The Times* a year later.[105] The Royal Geographical Society consequently received offers of support from other British bodies, such as the Institution of Mining and Metallurgy.[106] Suitably encouraged, the Institution addressed the Foreign Office with the suggestion of seriously investigating the mineral deposits on Spitsbergen and offered its assistance in such investigation if the Government desired it.[107] The Government, however, desired no such thing.

A claim dispute flared up between the Scottish Spitsbergen Syndicate and the Northern Exploration Co. The latter had produced a map of its areas, some of which Bruce and J. V. Burn Murdoch noticed to overlap with the Scottish claims.[108] Maples had offered the explanation that these parts were claimed by the AB Isfjorden-Belsund. Yet, the Northern Exploration Co. had interests in the Swedish company. So the syndicate's lawyers pointed out that the Scots had been careful to exclude all disputed territories before intimating its claims to the Foreign Office, thereby voluntarily renouncing the land between Recherche Bay, Cape Lyell, and Dunder Bay, and they drew attention to the fact that the Foreign Office had recognised its claims in Temple Bay, Sassen Bay, and Klaas Billen Bay, which rendered the Swedish counterclaims void. It is not known how the matter was resolved. By November, the Scottish Spitsbergen Syndicate had another reason to be suspicious of the Northern Exploration Co., but the latter asserted that 'it is absolutely untrue that we have sold, or attempted to sell, any of our Spitsbergen properties to a Company in Holland, or to any Dutchman.'[109]

Elsewhere, Bruce had persuaded the Chamber of Commerce of Dundee to address the Foreign Secretary. This occurred in December 1916, Dundee suggesting that 'the question of bringing Spitsbergen under the influence of Great Britain is worthy of investigating by His Majesty's Government.'[110] By mid-January

---

[104] Foreign Office (1917) *Letter to the President (RGS), 7 March*, RGS/CB8/Spitsbergen, Annexation of, RGS/IBG Archives, London.

[105] *The Times* (1918) 'British claims in Spitzbergen', 13 March, p. 5.

[106] Trewartha James, W. H. (1918) *Letter to A. Hinks, 12 April*, RGS/CB8/Spitsbergen, Annexation of, RGS/IBG Archives, London; Secretary (RGS) (1918) *Letter to the Secretary (IMM), 24 April*, RGS/CB8/Spitsbergen, Annexation of, RGS/IBG Archives, London.

[107] Dermid, C. M. (1918) *Letter to A. J. Balfour, 29 May*, RGS/CB8/Spitsbergen, Annexation of, RGS/IBG Archives, London.

[108] Aitken and Methuen (1916) *Letter to Maples, 20 December*, William Speirs Bruce Collection MS 101/103, Scott Polar Research Institute, Cambridge.

[109] Maples, J. R. (1917) *Letter to R. N. R. Brown, 6 November*, William Speirs Bruce Collection MS101/103, Scott Polar Research Institute, Cambridge.

[110] Keiller, G. C. (1916) *Letter to Balfour, 27 December*, William Speirs Bruce Collection MS101/103, Scott Polar Research Institute, Cambridge.

1917, the Scots were contemplating another brochure.[111] It would firstly serve to update those who had complained about not having had any information about the company. Secondly, it could be forwarded to several Admiralty officials, Foreign Office officials, and others. In addition, Bruce wondered if Brown could not have a word with the Colonial Office.[112] Besides approaching the Chamber of Commerce in Dundee and Edinburgh, Bruce now sought an introduction at the Chamber of Commerce in Glasgow.[113] Despite his best efforts, his rigorous campaigning bore no immediate fruits.

The unpopular Treaty of Brest-Litovsk between Russia and Germany on March 3, 1918, which became such a powerful tool in the rhetoric of the Northern Exploration Co., only seemed to distract the Scots temporarily. The Foreign Office had called for the recognition of Russian claims on Spitsbergen, which Bruce considered to be ill-advised.[114] He related,

> I do not believe the Russians have claims anything approaching the value of £220,000 and even more than a year ago we were warned that any private offer of purchase from Russians of Spitsbergen estates would almost certainly be with German money. In 1912 Dr Rudmose Brown and I met Dr Nansen and his son in Advent Bay, Dr Nansen told us the Americans were then inclined to sell, but he depreciated the American coal [...] One wonders now if Nansen had business objects in view and whether he and his friends are now shareholders in the Norwegian Company which has bought these same tertiary American claims. I think I am right in saying that it was Dr Nansen in 1912 who told Dr Brown and myself that there were Russian officers at that time in Spitsbergen and that their work was not of commercial character. One now associates those officers with the present officers named in the Foreign Office letter. I strongly suspect German influence at the back of the Russian offer.[115]

Rather than seeing the Treaty of Brest-Litovsk as an opportunity, Bruce dwelt on how easy it was for claims to be depreciated by so-called experts for personal gain. His account also demonstrated how observed actions could be interpreted to fit individual argumentation. One nation was seemingly as guilty of it as the next.

In June 1918, the Northern Exploration Co. revived communications with the syndicate. Colonel H. C. King, who was supposedly not a stakeholder but a

---

[111] Bruce, W. S. (1917) *Letter to R. N. R. Brown, 19 January*, William Speirs Bruce Collection MS101/103, Scott Polar Research Institute, Cambridge.
[112] Bruce, W. S. (1917) *Letter to R. N. R. Brown, 24 January*, William Speirs Bruce Collection MS101/103, Scott Polar Research Institute, Cambridge.
[113] Bruce, W. S. (1917) *Letter to J. S. Samuel, 7 February*, William Speirs Bruce Collection MS101/103, Scott Polar Research Institute, Cambridge.
[114] Bruce, W. S. (1918) *Letter to A. N. G. Aitken, 2 April*, William Speirs Bruce Collection MS101/103, Scott Polar Research Institute, Cambridge.
[115] Bruce, W. S. (1918) *Letter to A. N. G. Aitken, 5 April*, William Speirs Bruce Collection MS101/103, Scott Polar Research Institute, Cambridge.

personal friend of Davis, informed Bruce that the English would send an expedition to Spitsbergen within a week and presented the Scots with a chance to partake. King claimed the expedition to have support of the Foreign Office and the Admiralty and that Shackleton would be the leader. Since the syndicate had not heard about the planned expedition before, this was a lot to deliberate within the short period.

The first question was whether King was a responsible agent for the Northern Exploration Co. Secondly, the syndicate had previously expressed doubt as to the company's methods regarding the claims by Earl Morton and the Swedish claim-jumping. The current proposal was 'to seize every piece of available land [...] whether it is valuable or whether it is not, as long as it comes under British rights or claims.'[116] King asked Bruce if the Scots would contribute £10,000 towards the expedition or give the Northern Exploration Co. the rights to investigate their properties for three years with an option to take them over for 10,000 shares in the company.[117] It seemed that the Northern Exploration Co. had not raised the necessary funds yet. In addition, the Earl's lawyer stated that the Morton group had protested to the Foreign Office and elsewhere against Mansfield selling their claims and that the group still maintained its rights. Bruce feared that 'great caution is required in having anything to do with N.E.C. projects.'[118] Besides, the offer of £10,000 if within three years any investigation showed satisfactory results was preposterous. It would merely tie their hands for that length of time.[119]

The Scottish Spitsbergen Syndicate was under pressure to renew its own operations or run the risk of losing out. Bruce realised that 'it is obviously too late for an expedition this year but it is not too early to consider a proposal for one next year.'[120] He foresaw difficulties in raising the costs and proposed that money should in the first instance be spent on good mining geologists. He confided that he also wished for a solid Glasgow managing director to tackle the tasks in hand. 'Usher is good, so is Whitson and Aitken and I think these three men are as good men as we will find in Edinburgh and a long way round, but all three are wrapped up in other important business affairs [...]. We need a Glasgow sprinter, a real Glasgow sprinter badly.'[121] J. V. Burn Murdoch was glad to hear that 'the Directors

---

[116] Bruce, W. S. (1918) *Letter to R. N. R. Brown, 5 June*, William Speirs Bruce Collection MS101/103, Scott Polar Research Institute, Cambridge.
[117] Bruce, W. S. (1918) *Letter to R. N. R. Brown, 6 June*, William Speirs Bruce Collection MS101/103, Scott Polar Research Institute, Cambridge.
[118] Bruce, W. S. (1918) *Letter to R. N. R. Brown, 7 June*, William Speirs Bruce Collection MS101/103, Scott Polar Research Institute, Cambridge.
[119] Bruce, W. S. (1918) *Letter to R. N. R. Brown, 14 June*, William Speirs Bruce Collection MS101/103, Scott Polar Research Institute, Cambridge.
[120] Bruce, W. S. (1918) *Letter to A. N. G. Aitken, 3 July*, William Speirs Bruce Collection MS101/103, Scott Polar Research Institute, Cambridge.
[121] Bruce, W. S. (1918) *Letter to A. N. G. Aitken, 19 August*, William Speirs Bruce Collection MS101/103, Scott Polar Research Institute, Cambridge.

feel the need of a substantial push and of getting the right people in. My feeling is that with all respect and personal affection for our Directors, they are not the kind of businessmen for the execution of the particular kind of business immediately before them.'[122] Yet, it remained to be seen whether the right people could also get on with each other. Bruce did not want to be expedition leader himself, but he feared that no one else would do a good job. He wrote to Brown, 'Like yourself I find Spitsbergen business on the top of my ordinary work far too much for me, but who will follow things up the same as we will.'[123]

In November 1918, the war ended without a noticeable effect on the proceedings. Bruce, who had been invited to lunch at the Chamber of Commerce in Aberdeen[124], now requested a recommendation for the release of Hannay from service at the Industrial and Scientific Department.[125] By December 23, relations with the Northern Exploration Co. had improved somewhat. Neither firm seemed too confident that the Foreign Office would make the most favourable decisions regarding Spitsbergen's sovereignty. Maples and Bruce, therefore, discussed the following safeguards in case of a Norwegian annexation,

1) Norwegian Mining Laws must not apply.
2) Any dispute between British and Norwegian subjects must be heard and adjudicated upon by British and Norwegian judges in equal proportion.
3) No harbour dues to be imposed by Norwegians on vessels visiting British territory.
4) Rates and taxes on British territories to be agreed on by British owners.
5) No law or regulation made by Norwegian authority shall have force in British territory except with the consent of the owners of the territory.[126]

This was the first time the firms acknowledged the eventuality of a Norwegian take-over. A Scottish expedition to confirm and develop their claims was therefore imperative. It was resolved that the present syndicate should be dissolved and formed anew.[127] On January 8, 1919, Aitken and Methuen called an extraordinary general meeting, at which the resolution was submitted for passing.[128] It entailed a

[122] Burn Murdoch (1918) *Letter to W. S. Bruce, 20 September.* William Speirs Bruce Collection MS 101/103, Scott Polar Research Institute, Cambridge.
[123] Bruce, W. S. (1918) *Letter to R. N. R. Brown, 28 October.* William Speirs Bruce Collection MS 101/103, Scott Polar Research Institute, Cambridge.
[124] Aberdeen Chamber of Commerce (1918) *Letter to W. S. Bruce, 26 November.* William Speirs Bruce Collection MS 101/103, Scott Polar Research Institute, Cambridge.
[125] Bruce, W. S. (1918) *Letter to Sir William McCormick (Ind. Sci. Dept), 28 November.* William Speirs Bruce Collection MS 101/103, Scott Polar Research Institute, Cambridge.
[126] Bruce, W. S. (1918) *Letter to A. N. G. Aitken, 23 December.* Robert Neal Rudmose Brown collection, MS 356/30/3, Scott Polar Research Institute, Cambridge.
[127] Aitken (1918) *Letter to R. N. R. Brown, 24 December.* William Speirs Bruce Collection MS 101/103, Scott Polar Research Institute, Cambridge.
[128] *Special resolution* (1919) Scottish Spitsbergen Syndicate, Norwegian Polar Institute, Tromsø.

scheme of reconstruction during which a liquidator would be appointed for the purpose of winding up the syndicate and transferring its assets to a new firm under the same name. The voluntary liquidation of the Scottish Spitsbergen Syndicate Limited was announced in the *Gazette* on February 5, 1919.[129]

8.6 Formation of the second syndicate and chronological overview

Concurrent to the liquidation of the old, the new Scottish Spitsbergen Syndicate was incorporated.[130] It stayed a private company, aiming to carry out prospecting, mining, and just about any other business on Spitsbergen and elsewhere.[131] Its nominal capital was £100,000 divided into 10,000 shares of £10 each. Its original subscribers were William Gordon Burn Murdoch, Charles Hanson Urmston, and Thomas Leslie Usher. The three directors of the old syndicate had strong family ties and now formed the core of the new board. Rottenburg had resigned, so the directorate was offered to the aforementioned Henry Moubray Cadell.[132] Cadell proved himself unpopular with Bruce, being under the impression that the latter had been to Spitsbergen only once. Bruce, who assisted the new board as a consultant, confided in Brown, 'I am afraid that this Councillor of the [Royal Scottish Geographical Society] does not know more about Spitsbergen than the Editor who [...] associated my name with Ponting. I don't take this as flattery!'[133] Cadell simultaneously had a disagreement with the ship owner and broker Richard Mackie, who shortly afterwards handed in his resignation.[134] By mid-April, the board had settled into the constellation of Burn Murdoch, Cadell, the chartered accountant W. James Maxtone Graham, Urmston, and Usher.[135]

As with the Northern Exploration Co., milestones in the history of the new syndicate (Fig. 8.7) included the Paris Peace Conference, the post-war boom-and-bust, and the Great Depression, which the English firm did not survive. The Scots persisted throughout the Second World War, the nationalisation of the British coal industry, and the dawn of the Cold War. The Spitsbergen Treaty was signed and

---

[129] *Edinburgh Gazette* (1919) 'The Scottish Spitsbergen Syndicate Limited', 7 February, p. 784.
[130] *Certificate of incorporation* (1919) Scottish Spitsbergen Syndicate NAS02024 BT2-10219, National Archives of Scotland, Edinburgh.
[131] *Memorandum of association* (1919) Scottish Spitsbergen Syndicate NAS02024 BT2-10219, National Archives of Scotland, Edinburgh.
[132] Scottish Spitsbergen Syndicate (1919) *Letter to H. M. Cadell, 3 February*, Cadell of Grange papers, Acc. 5381, National Library of Scotland, Edinburgh.
[133] Bruce, W. S. (1919) *Letter to R. N. R. Brown, 6 March*, William Speirs Bruce Collection MS 101/103, Scott Polar Research Institute, Cambridge.
[134] *Minutes of the meeting, 19 March* (1919) Cadell of Grange papers, Acc. 5381, National Library of Scotland, Edinburgh.
[135] *Directors* (1919) Scottish Spitsbergen Syndicate NAS02024 BT2-10219, National Archives of Scotland, Edinburgh.

ratified, and the Danish Commission settled the claim disputes. However, the ownership of treaty properties expired after a ten-year period in 1937. A review of the situation was once again disrupted by war, after which the rivalry between Norway and Russia became more pronounced. As an alternative to British coal price movements, coal availability for overseas shipments and bunkers is taken as a proxy to show the destabilisation of the coal industry by the latest war and its decline into an unprofitable sector taken into state control. With British exports being out of the picture, one could assume a heyday for Spitsbergen coal. In 1935, the Scottish Spitsbergen Syndicate was still in a position to reject a Norwegian offer for its possessions. In 1947, C. W. B. Urmston became involved, staged an expedition, and formed a subsidiary company, which eventually obtained the syndicate's assets. The syndicate's voluntary liquidation was complete in 1953.

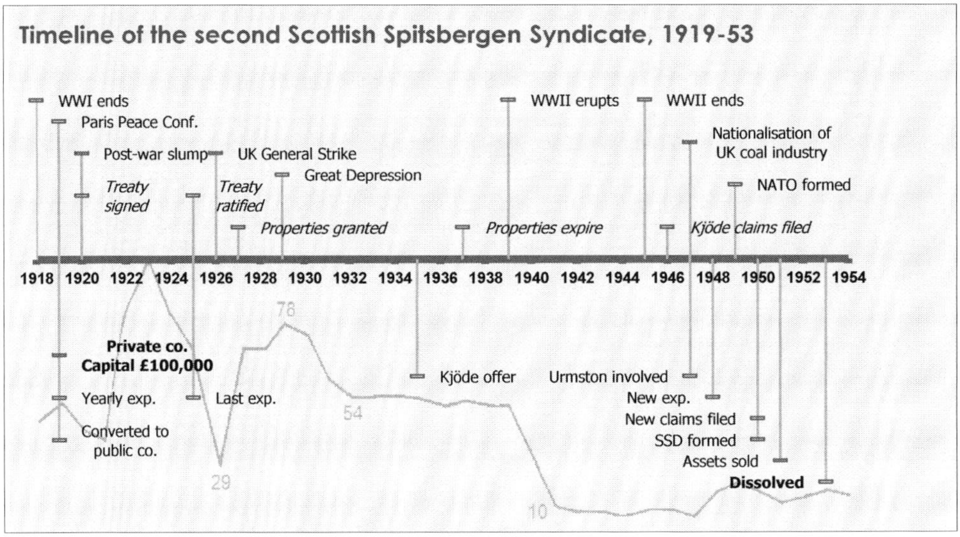

8.7 Timeline of the second Scottish Spitsbergen Syndicate, 1919-53. Events below the bar are company-specific. Above the bar, events in italics are relevant to Spitsbergen, while others are thought to have defined the historical context. The grey line indicates UK coal availability for overseas shipment and bunkers, whereby the bar denotes 100 million metric tons, and the bottom edge is 0. Actual values have intermittently been added for clarity. (Source: UK Department of Energy and Climate Change (2012) 'Historical coal data: coal availability and consumption, 1853 to 2011'. Available at: http://webarchive.nationalarchives.gov.uk/20121217150421/http://decc.gov.uk/en/content/cms/statistics/energy-stats/source/coal/coal.aspx (Accessed: 29 July 2013); Chart: F. Kruse.)

8.7 The global network of the second syndicate

8.7.1 Economic actors

The syndicate's reconstruction stipulated that it raised £30,000 to meet the costs of an expedition in spring 1919.[136] The private company could only raise the cash among its subscribers, their number still being restricted to 50. In February 1919, merely £10,000 had been subscribed to.[137] Nonetheless, the syndicate already attended to practical matters and searched for a qualified organisational secretary.[138] In April, only £13,150 had been raised, and the estimated costs of the expedition were cautiously adjusted to £20,000.[139] By June, £26,390 could be mustered, and the financial situation looked more promising.[140] Directors Burn Murdoch, Urmston, and Usher belonged to the 30 original subscribers of the old syndicate, but only Burn Murdoch acquired additional shares along with other newcomers to the syndicate. Directors Cadell and Maxtone Graham invested £500 each. Cadell later recounted that his pre-war opinion of Spitsbergen, which others may have shared, had been that the coal was not lucrative and that the oil shale would not pay.[141] After the war, there had been fresh reports of the islands' wealth. They were still exaggerated, but good coal was being mined at Advent Bay, and the £1 shares of the Northern Exploration Co., which had a capital of £500,000, were selling at £3. The prospects of the Scottish Spitsbergen Syndicate with a capital of £100,000 were bright, so Cadell eventually put £1,000 into it, joined the board of directors, and organised an expedition.

Although the money for the expedition could be raised, the syndicate must have realised that it could not squeeze its shareholders again in the future. On August 4, 1919, therefore, the board resolved to convert to a public company.[142] Within a fortnight, the resolution was sanctioned, and the division of the £10 shares into more attractive £1 shares was public knowledge.[143] A month later, another 31

---

[136] *Scheme of reconstruction* (1919) Scottish Spitsbergen Syndicate, Norwegian Polar Institute, Tromsø.
[137] Scottish Spitsbergen Syndicate (1919) *Letter to H. M. Cadell, 3 February*, Cadell of Grange papers, Acc. 5381, National Library of Scotland, Edinburgh.
[138] Bruce, W. S. (1919) *Letter to R. N. R. Brown, 19 February,* William Speirs Bruce Collection MS 101/103, Scott Polar Research Institute, Cambridge.
[139] W. S. Bruce papers, box 10/113, National Museums of Scotland Library, Edinburgh.
[140] *Return of allotments from 27 May to 9 June* (1919) Scottish Spitsbergen Syndicate NAS02024 BT2-10219, National Archives of Scotland, Edinburgh.
[141] *Extracts from H. M. Cadell's chronicle, Vol. 4 1910-1925*, Cadell of Grange papers, Acc. 5381, National Library of Scotland, Edinburgh.
[142] *Special resolution* (1919) Scottish Spitsbergen Syndicate NAS02024 BT2-10219, National Archives of Scotland, Edinburgh.
[143] *The Times* (1919) 'City news in brief', 20 August, p. 18.

investors had taken up shares worth £23,840.[144] Among them was Urmston, whom the sudden public interest may have encouraged to purchase an extra 400 shares. James Butler Bolton parted with £100 and became the most recent director.[145] Their investments were a pittance compared to £5,000, with which each of a spinster from Essex, a gentleman from Surrey, and two gentlemen from London entrusted the syndicate.

The decision to convert to a public company was accompanied by a memorandum.[146] Its contents were also published in *The Times*.[147] The statement set the syndicate back £12, but it was value for money because it reached the large readership of the influential paper. The readers were made aware of the syndicate's financial and operational structure. The statement fell back on the expeditions by the well-known Arctic and Antarctic explorer Bruce and the eminent mining and geological experts associated with him. The voyages had resulted in territorial claims of about 1,650 square miles, approximately one-twelfth of the archipelago. The Foreign Office 'has informed the Syndicate that its claims are officially recognised and will be protected. No alteration in the political status of Spitsbergen, which is now under consideration by an Inter-Allied Commission, can in the opinion of the Board affect the Syndicate's Mining, Fishing, Hunting, or other rights.'[148] The claims comprised coal, gypsum, oil shale, and possibly iron ore. An expedition, assisted by the Admiralty, was currently making a thorough survey of the properties, and a full report was expected in October. One preliminary result was the important discovery of a large coalfield workable all year round and easily accessible from Britain. Lastly, 'as several Companies are operating in Spitsbergen it may be well to mention, in order to avoid confusion, that the Scottish Spitsbergen Syndicate, Limited, has no connection with any other.'[149] The particulars pertaining to the syndicate were thus in the public realm, where it in all probability hoped to find additional sponsors.

It is not known when the shares of the Scottish Spitsbergen Syndicate first opened on the London Stock Exchange. On January 28, 1920, they were quoted at

---

[144] *Return of allotments from 27 August to 19 September* (1919) Scottish Spitsbergen Syndicate NAS02024 BT2-10219, National Archives of Scotland, Edinburgh.

[145] *Directors* (1919) Scottish Spitsbergen Syndicate NAS02024 BT2-10219, National Archives of Scotland, Edinburgh.

[146] *Memorandum for shareholders* (1920) R. N. R. Brown S.S.S. Correspondence, MS 356/95, Scott Polar Research Institute, Cambridge.

[147] *The Times* (1919) 'Miscellaneous financial. The Scottish Spitsbergen Syndicate, Limited', 17 September, p. 20.

[148] *The Times* (1919) 'Miscellaneous financial. The Scottish Spitsbergen Syndicate, Limited', 17 September, p. 20.

[149] *The Times* (1919) 'Miscellaneous financial. The Scottish Spitsbergen Syndicate, Limited', 17 September, p. 20.

45s, or £2 5s.[150] Other details occasionally appeared in the leading papers, but on the whole, publicity was at a minimum during the preparation of the follow-up expedition until the notices of another Scottish voyage surfaced in spring 1920. On May 5, an article related that the expedition would comprise three vessels and a staff of over 50 members.[151] Bruce was ill; it would be led by Mathieson. The expedition left on May 17.

Only after the expedition had sailed were further news issued. Since the sovereignty of Spitsbergen had been decided in favour of Norway, the syndicate was giving its full attention to the matter of proving its claims before the Danish Commission, which would subsequently grant complete titles.[152] The directors' report and the accounts had been submitted for public inspection. They indicated that the cost of the expedition in 1919 had been £24,974, which was well within the estimates, and that the cash balance of the syndicate was currently £46,964. The report preceded the syndicate's first annual general meeting in May. Its estates on Spitsbergen were its principal assets at a value of £15,000.[153]

The syndicate prided itself in having sent 'one of the largest, if not the largest and most important expedition which had so far visited Spitsbergen.'[154] The directors 'were aware of extensive coal outcrops occurring on our properties but we were hardly prepared for such a very favourable report as was brought home by our experts.'[155] Spitsbergen coal would find a ready market in Britain, and a price of £5 to £6 per ton could be obtained in France and Italy, 'where at present inferior Welsh coal is selling at 85s to 90s per ton.'[156] In light of the coal famine in Norway, an even higher price could be possible at the nearer market. In addition, iron ore, gypsum, oil, and copper were some of the resources that could be found on the claims, which with the recent acquisition of Edge Island had increased to nearly 3,000 square miles. The shareholders were reminded that, 'the functions of this company are to continue to explore and prove our minerals, and to grant mining and other concessions to subsidiaries, who will work them, and with whom we shall in most cases be partners.'[157] The syndicate would not dabble in active mining.

After the return of the 1920 expedition, the syndicate was shaken by recent reports regarding the coalfields of Spitsbergen. These implied that coal was not found on the Scottish claims. To rectify this grave and damaging mistake, the directors contacted several publications. While it did not intend to criticise the

[150] *The Times* (1919) 'Stock exchange prices', 28 January.
[151] *The Times* (1919) 'Coal from the Arctic', 5 May, p. 18.
[152] *The Times* (1919) 'Company results', 21 May, p. 25.
[153] *The Times* (1920) 'The Scottish Spitsbergen Syndicate, Limited', 26 May.
[154] *The Times* (1920) 'The Scottish Spitsbergen Syndicate, Limited', 26 May.
[155] *The Times* (1920) 'The Scottish Spitsbergen Syndicate, Limited', 26 May.
[156] *The Times* (1920) 'The Scottish Spitsbergen Syndicate, Limited', 26 May.
[157] *The Times* (1920) 'The Scottish Spitsbergen Syndicate, Limited', 26 May.

reports of other companies, it wished to point out that its properties had been 'more thoroughly explored by expert geologists than those of any other company (British or otherwise) operating in Spitsbergen, and that they contain coal-bearing areas at least as important as those found on the properties of any other Spitsbergen company.'[158] For the benefit of its shareholders, the syndicate would soon prepare a summary of the reports of the scientific staff.

At the end of 1920, the syndicate presented the highly satisfactory results of its second important expedition.[159] A first coalfield at Ebba Valley to the north of Adolf Bay extended over three miles and comprised two coal seams. The lower seam was about four feet thick, of which about a quarter was workable. This large store of coal could be doubled if worked below sea level. The coal was hard, clean-looking, with a semi-anthracite lustre, and of excellent quality. A second coalfield at Bruce City to the east of Klaas Billen Bay covered an area of roughly 18 square miles. It also contained two coal seams. The Upper or Carron seam was up to three feet four inches thick. It was separated from the Lower seam by about 60 yards of mainly sandstone. The Bruce City coalfield held roughly 90 million tons of coal. The seams, especially the lower one, were likely to increase in thickness and continuity and were found under the most favourable geological and working conditions. The board had visited Spitsbergen to gather practical information for the future administration of the properties and dispelled the myth that the islands were barren and wind-swept. The directors had ascertained that coal was finding a ready market in Norway at prices between £10 and £12 per ton.

At the time of the report, however, the post-war depression had set in. As with the Northern Exploration Co., the Scottish shares had taken a turn for the worse (Fig. 7.27). Nor was it only the British firms, who suffered the decline. In a letter to Cadell, the manager of Store Norske could understand that the mine owner should be very interested in the great change in the market since the two men met in summer 1920. Karl Bay cautioned, 'How far prices may go it is difficult to say today, but should the market decline to such an extent that we shall face the rates from before the war and keep about the mining expenses today, none of us may be expected to produce and ship the coals with profit.'[160]

Meanwhile, the syndicate had released the report and accounts for the year to February 28, 1921.[161] The expedition in 1920 had cost £39,810, but considering that it had a larger staff and lasted twice as long as the expedition in 1919, this was not thought excessive. The cash balance was £12,970, and the

[158] 'Spitsbergen coal' (1920) *Mining Journal*, 131, p. 842.
[159] *The Times* (1920) 'The Scottish Spitsbergen Syndicate, Limited', 18 November, p. 18.
[160] Bay, K. (1921) *Letter to H. M. Cadell, 16 March,* Cadell of Grange papers, Acc. 5381, National Library of Scotland, Edinburgh.
[161] 'Scottish Spitsbergen Syndicate, Ltd.' (1921) *Colliery Guardian*, 121, p. 1615.

stores had been valued at £6,200. The reserve account had been increased to £1,547 through the issue of option shares. The report customarily preceded the annual general meeting.[162] On June 3, Maxtone Graham presided and outlined that two expeditions at a total of £65,000 had proven the existence of coal measures to such an extent as to warrant a subsidiary company, the preparations of which were well advanced. He expected an improvement in the general conditions and emphasised that the artificial conditions in Britain's coal industry were largely brought about by governmental control. They did not affect Spitsbergen, but as European markets settled, the demand for coal from the archipelago was bound to rise again. When it did, the Scots would draw on the many applications from British miners wanting to work up north.

The directors' reports and the ensuing annual general meetings continued to be the main source of information for the shareholders. The report for the year ending February 28, 1922, explained that the particulars of the expedition in 1921 had not been made public because the directors had entered into negotiations concerning the sale of Gips Valley shortly after its return.[163] Lengthy discussions with influential groups, however, did not result in a definite sale. Another expedition would be sent in 1922 to continue the investigation of the properties.

According to the report for the year to February 1923, the expedition of 1922 had cost £4,650.[164] It had satisfied the board and its technical advisers that Gips Valley, too, contained a valuable coalfield that merited a large colliery. In Prof. Robert Wilson Dron's opinion, which stood in stark contrast with Bay's foreboding above, 'a colliery well equipped and well managed should be able to produce coal f.o.b. Spitsbergen at a lower rate than can be done on the Forth or the Tyne. So far as competition with the other collieries in Spitsbergen is concerned, the fact that Gips Valley coal is a coking coal is very important.'[165] The negotiations referred to previously had still not come to a conclusion, owing to the market being unstable and the treaty not having been ratified.

The financial slump forced the syndicate to mark time. The report made out to February 1924 presumably read like the one to February 1925. The latter stated that another expeditionary party had been sent in 1924 'to inspect hutments and overhaul gear and other property and report on conditions in general.'[166] Apart from this small cost, there had been no financial change. While negotiations were still

[162] *The Times* (1921) 'Company meeting', 6 June, p. 16; *Financier* (1921) 'Scottish Spitsbergen Syndicate Limited', 6 June; *Scotsman* (1921) 'Company meeting', 6 June.
[163] 'Scottish Spitsbergen Syndicate, Ltd.' (1922) *Colliery Guardian*, 123, p. 1309.
[164] 'Scottish Spitsbergen Syndicate, Ltd.' (1923) *Colliery Guardian*, 126, p. 483.
[165] 'Scottish Spitsbergen Syndicate, Ltd.' (1923) *Colliery Guardian*, 126, p. 483.
[166] 'Scottish Spitsbergen Syndicate, Ltd.' (1925) *Colliery Guardian*, 129, p. 1583.

ongoing, the agent in Norway had informed the directors that arrangements were being made for Norway's formal take-over in summer 1925.

8.7.2 Political actors

The reconstruction of the Scottish Spitsbergen Syndicate fell into the momentous period marked by the Paris Peace Conference and post-war optimism. Yet, despite a surprising amount of documentary evidence regarding the syndicate in general, sources of any politically motivated activities are rare. Following an interview between Brown and the Foreign Office on December 31, 1918, the syndicate voiced its concerns in a letter to the British delegate in Paris.[167] It maintained that the protection of the British Government continued to be the weightiest on Spitsbergen. If the Government could not annex the archipelago, there should be a minimum of safeguards. These safeguards included: 1) the absolute, valid, and perpetual title to all territories claimed, 2) the freedom to work and develop the claims, 3) no tax discrimination, 4) no harbour or other dues, 5) no Norwegian mining laws, 6) no Norwegian civil laws, 7) no import tax, 8) free navigation in Icefiord and Stor Fjord, and 9) no Norwegian laws on British territory, unless consented. Furthermore, the syndicate assumed that its claims extended to the customary three-mile limit from the shore and again pointed out that wildlife on the islands was in need of protection. The Scots were not as adamant as their English counterpart about being present in Paris, nor did they threaten the Foreign Office in order to influence foreign policy. They had communicated their concerns and remained patient, merely pointing to an unresolved claim dispute in April.[168] Later in the year, contact between the syndicate and the Foreign Office regarded the new Scottish claims made during the 1919 expedition.

Neither did the signing of the Spitsbergen Treaty on February 16, 1920, seem to faze the board. Privately, however, Cadell was displeased about the turn of events. He aired his frustration in an academic publication, writing that 'the Norwegians must be heartily congratulated on the valuable virgin territory that they have so easily acquired without taking any part in the Great War.'[169] Brown also vented his anger. He maintained that the Spitsbergen Treaty removed the last *terra nullius* of economic importance from the map of the world and gave Norway its first overseas possession. As such, '[it] exhibits the anomaly of a State which remained

[167] Aitken and Methuen (1919) *Letter to E. Howard, 13 January*, British Legation 1919 FO 608/123 Files 427/3/1 (start- pp. 17998), National Archives, Kew.
[168] Scottish Spitsbergen Syndicate (1919) Letter *to the Foreign Office, 23 April,* British Legation 1919 FO 608/123 Files 427/3/1 (start- pp. 17998), National Archives, Kew.
[169] Cadell, H. M. (1921) 'Coal mining in Spitsbergen', *Transactions of the Institution of Mining Engineers,* 40, pp. 119-42.

neutral throughout the war gaining considerable territorial aggrandizement at the hands of the Peace Conference.'[170] He asserted that 'in view of Europe's gift to Norway, it is not unreasonable to suggest that Norway should bear the cost of [the arbitration of claim disputes], which, after all, is forced on the mining companies in spite of the indisputable nature of most of their titles.'[171] Regarding the new legal status of Spitsbergen, the board appeared to place its trust in the assurance from the Foreign Office that its claims would be fully and adequately protected.

Five years elapsed. The Scottish Spitsbergen Syndicate was anxiously waiting for the Norwegian Government to ratify the Treaty.[172] Only then could the Danish Commission deal with the arbitration of the claims. The delay had badly hampered the directors' negotiations to form a subsidiary company and to develop the mineral resources. They lamented that companies of other nations enjoyed governmental aid. Those Governments had realised their dependency on British coal and had studied the possibilities of Spitsbergen to become self-sufficient. None of the coal from Spitsbergen had as of yet reached Britain. The treaty eventually came into force on August 14, 1925.

8.7.3 *Other allies*

Other allies of the Scottish Spitsbergen Syndicate continue to be elusive. The syndicate appears to have acted in isolation, or better, its directors seem to have done their networking among their own extensive professional and personal connections. Proof is rare. Perhaps it occurred verbally in the various Edinburgh and London clubs, of which the directors were invariably members. Their private files may provide relevant information, but these have not been consulted.

One James Campbell from Middlesbrough approached the syndicate, proposing to take miners on the 1919 expedition in order to look for iron ore, manganese ore, and the like.[173] The proposal envisaged two mining parties from Middlesbrough, but by June, it looked like only the mining engineer Captain H. Sandys would accompany the Scots.[174] His group was paying £4,000 for the privilege, and he offered his assistance to the expedition in return for being shown all there was to see.

---

170 Brown, R. N. R. (1920) 'Recent developments in Spitsbergen', *Scottish Geographical Magazine*, 36, pp. 111-6.
171 Brown (1920).
172 Scottish Spitsbergen Syndicate (1926) *Memorandum as to the company's properties in Spitsbergen, and their possible development*, Norsk Polarinstitutt 239, Regional State Archive, Tromsø.
173 Campbell, J. (1919) *Letter to the Scottish Spitsbergen Syndicate, 13 May*, MS356/95, Scott Polar Research Institute, Cambridge.
174 Arthur (1919) *Letter to R. N. R. Brown, 7 June*, MS 356/95, Scott Polar Research Institute, Cambridge.

In relation to the same expedition, Bruce recommended to convey Burn Murdoch to Tromsø early to prearrange matters there, which suggested that business ties existed in the Norwegian town.[175] Since no Norwegian agent was named, the Scots may have worked in conjunction with the British consul there.

8.7.4 Competitors

Global competition consisted chiefly of perceived trespassers on Scottish claims. In November 1919, therefore, the Foreign Office contacted the Norwegian Foreign Department to ascertain the details relating to a claim dispute with the Norwegian Exploration Co. at Bjona Haven.[176] By now, however, all stakeholders were anticipating the ratification of the Spitsbergen Treaty.

The Scots' feeling that they were hard done by was offset by the contempt, in which they held accusations made against them. After a Scottish representative met the Swede Einar Wirén on board the *Sigurd Jare*, he gave an account of their conversation. Wirén alleged that the syndicate only arrived in Ebba Valley in summer 1919 because the coal was more noticeable than on its property in Adolf Bay. He maintained that the Swedish Foreign Department held no records indicating a prior Scottish claim in Ebba Valley, and the Scot got the impression that 'their attitude is that the war is being used as a cloak to cover the annexing of their claims.'[177] Wirén had previously written to the board, offering to sell his company's house in Ebba Valley, but there had been no reply. The writer assumed that the Swedish company owning Ebba Valley was small because Prof. Erik Andersson Stensiö from Uppsala held a large portion of the shares. Wirén was a shareholder, as was Stensiö's younger brother. They seemingly did not have governmental support because the Swedish Coal Commission, scheduled to visit Spitsbergen, planned to bypass the area. The writer stressed, 'that the Swedish attitude on the matter is hysterical is shown by the fact that Wordie and Tyrrell in the course of their academic work, having applied to Uppsala for various publications, were informed that they would not be sent until the Ebba Valley question was settled.'[178] Yet, there had been no Swedish attempt to work the claim in 1919, supposedly because money was scarce.

---

[175] Bruce, W. S. (1919) *Letter to R. N. R. Brown, 31 May,* MS 356/95, Scott Polar Research Institute, Cambridge.

[176] Foreign Office (1919) *Letter to Ihlen, 13 November,* Norwegian Foreign Department Box 5373, National Archives of Norway, Oslo.

[177] *Phil Dr Einar Wirén* (1920) Scottish Spitsbergen Syndicate, Norwegian Polar Institute, Tromsø.

[178] *Phil Dr Einar Wirén* (1920) Scottish Spitsbergen Syndicate, Norwegian Polar Institute, Tromsø.

8.8 The local network of the second syndicate

8.8.1 Claims and natural resources

An advance party preceded the main expedition in 1919. It was headed by John Mathieson and was supposed to leave at the end of May. However, the party was delayed by difficulties such as obtaining permission from London to take food out of the country.[179] When it did depart, on June 16, it was reported to be a surveying expedition organised by the Scottish Oceanographic Laboratory; no mention of the syndicate.[180] Yet, Mathieson had 'to find all exposures I could, so that I might be able to point them out to Mr. Tyrrell on his arrival and at the same time to find something of commercial value.'[181]

Hindered by fog and ice, the ship entered Advent Bay on June 30.[182] Adolf Bay at the northern extremity of Klaas Billen Bay was full of ice, so Mathieson's party landed in an inlet on the eastern shore, which they christened Phantom Bay. Fast ice made landings difficult, and at times, the three men jumped over floes to reach the shore. During their surveys, they ceaselessly kept an eye out for coal. Other locations targeted over the next month included Ebba Valley, the Anser Islands at the junction of Klaas Billen Bay and Sassen Bay; Gips Bay and Gips Valley; the neighbourhood of Bjona Haven; and the Post Glacier. Mathieson thought Ebba Valley to be Scottish territory, on which the Swedes had encroached and built a house. They had partly opened a coal outcrop situated about two miles from the shore at a height of roughly 500 feet. During the production of a large-scale map of the area (Fig. 8.8), which lacked the location of the Swedish hut and workings, the Scots observed that the coal appeared to have slipped, and A. F. Campbell recommended opening an outcrop at another site. In addition to Ebba Valley, the men mapped a district to the south of Nordenskiöld Glacier, again paying particular attention to an outcrop of coal.

The main expedition, too, was delayed by several weeks. Only in mid-July did the *Scotsman* announce the forthcoming departure of 'an expedition of more than usual magnitude.'[183] The article portrayed a patriotic endeavour supported by the Admiralty; the application to the Treasury had received the backing of the Foreign Office. On July 15, the party finally sailed aboard the RY *Petunia,* a mine

[179] Bruce, W. S. (1919) *Letter to R. N. R. Brown, 4 June,* MS 356/95, Scott Polar Research Institute, Cambridge.
[180] *The Times* (1919) 'Expedition to Spitsbergen', 17 June, p. 15.
[181] Hoel (1967a) p. 1055.
[182] Mathieson, J. (1919) *Diary,* Scottish Spitsbergen Syndicate Papers 1909-51 Vol. 1, Scott Polar Research Institute, Cambridge.
[183] *Scotsman* (1919) 'Scottish Spitsbergen expedition', 12 July.

sweeper and convoy sloop with 13 bulkheads built during the war.[184] According to the *Scotsman*, 'no expedition to Spitsbergen has ever set out with a better equipment of stores, mining materials and scientific instruments, or with a larger or better trained staff of experts on board.'[185] The article listed key staff, who would carry out extensive prospecting and preliminary development work.

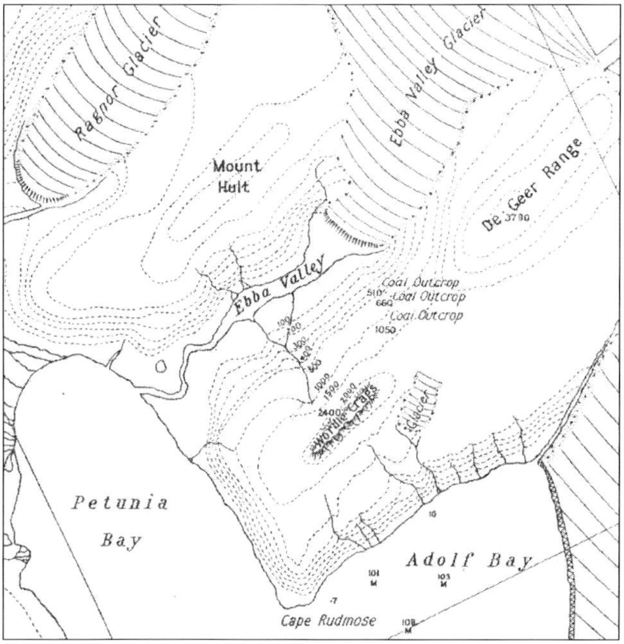

8.8 Detail of a map showing three coal outcrops in Ebba Valley. (Source: Cadell, H. M. (1920) 'Spitsbergen in 1919', Scottish Geographical Magazine, 34, pp. 1-10.)

On July 26, the main expedition reached Prince Charles Foreland and landed a prospecting party of seven in Richard Lagoon.[186] Headed by the Glasgow geologist and mining engineer James Scott, the men erected a first tent camp on the north side of the lagoon near the mouth of the Mount Balfour stream.[187] After eight days of mist and heavy rain, they shifted the camp to the most westerly shore of the lagoon. The men also established an auxiliary camp on the west side of the island

[184] Cadell, H. M. (1919) *Diary*, National Library of Scotland, Edinburgh; *Extracts from H. M. Cadell's chronicle, Vol. 4 1910-1925*, Cadell of Grange papers, Acc. 5381, National Library of Scotland, Edinburgh; Brown, R. N. R. (1920) 'Recent developments in Spitsbergen', *Scottish Geographical Magazine*, 36, pp. 111-6.
[185] *Scotsman* (1919) 'Spitsbergen expedition, R.Y. Petunia leaves Leith', 16 July.
[186] Bruce, W. S. (1919) *Report to directors*, Scottish Spitsbergen Syndicate papers MS 311, Scott Polar Research Institute, Cambridge.
[187] Scott, J. (1919) *Report to the Scottish Spitsbergen Syndicate on the expedition to Prince Charles Foreland, 19/8/1919*, Scottish Spitsbergen Syndicate papers MS 311, Scott Polar Research Institute, Cambridge.

to the south of Miller Glacier, from where they studied an outcrop of iron ore. A change of staff occurred during a rendezvous with the *Petunia* at Point Napier on August 19. George Walter Tyrrell, chief geologist of the expedition, now arrived on the island and took charge of a reduced party.[188] His group relocated to the west coast of the Foreland. Two miners then returned to house No. 3, which had been assembled at Kenmore, for more supplies. On September 5, the men were picked up by the homebound *Phantom*.

After its initial call at the Foreland, the *Petunia* continued to Klaas Billen Bay and anchored in Adolf Bay on July 28, 1919.[189] The main expedition met up with the advance party, who had explored the area south of Nordenskiöld Glacier and discovered the Bridgeness coal outcrop along the River Carron, a tributary to the River Gerrit. According to A. F. Campbell, the mining tools and drilling plant were unloaded, and the drillers went to work on the first of three borehole locations previously marked out by the mining engineer (Fig. 8.9).[190]

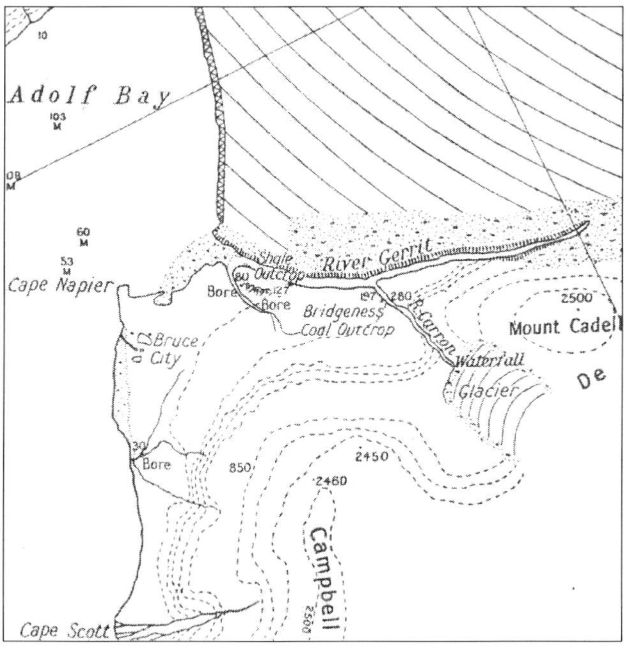

8.9 Detail of a map showing the Bridgeness coal outcrop and the three borehole locations near Bruce City. (Source: Cadell, H. M. (1920) 'Spitsbergen in 1919', Scottish Geographical Magazine, 34, pp. 1-10.)

[188] Tyrrell, J. W. (1919) *Report to directors*, Scottish Spitsbergen Syndicate papers MS 311, Scott Polar Research Institute, Cambridge.
[189] Bruce, W. S. (1919) *Report to directors,* Scottish Spitsbergen Syndicate papers MS 311, Scott Polar Research Institute, Cambridge.
[190] Campbell, A. F. (1919) *Report to directors*, Scottish Spitsbergen Syndicate papers MS 311, Scott Polar Research Institute, Cambridge.

Bore No. 1 was situated about one and a half kilometres to the south of the imminent Bruce City. According to Hoel, Campbell expected to find coal at a depth of 100 metres, but after the hole was sunk without coal being found, he calculated that the seam lay instead at 290 metres, which could not be reached with the equipment available.[191] Bore No. 2 lay to the south of the River Gerrit. It was intended to investigate a layer of quartzite and the strata beneath. After about six metres, it was discontinued. Bore No. 3 was roughly half a kilometre further inland. Coal here had been expected at circa 47 metres, but it had not been encountered when drilling stopped at 50 metres on September 4. Campbell assumed the coal to be associated with a dark shale. The miners had already made a cutting into such an outcrop along the River Gerrit, but it took them too long to reach solid rock here. They moved into the river bed instead, mining only six feet before coming across a nine-inch seam. The results suggested a fault to the north and west of bore No. 3. Nonetheless, Campbell recommended the future investigation of the area to the south of the River Gerrit, unless, of course, the analysis of the coal was unfavourable. While prospecting commenced, a site was chosen for the pre-fabricated houses No. 1 and No. 2. This marked the beginning of Bruce City and gave rise to the name of the Bruce City coalfield.

The season's programme also entailed the exploration of Stor Fjord. On August 2, Bruce, Mathieson, Cowan, Tyrrell, and Wordie landed at Mount Keilhau to the east of the South Cape, where two sets of supposedly Norwegian footprints warranted a comment.[192] As ice conditions improved, the ship was able to reach Agardh Bay. This allowed them to land stores for a geological party working in the direction of Sassen Bay. The stores were deposited next to a Scottish claim board. The men stayed overnight in Mohn Bay. Thereafter, Mathieson briefly surveyed Changing Point on Barents Island. On the return journey, the party temporarily sought shelter at Cape Lee on Edge Island, which Bruce took the opportunity of claiming on August 7.[193] Wordie noted a successful trip, yet the ice could be deceptive: a ship entering the sound in July may have six weeks without fear of being frozen in; August may already be too late.

The ice conditions in Stor Fjord, however, would only be of interest if the region could be exploited economically. In 1909, Bruce and Hannay had reported a Broxburn-like odour on the east coast, Broxburn being a Scottish village with an

---

[191] Hoel (1967a) p. 1061.

[192] Wordie, J. M. (1919) *East Spitsbergen in 1919*, Gen 1652 107/16, Centre for Research Collections, Edinburgh University, Edinburgh.

[193] Foreign Office (1919) Letter to the Norwegian Foreign Ministry, 13 November. Norwegian Foreign Ministry Box 5373, National Archives of Norway, Oslo.

early oil works. Wordie traced this odour in August 1919.[194] It was located at a large glacier immediately to the south of Hayes Glacier in Mohn Bay. Wordie named it Usher Glacier. At a point roughly two miles from the coast, a torrent gushed out of an ice cave. The strong bituminous smell around the ice cave was intensified by the moist atmosphere and seemed to be coming from below the glacier. The Scot gathered fragments of black shale, which reeked when he broke them. He concluded that the occurrence of any oil needed to be looked for higher up the glacier.

Having visited Prince Charles Foreland on August 19, Bruce and Wordie arranged their early return to Britain with the assistance of Store Norske.[195] They delivered some preliminary results, while the expedition laboured on until the beginning of September. Brown had taken over command, and it fell to him to coordinate the departure. On September 1, both the *Petunia* and the *Phantom* sailed for Advent Bay to coal.[196] Campbell and 15 men remained at Klaas Billen Bay, where bore No. 2 had been completed and bore No. 3 had reached a depth of 22 fathoms, although no coal had been discovered. In Gerrit Valley, an adit had been driven 60 feet into a coal seam two feet and three inches thick. A two-ton sample had been loaded. On Brown's orders, all participants embarked on the *Petunia* and left Klaas Billen Bay on September 5.[197] In Green Harbour, they were reunited with the party from the Foreland and departed for Norway.

A short report subsequently summed up the Scottish properties and expressions of ownership.[198] Bruce had claimed the north of Edge Island, enlarging the syndicate's territory by 1,330 square miles. This acquisition marked its greatest extent of 2,980 square miles and was accompanied by an updated claim map (Fig. 8.10). The syndicate had placed claim boards at Bjona Haven, the mouth of Sassendal, the Anser Islands, the head of Klaas Billen Bay, the north shore of Adolf Bay, at Napier Point and to the west of Richard Lagoon on Prince Charles Foreland, at Changing Point on Barents Island, and at Cape Lee on Edge Island. It had strengthened the claim of the Bruce City coalfield through the construction of two pre-fabricated houses. A third had been built at Kenmore, and a fourth at Bjona Haven. At Bruce City and Bjona Haven, large Union Jacks had been painted onto

---

194 Wordie, J. M. (1919) *Note on the location of the Broxburn-like odour reported by Dr Bruce and Mr Hannay in 1909, based on its re-discovery by Mr Wordie in August 1919*, Gen 1652 103-5, Centre for Research Collections, Edinburgh University, Edinburgh.

195 Hoel (1967a) p. 1060.

196 Brown, R. N. R. (1919) *Report*, Scottish Spitsbergen Syndicate papers MS 311, Scott Polar Research Institute, Cambridge.

197 Bruce, W. S. (1919) *Report to directors*, Scottish Spitsbergen Syndicate papers MS 311, Scott Polar Research Institute, Cambridge.

198 *Report of claims, claim boards, and houses* (nd) Scottish Spitsbergen Syndicate papers MS 311, Scott Polar Research Institute, Cambridge.

the house walls.[199] A text in both English and Norwegian informed any visitors that the house and its contents belonged to the Scottish Spitsbergen Syndicate and asked to kindly close the door, 'Luk døren'.

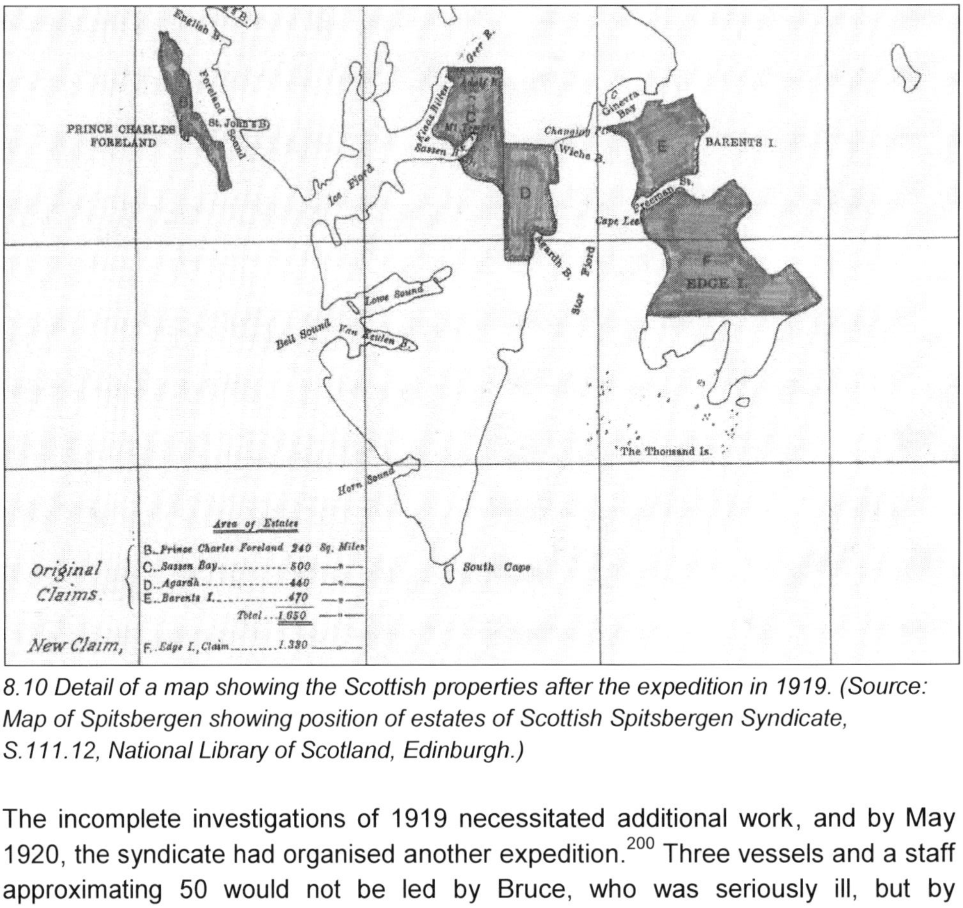

8.10 Detail of a map showing the Scottish properties after the expedition in 1919. (Source: Map of Spitsbergen showing position of estates of Scottish Spitsbergen Syndicate, S.111.12, National Library of Scotland, Edinburgh.)

The incomplete investigations of 1919 necessitated additional work, and by May 1920, the syndicate had organised another expedition.[200] Three vessels and a staff approximating 50 would not be led by Bruce, who was seriously ill, but by Mathieson. An advance party of borers headed by A. F. Campbell left Edinburgh aboard the Norwegian sloop *Autumn* as early as May 4.[201] The main party followed on May 17 on the *Easonian* and on the *Lady of Avenal*. The syndicate had found it most economical to purchase the wooden vessels furnished with auxiliary power

[199] Hoel (1967a) p. 1062. A photograph of the painted flag at Bruce City exists in the Photo Library (np002354) of the Norwegian Polar Institute in Tromsø.
[200] *The Times* (1920) 'Coal from the Arctic', 5 May, p. 18.
[201] *Linlithgowshire Gazette* (1920) 'Spitsbergen expedition, departure from Granton', 21 May; *The Times* (1920) 'The Scottish Spitsbergen Syndicate, Limited', 26 May.

and to refurbish them for their purposes.[202] Even after their departure, the directors maintained that it was 'not desirable to make public the full details of the elaborate programme [...]: suffice it to say that the work of our experts will be mainly directed in proving the coal and iron, in addition to which special attention will be given to the existence of oil and copper, of which valuable minerals we have distinct indications.'[203] The directors did disclose, however, that they contemplated the development of tourism and the construction of hotels and health resorts.

Delayed by drift ice, the *Autumn* was only able to enter Klaas Billen Bay on June 6.[204] The *Easonian* had caught up and arrived already on June 11. More ice blocked the way north, so the men disembarked at Cape Ekholm, which provided the opportunity for prospecting work on the lower slopes of the Campbell Range and on Mount Tyrrell and Mount Tjosaas.[205] Yet, A. F. Campbell was eager to begin the season's drilling programme. He organised a series of sledging trips that conveyed 12 tons of material to Bruce City approximately five miles to the north over a period of ten days.[206] The large amount of snow on land demanded the greatest exertion from the men. The *Lady of Avenel* was the last to arrive on June 20; the ice would not budge until July 14.

Campbell continued the sinking of bore No. 3.[207] It cut the Carron seam at a depth of 48 fathoms and proved each of the two main horizons to be two feet ten inches thick before being terminated at about 61 fathoms.[208] Since ice too weak for sledging and too strong for a ship to break through still bared the route to the northern shore of Adolf Bay, the expedition leaders discussed the most practical position for another borehole.[209] They chose a spot in a valley to the east of Cape Ekholm 2.8 kilometres from the sea and as close to the Ekholm River as possible. They estimated that coal here lay between 110 metres and 125 metres. So drilling equipment, a cabin, and provisions were brought to the location, using a four-wheeled cart pulled by up to ten men. When drilling was already in progress, a re-assessment suggested that the coal actually lay much deeper. The drillers could, in any case, not penetrate the thick overburden, and the work was suspended.

Meanwhile, on the basis of bore No. 3 having been inconclusive, the site for bore No. 5 had been selected. This lay near the River Gerrit about 400 metres

---

202 *The Times* (1920) 'The Scottish Spitsbergen Syndicate, Limited', 26 May.
203 *The Times* (1920) 'The Scottish Spitsbergen Syndicate, Limited', 26 May.
204 Hoel (1967a) p. 1062.
205 Campbell, R. (1920) *Geological notes on Spitsbergen 1920*, Robert Campbell Collection MS 502/1, Scott Polar Research Institute, Cambridge.
206 Hoel (1967a) p. 1063.
207 Hoel (1967a) p. 1063.
208 Campbell, A. F., Campbell, R., and Mathieson, J. (1920) *Joint report on Bruce City coal field Spitsbergen,* Robert Campbell Collection, Scott Polar Research Institute, Cambridge.
209 Presumably bore No. 4, which does not find mention in the sources.

to the northeast of the former.[210] Drilling began on June 28 and was completed on September 1. Campbell had taken five coal samples from this borehole (Fig. 8.11), before it was terminated at 117 metres.

	Thickness		Depth			Box
	Ft.	Ins.	Fms.	Ft.	Ins.	
Coal	1	5	19	3	8	1
Coal	2	10	23	1	10	2
Coal	2	0	47	3	0	3
Coal	1	3	50	4	11	4
Coal	1	6				
Dirt rib		3	53	0	0	5
Coal	1	4				

8.11 Details of samples taken from bore No. 5. (Source: Campbell, A. F. (1920) Details of samples taken, Scottish Spitsbergen Syndicate papers, Scott Polar Research Institute, Cambridge.)

Both bore No. 3 and bore No.5 revealed an upper and a lower coal horizon in the Bruce City coalfield.[211] The coalfield's outcrops belonged exclusively to the upper horizon and were confined to the bed of the Carron River. The Scots examined these exposures with the greatest care. Two staff initially dug several shallow trenches; larger excavations were made by the miners. Although coal seams and coal blaes (shale) were frequent, they were thin and impersistent, implying that last season's successes in the mine at the Bridgeness outcrop had been an exception.

On July 14, the fast ice finally budged, and the ships were able to reach Bruce City. Tyrrell and four miners proceeded to the northern shore of Adolf Bay to continue the geological investigations and to open up the coal discovered in the previous year.[212] Tyrrell initially marked out a borehole location approximately 1.6 kilometres to the west of Nordenskiöld Glacier and 270 metres from the beach, but it was not possible to sink a borehole here. An alternative location was chosen 270 metres to the west of the glacier and only 60 metres inland. The drilling equipment from Cape Ekholm was now transferred to the site of bore No. 6. The operations lasted from August 4 until August 31. Coal was encountered on three occasions, proving the upper and lower horizon. A. F. Campbell again collected samples (Fig. 8.12) prior to bore No. 6 being terminated at 50 metres.

Bore No. 6 emphasised the need for further work between Bruce City and the Bridgeness mine. R. Campbell and A. F. Campbell selected a site for another

[210] Hoel (1967a) p.
[211] Campbell, R. (1920) *Geological notes on Spitsbergen 1920*, Robert Campbell Collection MS 502/1, Scott Polar Research Institute, Cambridge.
[212] Hoel (1967a) p. 1064.

borehole approximately 300 yards south of bore No. 3 and had the plant from bore No. 5 erected there. This was deemed as far in the down-dip direction of the strata as drilling would be able to reach the base of the Culm formation with the machinery presently available. R. Campbell recommended the use of heavier plant in the future, which should preferably be landed at the Ekholm site.

	Thickness		Depth			
	Ft.	Ins.	Fms.	Ft.	Ins.	
Coal	2	3	13	2	5	Wooden box
Coal	1	2	17	0	8	Tin box
Coal	1	5	18	1	2	Tin box

8.12 Details of samples taken from bore No. 6 (Source: Campbell, A. F. (1920) Details of samples taken, Scottish Spitsbergen Syndicate papers, Scott Polar Research Institute, Cambridge.)

After the investigations at the northern shore of Adolf Bay were completed, Tyrrell and his miners progressed to Ebba Valley. He was charged with the continuation of geological studies to prove the Ebba Valley coalfield.[213] By now, the Scots had absorbed the territory into their claims. This was demonstrated by both a large claim board dated back to the year 1909 and an even larger flag post that presumably sported a Union Jack (Fig. 8.13).

8.13 The Swedish hut in Ebba Valley was visibly incorporated into the Scottish claim, using a claim board and a flag post. (Source: Sheila Watt private archive, Edinburgh.)

When they were no longer working on a landslide but had found an *in-situ* outcrop, R. Campbell conducted further trials. At a height of approximately 268 metres, an adit was driven into a coal seam and over 20 tons of coal were extracted. Since the so-called Ebba Valley mine lay roughly 4.8 kilometres from the coast and its

[213] Hoel (1967a) p. 1064.

accessibility was terrible, only small samples could be taken away. A. F. Campbell recorded these to have been collected at the working face 43 feet in. His record also indicated the seam to have been a total of 3 feet 11 inches thick. Although the seam corresponded to the Bridgeness coal at the River Carron, its quality was much poorer.[214]

Meanwhile, Wordie had advanced the investigation on Prince Charles Foreland.[215] His party of six had anchored off the mouth of Richard Lagoon on June 26.[216] It had been difficult to enter the lagoon, so the men landed on the spit and dragged the boat across.[217] Their camp lay on the west side and comprised two bell tents. An outside galley was built from turf in the style of a grouse butt, and driftwood provided the necessary fuel. At first the wood was covered in snow, but later there was plenty. Wordie related little about the task of the group, but he was enthusiastic about their reindeer sleeping bags. According to Hoel, the party's mission had been to ascertain if the iron ore discovered the previous summer could be found elsewhere on the island, but they were unsuccessful. The men left the Foreland on July 22. A. F. Campbell later concluded that any operations in the area would be subject to major technical difficulties that would render the exploitation of mineral resources financially irresponsible.

The busy summer of 1920 also witnessed the visit of the syndicate's board. With the exception of W. G. Burn Murdoch, the directors were accompanied by Bruce, Brown, and a few others. Cadell sighted Spitsbergen, 'a dismal prospect, cold, inhospitable and bleak to the last degree', on July 21.[218] After calling at a small, smelly whale oil factory, he resolved that 'the place was not quite so pleasant at first sight as I expected.'[219] On July 23, the distinguished guests arrived at Bruce City. The *Autumn* took them to Ebba Valley, where Tyrrell was opening the coal, but Cadell could not walk as far as the workings. Further excursions led the director to the Bridgeness outcrop at the River Carron and around Klaas Billen Bay, from where he was able to observe Mount Cadell named in his honour. He found the hut at Bjona Haven in good order, and on July 30, the group left for Advent Bay, where they received a tour of Store Norske's workings. They were 'very much impressed [...] at the provisions by the Management of the Colliery for the social and moral welfare of the community, including clergy, doctors, nurses

---

214 Hoel (1967a) p. 1065.
215 Wordie, J. M. (1921) 'Present-day condition in Spitsbergen', *Geographical Journal*, 58 (1), pp. 25-45.
216 Hoel (1967a) p. 1065.
217 Wordie (1921).
218 Cadell, H. M. (1920) Diary, Cadell of Grange papers, Acc. 5381, National Library of Scotland, Edinburgh.
219 Cadell, H. M. (1920) Diary, Cadell of Grange papers, Acc. 5381, National Library of Scotland, Edinburgh.

and a well-equipped hospital.'[220] They undoubtedly thought this befitting of a mining company, keeping in mind that they focused on exploration. After more sightseeing in Recherche Bay and Braganza Bay among other places, the men were back in Norway on August 7.

Wordie, Tyrrell, Mathieson, and some workers continued into Stor Fjord aboard the *Autumn.*[221] The sound had not been navigable until August 8, and although fog and ice were common here, 1920 was a good year. Wordie even thought the water too open since heavy swells hampered the landings. The men visited Agardh Bay, Mohn Bay, various points on Barents Island, the northern part of Edge Island, and Whales Bay. The main objective was to revise the Admiralty chart for this area, and they ran some soundings that disproved the ancient Dutch charts. Hoel later related that Wordie discovered a thin seam of good-quality coal and Tyrrell two layers of phosphate-bearing limestone in Agardh Bay, while a study of Mohn Bay had led to the same conclusions regarding the oil smell from beneath Usher Glacier.[222] On the east-lying islands, coal had been found in four locations, none thicker than a mere five centimetres.

Meanwhile, A. F. Campbell and R. Campbell had ventured into Gips Valley.[223] On August 17 and 18, they studied a layer of coal approximately 19 kilometres from the coast. On the west side of the valley, it lay at a height of about 64 metres, on the east side at 79 metres. The seam was thought to correspond to that in Ebba Valley, but it could not be sufficiently examined because the men only had light tools at their disposal.[224] They therefore secured a 'sample representing about one foot of coal in tin box (oatmeal).'[225] In addition, a large sample of gypsum was shipped from gypsum workings at the shore (Fig. 8.14).

Based on the expeditions in 1919 and 1920, the board decided to limit future activities on Spitsbergen to the exploration of the Bruce City coalfield and Gips Valley.[226] It would also drill additional boreholes to prove the Ebba Valley coalfield. It dispatched a party of ten under the leadership of John Mathieson from Edinburgh on May 31, 1921.[227] The journey led via Newcastle and Bergen to Tromsø, where the *Autumn* had again been chartered. The cutter left Norway on June 9, and her route was ice-free until she reached Cape Ekholm, where she moored along the ice edge. On June 14, the men embarked on the familiar five-

---

[220] Hoel (1967a) p. 1066.
[221] Wordie, J. M. (1921) 'Present-day condition in Spitsbergen', *Geographical Journal*, 58 (1), pp. 25-45.
[222] Hoel (1967a) pp. 1065-6.
[223] Hoel (1967a) p. 1065.
[224] Campbell, R. (1920) *Geological notes on Spitsbergen 1920*, Robert Campbell Collection MS 502/1, Scott Polar Research Institute, Cambridge.
[225] Campbell, A. F. (1920) *Details of samples taken*, Scottish Spitsbergen Syndicate papers, Scott Polar Research Institute, Cambridge.
[226] Hoel (1967a) p. 1067.
[227] *Financier* (1921) 'Scottish Spitsbergen Syndicate Limited', 6 June.

mile journey to Bruce City, pulling a heavy sledge loaded with tools, supplies, and other equipment. The trip there and back lasted a whole day.

8.14 Working gypsum in Gips Valley. (Source: Sheila Watt private archive, Edinburgh.)

8.15 Scottish expedition members in front of the hut in Gips Bay in 1921. (Source: Sheila Watt private archive, Edinburgh.)

The *Autumn* departed for Gips Bay late the next day.[228] After landing equipment and supplies, the party erected a hut on the raised beach (Fig. 8.15). Thereafter, they chose a suitable camp site some distance into Gips Valley. Over the next six days, the men moved nearly three tons of stores to the camp. Their subsequent survey work was later summarised in Mathieson and Finlay's topographic map of Gips Valley.[229] It depicted the hut at the shore and two camps. The first camp lay about five miles inland in Aitken Valley at an altitude nearing 200 feet. The second camp lay almost nine miles from the shore south of the Gips River at a height of

[228] Hoel (1967a) p. 1068.
[229] Mathieson, J. and Finlay, T. R. (1921) *Map of Gips Valley*, Norwegian Polar Institute, Tromsø.

about 140 feet. Its proximity to a series of coal outcrops circa ten miles inland singled it out to be the 1921 base camp. According to Hoel, the outcrops could be traced over a considerable distance and were found to dip at an angle of 2.5 degrees towards west-south-west.[230] As such, coal could be expected to lie 600 metres below sea level at the coast. Only 3.2 kilometres down the valley, however, it should be found at no more than 90 metres.

On August 8, Prof. Robert Wilson Dron, a mining engineer and gypsum specialist, arrived in Klaas Billen Bay.[231] He examined the trials in the Bruce City coalfield and in Gips Valley. The Bruce City coal lay 'at very even and undisturbed gradients and there is no difficulty in tracing the general sequence of the strata along the coast line and to the entrance of Gips Valley and thence along the two sides of the Valley to the point where the coal comes to the surface.'[232] On the basis of trials in Gips Valley (Fig. 8.16), he concluded that 'the coal proven [...] is a seam 3 feet thick of good coking quality. There are geological reasons to anticipate that this seam will improve towards the dip and in addition other seams may be found. I advise that boring and proving operations be carried out at a point about a mile west from the openings made this year. If these operations give the results which I expect a good workable coalfield extending to many square miles will be proven.'[233] He added that 'gypsum occurs in practically inexhaustible quantities in this region. There are two main beds of gypsiferous rock, in one of which occurs 263 feet of solid Gypsum. The mineral occurs in suitable localities at or near the seashore with good loading and transport conditions.'[234]

8.16 Trial workings in the coal seams in Gips Valley in 1921. (Source: Sheila Watt private archive, Edinburgh.)

[230] Hoel (1967a) p. 1068.
[231] Hoel (1967a) p. 1069.
[232] Dron (1921), cited in Hoel (1967a) p. 1069.
[233] Dron (1921), cited in Hoel (1967a) p. 1069.
[234] Dron (1921), cited in Hoel (1967a) p. 1070.

Between August 10 and 24, the mining engineer James Cooper and the geologist T. R. Finlay were active in Ebba Valley.[235] They inspected the workings of earlier years and confirmed that the coal at this location was similar to that in Gips Valley. They tried to discern the lower horizon closer to the shore but were unsuccessful. However, they correlated an outcrop of the upper horizon on the west side of the valley with a layer of coal on the east side. They continued to the northern shore of Adolf Bay, where additional studies underlined their view that a new borehole approximately 360 metres to the west of bore No. 4 would reach the lower layer. Presumably after collecting a sledging party of the first Oxford University Expedition from their base at Bruce City[236], the syndicate's voyage concluded on September 12. Within three days, the men were back in Tromsø and reached Edinburgh on September 22.

The expedition in 1922 entailed the syndicate's last major programme and concluded four intensive field seasons. Its 15 participants, again led by Mathieson, left Edinburgh on June 10.[237] In Tromsø, their departure was delayed by taking on supplies for the Anglo-Russian Grumant Co., whose warehouse had been destroyed by an avalanche. The journey commenced on June 19, and within three days, the ship lay outside a wide belt of pack ice. The ice blocked the way until June 25, when the Store Norske coal boat could break a lead through the floes, which the *Autumn* followed in. Another three days were spent discharging the supplies for the Anglo-Russian Grumant Co. in Coles Bay, unloading equipment at Gips Bay, and inspecting Bruce City, before operations in Gips Valley commenced in earnest on June 29.

Based on Prof. Dron's recommendations, a borehole was positioned a little less than nine miles inland. At this site, the upper coal horizon was expected at approximately 90 metres.[238] A hut and a tent camp were erected, which acted as the base (Fig. 8.17). On July 17, sinking of No. 1 bore commenced.[239] Rockhead was encountered at 26 feet and proven to 31 feet before the hole was securely lined and the shot-cutter employed on July 27. The work initially proceeded without major difficulties to 38 fathoms. With increasing depth, however, it was plagued by frequent problems and stoppages. On August 11, the core jammed at 46 fathoms 4 feet 8 inches. After efforts to free it failed, the hole was abandoned on August 12 without having reached the coal.

[235] Hoel (1967a) p. 1068.
[236] *The Times* (1921) 'Exploring the unknown. The Spitsbergen expedition. Wonders of bird life. Scenery and bagpipes.' 7 September, p. 9.
[237] Hoel (1967a) p. 1070.
[238] Hoel (1967a) p. 1071.
[239] Hoel (1967a) p. 1071.

8.17 Hut and tent camp in Gips Valley in 1922. (Source: Photo Library np001634, Norwegian Polar Institute, Tromsø.)

The summer had advanced considerably, so subsequent boreholes needed to target shallower depths. The site of the second borehole was therefore selected about two kilometres closer to the outcrop.[240] Bore No. 2 commenced on August 14. By September 7, the drillers reached a depth of 96 ½ feet without finding coal. The fact that a sudden gale blew over the sheer legs, breaking the crane handle, and that the hole was in turn abandoned to save time only found mention in the engineer's report to the syndicate, not in his later scientific article.[241] A total of seven bores was sunk in 1922.[242] Coal was found in none of them. Yet, they confirmed Dron's opinion that the layer of gypsum was extensive and without faults.[243] Upon leaving Gips Valley, the boring plant was properly greased and stored in the engine hut at No. 2 bore, while a few rods and the chisel were also left at bore No. 7.[244]

On Prince Charles Foreland, the snow had melted more extensively than in previous years, and the iron ore could be traced over larger distances. Yet, Mathieson retained the view that transport would be difficult and that the harbour facilities were poor. An analysis of the samples would discern if the iron ore was worth any effort and expenditure. Some work was done on a copper vein, and with the exception of Black Hill, no pyrites were found in the pieces of iron-stained

[240] Hoel (1967a) p. 1071.
[241] Dron, R. L. W. (1922) *Report on work done for the Scottish Spitsbergen Syndicate*, Scottish Spitsbergen Syndicate papers MS 311, Scott Polar Research Institute.
[242] Dron (1922).
[243] Hoel (1967a) p. 1071.
[244] Dron (1922).

chalybite (siderite) lying on the surface.[245] The expedition left Spitsbergen on September 17. After a stormy crossing, it arrived in Tromsø on September 20. By September 28, the men had returned to Edinburgh.

8.8.2 Manifestations and technologies

The rough cost estimate for the 1919 expedition listed the items intended for use on shore and probable additional expenses.[246] These items included huts (£930), hut fittings (£500), and tents and outfits (£320), which hinted at the structures to be put up. Borers (£450), miners (£901), and geological implements (£350) suggested the tasks to be carried out. The huts were reportedly constructed by Messrs Thomson and Balfour's Victoria Sawmills in Bo'ness.[247] They were strong structures measuring 23 feet by 15 feet, and they were lined, the spaces between the lining and the outside walls packed with sawdust. It was hoped that this would keep out the cold. Bruce personally inspected the huts and suggested the addition of a porch to allow for double doors to be fitted. Each hut could accommodate six men. However, the *Petunia* soon proved very clumsy, used too much coal, and was ill-adapted for carrying stores, so two pre-fabricated huts were left behind in Leith.[248] Of the four huts taken, No. 1 and No. 2 were erected at Bruce City in Klaas Billen Bay.[249] They were lined with bunks and had tables and two stoves each. It is perceivable that the two remaining huts were delivered in 1920, since a photograph taken in early 1921 already showed four buildings at Bruce City. House No. 3 had been put up at Kenmore. It was lined, except its partition, and it had one large stove. House No. 4 at Bjona Haven was unlined and without any bunks, tables, or stoves.[250] Although this hut was shown in Mathieson's map of 1919, no image could as of yet be found. All hinges and locks were oiled, and each hut displayed the unambiguous 'SSS' stencil of ownership.

In 1919, Messrs Kyle of Cumnock in Ayrshire reportedly undertook the boring.[251] Meant was the firm Andrew Kyle & Co. of nearby Galston. According to Cadell, the main expedition comprised two boring parties.[252] The mining foreman

---

[245] Dron (1922).
[246] *Rough estimate of cost of expedition* (nd) Scott Polar Research Institute, Cambridge.
[247] *Bo'ness Journal* (1919) 'Bridgeness miners for Spitsbergen', 20 June.
[248] *Extracts from H. M. Cadell's chronicle, Vol. 4 1910-1925*, Cadell of Grange papers, Acc. 5381, National Library of Scotland, Edinburgh.
[249] Scottish Spitsbergen Syndicate papers MS 311, Scott Polar Research Institute, Cambridge.
[250] Archaeological fieldwork has shown a house labelled No. 4 to stand at Kenmore.
[251] *Linlithgowshire Gazette* (1919) 'The Scottish Spitzbergen expedition, coal-winning in the Arctic islands', 19 September.
[252] *Extracts from H. M. Cadell's chronicle, Vol. 4 1910-1925*, Cadell of Grange papers, Acc. 5381, National Library of Scotland, Edinburgh.

listed four drillers among the staff.[253] The drillers set to work at the first of three locations.[254] Their machinery comprised a light drill mounted on a small cart. It was powered by a belt drive connected to an engine in a small engine shed. Miller did not recollect any problems in the frozen overburden, but he remembered that diamond drilling was difficult because the water for clearing out the debris turned to ice and closed the borehole.[255] This was overcome by bending a copper tube into a spiral, connecting it to the water supply, and placing it over a coal fire. It fed hot water down the hole, but it had to be manned continuously. A. F. Campbell recommended using steam-driven rather than oil-driven machinery and including an apparatus that would heat the circulating water during drilling in order to keep the borehole from freezing.[256] He also suggested the transition from diamond drilling to using abrasive sands, which would enable the use of drill bits with a larger diameter. He highlighted the necessity of allocating a crew of eight to each drilling rig, divided into two teams of four on 12-hour shifts. Drilling could then be continuous in order to prevent the borehole from closing up.

8.18 The Fordson tractor was reliably employed under difficult conditions in 1922. (Source: Photo Library np001633, Norwegian Polar Institute, Tromsø.)

In 1922, a Fordson tractor (Fig. 8.18), inexpensive compared with a railway, solved the problems of transportation in Gips Valley.[257] Since the tractor weighed over a ton, landing it presented a challenge. This was resolved by the captain of the

---

[253] Miller, G. W. (1978) *Expedition to Spitsbergen 1919* [Audio cassette], Callender House Museum, Falkirk.
[254] Campbell, A. F. (1919) *Report to directors, 28 July*, Scottish Spitsbergen Syndicate papers MS 311, Scott Polar Research Institute, Cambridge.
[255] Miller, G. W. (1978) *Expedition to Spitsbergen 1919* [Audio cassette], Callender House Museum, Falkirk.
[256] Hoel (1967a) p. 1061.
[257] Dron, R. L. A. (1924) 'Boring in frozen strata', *Transactions of the Institution of Mining Engineers*, 67, pp. 186-91.

Autumn, who had two lifeboats strapped together and planks placed across them.[258] The makeshift raft was sturdy enough to take the weight of the tractor. At high tide, it was rowed a little distance up Gips River, where the vehicle could be landed on the river bank without mishap. The Fordson was primarily employed for haulage. It pulled a two-wheeled trailer repeatedly loaded with boring gear, provisions, huts, tents, bedding, coal, and petrol. After 15 days of negotiating rough terrain in the form of boulder-strewn glacial fans intersected by numerous streams, it succeeded in conveying the entire stock to the camp in Gips Valley.

Outboard motors were another important contribution easily overlooked in the technological evolution on Spitsbergen. Historical photographs commonly show rowing boats or sailing vessels, yet Bruce had already taken engines, presumably outboard motors with internal combustion engines, to be fitted onto such crafts in 1914. They also featured in the cost estimate for 1919. Motor boats were faster and less dependent on natural conditions. They therefore speeded up travel between the claims. It is likely that all three types coexisted for a while. The final transition from manual toil to routinely using motorised power is not well recorded. It notably occurred not long after the initial experimentation with motor boats at the turn of the century.[259]

The technological innovations of 1919 defined the development programmes of successive expeditions. The Scots considered boring to be the best means by which to prove the geological strata and any natural resources. In 1922, Messrs. Thomson of Dunfermline did the drilling.[260] Their plant comprised a hand-operated machine, which was driven by a belt from a five-horsepower National petrol-paraffin engine. The drilling rods were one inch thick and 12 feet long. The sheer legs were high enough to allow for a 24-foot length to be raised and lowered, which was done by a hand-crane attached to the sheer legs. The drill bits supplied were both shot cutters and diamond crowns. Although thick overburden was not expected, 30 fathoms of lining tubes to prevent the borehole from collapsing, and a set of chisels were available.

On July 17, 1922, No. 1 bore commenced in Gips Valley, and an early decision was made to treat the frozen overburden like solid rock.[261] The drillers initially employed the shot cutter, but circulating water raised the temperature in the borehole, causing the gravel to run and a whole host of associated problems. 'This seemed to prove conclusively that Spitsbergen surfaces, although hard frozen,

---

258 Hoel (1967a) p. 1071.

259 Redwood, B. B. (1906) 'Motor boats', *Journal of the Society of Arts*, pp. 512-20. Available at: http://www.lesliefield.com/other_history/a_short_history_of_motor_boating.htm (Accessed: 12 November 2012).; Encyclopædia Britannica (2012) [Online]. Available at: http://www.britannica.com/EBchecked/topic/394352/motorboat (Accessed: 12 November 2012).

260 Dron (1924).

261 Dron (1924).

cannot be bored through by any other means than by the chisel.'[262] The chisel repeatedly got stuck because ice had grown on the sides of the hole. The men therefore decided to use only salt water in the hole and that just enough water should be used to keep the cuttings from clogging the chisel. Rockhead was first encountered at 26 feet and proven to 31 feet before the hole was lined and the shot-cutter re-employed on July 27. The drillers drew on the experience of previous bores to keep the circulating water at a moderate temperature. Hence, the water from the pump passed through a small cast-iron boiler. However, the coal-fired boiler was inefficient because it consumed too much coal, which had to be transported up the valley, while most of the heat went out the chimney. It was felt that if the temperature of the outlet water from the hole was never allowed to fall beneath 45°F (7°C), there should not be any problems with ice down the hole. The work proceeded steadily to a depth of 38 fathoms. On August 11, the core jammed and could not be freed by passing water. The hole was abandoned.

8.19 Drilling in progress at No. 2 bore in Gips Valley (Source: Images for All, Royal Scottish Geographical Society, Perth.)

The site of No. 2 bore lay about two kilometres closer to the outcrop.[263] Operations commenced on August 14 (Fig. 8.19). At an initial speed averaging 3 feet 4 inches per 8-hour shift, the hole was chiselled and lined to 47 feet 5 inches.[264] As an alternative to the shot cutters, the diamond crowns were now tested, but as soon as the drillers realised that they were still in overburden, the crown had jammed,

[262] Dron (1924) p. 187.
[263] Hoel (1967a) p. 1071.
[264] Dron (1924).

losing one of its diamonds.[265] Chiselling with a 6-inch cutting tool continued to 66 feet. The tool was then exchanged for a smaller chisel, which significantly reduced the average speed. On September 7, the borehole reached a depth of 96 ½ feet, without encountering coal, when a gust of wind blew over the sheer legs.[266] Another five boreholes were attempted but with little success.[267]

8.8.3 Employees

According to the *Scotsman*, the 1919 expedition compensated for its delay by increasing the staff of experts, including a large scientific staff.[268] In successive years, the scientific rather than the technical personnel continued to be the show horse of the Scottish work on Spitsbergen. Newspapers repeatedly highlighted academic titles and professional affiliations. University appointments and fellowships at institutions such as the Royal Geographical Society, the Royal Scottish Geographical Society, the Royal Society of Edinburgh, the Geological Society, the Institute of Mining and Metallurgy, and the Institute of Civil Engineering were important. The syndicate itself mostly referred to staff in relation to the costs attached to each person, but personal communications between the men revealed additional details. The workforce would be practically invisible, were it not for some exceptional photographs and documents. The staffing attached in Appendix 2 is neither complete nor was every name on the list accounted for in the field, but it provides the foundation for this section and for future research.

Staffing weighed heavily on Bruce's mind because he wanted to know his legacy looked after. He informed Brown that he had been asked to be the organisational secretary to oversee boats, stores, and the like, which he had angrily declined.[269] He had recommended Frank B. Napier as a practical man and skipper, and he had suggested James Mann Wordie. Wordie's attendance of Shackleton's *Endurance* expedition was not mentioned; perhaps his polar competence was obvious. He was thought capable of replacing Bruce and Brown in later years. For the time being, however, Bruce felt that the two veterans had certainly earned the right to a decent salary and should insist on it.[270] By May 21, they had not heard of their employment, so Brown reminded the syndicate of his

---

[265] Synthetic diamonds were not invented until the 1950s.
[266] Dron, R. L. W. (1922) *Report on work done for the Scottish Spitsbergen Syndicate*, Scottish Spitsbergen Syndicate papers MS 311, Scott Polar Research Institute.
[267] Dron (1922).
[268] *Scotsman* (1919) 'Scottish Spitsbergen expedition', 12 July.
[269] Bruce, W. S. (1919) *Letter to R. N. R. Brown, 19 February,* MS 356/95, Scott Polar Research Institute, Cambridge.
[270] Bruce, W. S. (1919) *Letter to R. N. R. Brown, 7 April,* MS 356/95, Scott Polar Research Institute, Cambridge.

offer to work for £500 on condition of the expedition returning before the end of September.[271] Within a day, both men were appointed. The part of organisational secretary fell to Lieutenant Chester M. Scott, and Napier captained the *Petunia*.[272]

Although Bruce had hoped to obtain Hannay as leading geologist, the latter was unavailable. George Walter Tyrrell was chosen in his stead. Bruce also proposed to take Mathieson despite Brown's criticism of his survey work.[273] He thought of Mathieson as supercargo.[274] Nonetheless, the surveyor agreed to £60, although he considered himself worth £70.[275] Not all staff were handpicked; a call had presumably gone out to in search of geologists. Many applied, but most were turned down.

Mathieson captured engaging impressions of the advance party in 1919 in his diary. When A. Fleming Campbell and George M. Cowan shot two ducks in Klaas Billen Bay, for instance, Mathieson lamented their deaths. Similarly, James Scott wrote compassionately about the shore party in his care on the Foreland. It included Captain H. Sandys, the junior geologist George Ross, the miners Laffay, Shanks, Stanners, and the cook Bell. Bell proved to be particularly apt at shooting and preparing eider ducks, a skill also praised by Tyrrell. When mist and heavy rain disrupted their work, Scott put the well-being of his men before any explorative tasks.[276] On the whole, he was optimistic about the Scottish prospects of working Arctic coal without special difficulties. Tyrrell recounted his pertinent remark, 'What is four miles of glacier compared with 200 miles of African jungle that we have to contend with in Nigerian tin mining.'[277]

Regarding the workforce in 1919, the *Bo'ness Journal* excitedly reported that 13 local miners from Cadell's Bridgeness Co. had volunteered to travel north.

---

[271] Brown, R. N. R. (1919) *Letter to Aitken, 21 May*, MS 356/95, Scott Polar Research Institute, Cambridge.

[272] Bruce, W. S. (1919) *Letter to R. N. R. Brown, 21 May,* MS 356/95, Scott Polar Research Institute, Cambridge.

[273] Bruce, W. S. (1919) *Letter to R. N. R. Brown, 7 April,* MS 356/95, Scott Polar Research Institute, Cambridge.

[274] Bruce, W. S. (1919) *Letter to R. N. R. Brown, 14 April,* MS 356/95, Scott Polar Research Institute, Cambridge.

[275] W. S. Bruce papers, box 10/113, National Museums of Scotland Library, Edinburgh.

[276] 'We would have moved camp earlier in the week but for this rain. Under each tent was a small area of dry ground which was the only dry ground at that time in the neighbourhood. [...] I began to consider the advisability of turning out all hands in the rain to build a shelter of the moss in which a fire of driftwood might be kindled so that some clothes might be dried. But there was not enough moss to build such a shelter. I therefore disposed things so that we might conserve as much dry clothing as possible and trust to the weather clearing. With this in view I sent all the others to their tents with orders to look on keeping dry as their principal business while I cooked the meal and took it round to them .' (Scott, J. (1919) *Report to the Scottish Spitsbergen Syndicate on the expedition to Prince Charles Foreland, 19/8/1919*, Scottish Spitsbergen Syndicate papers MS 311, Scott Polar Research Institute, Cambridge.)

[277] Tyrrell, G. W. (1919) *Letter to H. M. Cadell, 30 September*, Cadell of Grange papers, Acc. 5381, National Library of Scotland, Edinburgh.

They took part 'more as a sporting venture and for the sake of experience.'[278] The foreman of the Carriden mine, George W. Miller, was in charge of the group. At 90 years of age, he later recorded his memories of the voyage. Cadell, being the only mine owner on the board, had been asked to provide the labour. Incidentally, production at one of his mines had stopped, and while a new shaft was being sunk, there were men to spare. There was no shortage of volunteers, and two local army cooks could be got on demob. A week spent in Edinburgh seemed like a holiday to them. Millar was amazed to see the staff swell to 85, including an official piper and 'several whose occupations in the expedition always eluded me.'[279]

After another week in Tromsø, where James Uncles played the bagpipes around the town, Miller's first sight of Spitsbergen unveiled tall mountains rising out of bleak terrain. The party split. Some went with Brown and a geologist while Miller's lot aided by Uncles, who was a joiner by trade, constructed two houses at Bruce City and opened the Carron seam named after Cadell's connection with the Carron Co. The miners were engaged to work eight hours a day, but they soon decided on twelve hours around the clock, to which the syndicate agreed. They spent their time working and sleeping, occasionally playing games. When he had a chance, Miller sent a message via Green Harbour to his wife to let the families know the men were alright. Towards the close of summer, the syndicate contemplated leaving a small party to winter. Although a few volunteers were initially found, an early snowstorm convinced them that it was not such a good idea after all. The miners returned home in September, and while Miller's first view of Spitsbergen had been reserved, he now observed, 'It was harvest time and the reapers were in the field reaping the rich yellow corn. What a contrast it was from conditions where we had just come from.'[280]

On the whole, the 1919 expedition, which had been led by Bruce for a last time, was noteworthy for the staff's frequent commendations of the workforce. As such, Bruce commended the piper Uncles, Campbell the miners and drillers at Bruce City, especially the foreman Miller, and James Scott his shore party on the Foreland with special mention of Bell, the cook. The commendations were made to the board, but it is unclear what the purpose of the gesture may have been and whether it entailed any reward. Perhaps it acted as the official sanction of the pride the men may have felt regardless.

There were no such commendations in 1920. It was the first expedition under Mathieson's leadership, and the staffing was much less commented on. Napier's letter to Brown gave the impression that crew, staff, and workers had got

---

[278] *Bo'ness Journal* (1919) 'Bridgeness miners for Spitsbergen', 20 June.
[279] Miller (1978).
[280] Miller (1978).

on a lot less famously. The captain took the liberty of stating that 'this has been the worst five months I ever put in at sea with the exception of the short time you and the directors were on board.'[281] He did not see eye to eye with his mate, with whom he had to share a cabin, and fell out with him over presents made to the men that only served to cause a bout of self-importance. Furthermore, '[Mathieson] might be a good surveyor [...] but not a leader of a lot of men of the class we had. McDonald broke down over it and so did I before I got home; I hope you tear this up when you read this.'[282] Nor had he calmed down a fortnight later, when it transpired that there had also been problems with the provisions, especially the butter, and with the Norwegian workers.

1920 appears to have been the first and only occasion that the Scots had employed a partially Norwegian workforce. There are at least no sources to the contrary. This may have been a move towards cost reduction, the number of Norwegian workers that summer being nine.[283] None were used as watchmen in the subsequent winter or at any other time.

Cadell had a different opinion. He commended the hardy, unsophisticated workmen of northern Norway, who were well acclimatised to the long, dark winters. Spitsbergen had few or no natural facilities for sport or outdoor recreation, and he was afraid that British labour need hardly apply for a job here. On another occasion, Cadell praised the 65-year-old Mathieson, 'a Highlander who can speak Gaelic and a wonderfully able man for his years. Like Moses his eyes are not dim nor his natural force abated and he can walk all the younger men off their legs and climb the mountains like a deer without ever getting tired. I wish I could feel like that.'[284] In contrast, he did not believe Napier to be a very competent: the skipper had refused to sail into Horn Sound when the engine of the ship could not be relied on to get them out of there in unfavourable winds.

The limited programmes in 1921 and 1922 required less personnel. The scientific core vanished. The sources focus instead on Scottish mining experts and their assistants. The latter again included the sons of eminent professors. Cadell's continued correspondence with the manager of Store Norske now regarded the effect of the changing market on the workforce. Bay responded that 'we do not fear the consequences of reduced wages opposite to labourers at Spitsbergen, as with the approaching bad times on the working market our people in the north of

[281] Napier, F. B. (1920) *Letter to R. N. R. Brown, 2 November,* Frank Napier Collection MS 356/78/1, Scott Polar research Institute, Cambridge.
[282] Napier, F. B. (1920) *Letter to R. N. R. Brown, 2 November,* Frank Napier Collection MS 356/78/1, Scott Polar research Institute, Cambridge.
[283] Fountain, H. (1921) *Letter to W. M. Johannessen, 11 February,* Norwegian Foreign Department Box 5373, National Archives of Norway, Oslo.
[284] *Extracts from H. M. Cadell's chronicle, Vol. 4 1910-1925*, Cadell of Grange papers, Acc. 5381, National Library of Scotland, Edinburgh.

Norway will prefer going on living in Spitsbergen on lower wages than be starving in Norway.'[285] Despite this downturn, the syndicate relied on Scottish staff, and the men seemed content. In the next three years, only Mathieson and one or two companions travelled north to uphold the claims and lead negotiations with potential allies. Mathieson made a last journey in 1925. A revival of such journeys would not take place until after the Second World War.

8.8.4 Local allies

In 1920, the directors had the opportunity to mingle with other Spitsbergen stakeholders.[286] On July 28, they met Reynolds of the Royal Society Club, a London dining club, and Heron of the Northern Exploration Co. Reynolds and Heron were examining as much of the islands as possible with an option on the company's land if oil were discovered. The syndicate decided to facilitate their efforts on the Scottish claims, too, and gave them a letter of introduction to Mathieson. Heron had been with the Northern Exploration Co. in the previous year, and according to the men's account, the company was on the brink of collapse. Davis City in Lowe Sound was practically abandoned, and no coal was being worked; Calypso Bay was merely a store and a wireless depot; and Marble Island appeared to be the only workable asset at present. The directors later passed by Davis City and observed about 2,000 tons of coal in a stockpile, but nothing was being done there. The Norwegian Ingvald Svendsen, shareholder in the syndicate, said that the coal would not burn, that mining had ceased in any case, and that there was no export. The party later dined with Bevan in Recherche Bay, who assumed that the Norwegians were prejudiced against the Northern Exploration Co. because of its flamboyant anti-Norwegian policy and its refusal to employ Norwegian labour.

For an hour on July 31, the directors landed at the camp of the Anglo-Russian Grumant Co. situated in a ravine in Icefiord a few miles east of Coles Bay. They were received by the only man who spoke some English, 'a somewhat drunk steward who had overwintered with the Northern Exploration Co. at Lowe Sound last winter and intends to winter here this year.'[287] The mine lay at about 150 to 200 feet above sea level, and the approximately 1.4-metre seam was divided by a band of dirt. The adit had been driven for about 100 yards, and several hundred tons of

[285] Bay, K. (1921) *Letter to H. M. Cadell, 16 March*, Cadell of Grange papers, Acc. 5381, National Library of Scotland, Edinburgh.
[286] Brown, R. N. R. (1920) *Journal kept by R. N. R. Brown, Spitsbergen, 17 July to 19 August 1920*, Robert Neal Rudmose Brown Collection MS 356/2/3, Scott Polar Research Institute, Cambridge.
[287] Brown, R. N. R. (1920) *Journal kept by R. N. R. Brown, Spitsbergen, 17 July to 19 August 1920*, Robert Neal Rudmose Brown Collection MS 356/2/3, Scott Polar Research Institute, Cambridge.

coal had been extracted. 400 tons had already been sent to Tromsø that year. Loading was done at the foreshore, but the anchorage had no protection. The camp consisted of four huts and several tents. The huts were either on soft, boggy ground or on an unsecure foundation near the foot of the ravine. One at least was in a filthy state. Still, Brown noticed a copy of his recent book there. Great heaps of stores were being landed, which included two baths.

In 1921, Mathieson re-visited the Anglo-Russian Grumant Co.[288] He reported high-quality coal being broken out of a seam which comprised a band of sandstone. The mine opening lay a mere 90 metres from the shore, but the harbour facilities were extremely poor. The coal needed to be loaded onto flat-bottomed barges. According to the surveyor, this could be remedied by a railway along to the coast into the nearby Coles Bay. Hoel got the impression that Mathieson pointed this out because he could see an amalgamation between the Scots and the Anglo-Russian Grumant Co. The press releases concerning the mine had been good, and Mathieson urged the board to consider this option. It should at least not be dismissed until the coalfield at Grumant City had been investigated thoroughly.

Mathieson upheld the relations with the Anglo-Russian Grumant Co., but his report in 1922 was brief, stating merely that the mine was 'in excellent order.'[289] The syndicate had placed some of their equipment at the Anglo-Russian disposal, and in 1923, Mathieson and his companions arranged for their own motorboat to meet them at Advent Bay on their arrival. The arrangement was honoured, although the decrepit boat was replaced by a tugboat. After the Scottish claims had been inspected, Mathieson again went over the mine at Grumant City, from which he reported in the distinct language of practical mining.[290] He advocated the taking over of the mine by the syndicate.

A visit to the Anglo-Russian Grumant Co. was once more included in the syndicate's programme in 1925.[291] However, the Russian director would not arrive

---

[288] Hoel (1967a) pp. 1068-9.

[289] Mathieson (1922), cited in Hoel (1967a) p. 1072.

[290] 'We went over the whole of the mine and found it in very good working order. With a sum of £5000 to £10000 to supply one or two coal cutters, proper hauling gear and a better shoot for filling the barges, 40000 tons could easily be shipped each year. There are, on an average, 60 to 80 days each season during which loading can be carried on, and with the present two barges 1000 tons per day can be placed on board ship. The dumping ground for storing is scanty but with a little expense it could easily be extended and instead of using barges a pier and harbour could be erected at Coles Bay and a rope way made to the mine. My colleagues, Dr. Brown and Mr. Cowan, are entirely with me in thinking that the taking over of this mine by the S.S.S. and working it in conjunction with the carboniferous coal at Bruce City and Gips Valley would enable the latter to be developed at less expense and the Company would have the advantage of being able to place two kinds of coal, viz. carboniferous and tertiary and semi-anthracite coal, on the market, the latter being an excellent steam coal and the former a good coking coal which Sweden required badly.' (Mathieson (1923), cited in Hoel (1967a) p. 1073.)

[291] Hoel (1967a) pp. 1074-5.

until later in the season, and the site manager could not act without permission. There was nothing for Mathieson and Brown to do but to depart. Meanwhile, the syndicate's board had given some serious thought to a merger, and Mathieson returned to Spitsbergen post-haste to lead the discussions. Director Nachimson was still absent, but the manager Frost had been instructed to examine the Scottish claims. Shortly afterwards, Frost compiled a satisfactory report, and Nachimson was ready to take decisive actions towards a partnership. When he returned to England in the latter half of September, he would seek out the syndicate's board to agree on the particulars.

At the same time, however, the Anglo-Russian Grumant Co. had a contract with the Russian authorities to deliver 50,000 tons of coal to Murmansk each year.[292] That contract gave the Russians the right to examine the workings at Grumant City. While Mathieson was on site, a Russian professor and an agent of the Russian Government arrived. Consequently, an amalgamation between the Scottish Spitsbergen Syndicate and the Anglo-Russian Grumant Co. was never achieved. Instead, the Russians secured the coalfield at Grumant City.

8.8.5 Products and achievements

In mid-September 1919, Cadell met with Bruce and Wordie, who had returned prior to the main expedition.[293] The director thought the trip only partially successful.[294] The Admiralty had been difficult, and the expedition had departed too late to achieve its goals during the short Arctic summer. The Foreland party merely discovered some iron ore of secondary quality. The Stor Fjord group traced the oil shale, but it was neither worth much nor workable under the ice conditions. Although Edge Island had significantly added to the Scots' territory, Cadell viewed it as too exposed and of too little mineral value. The main party explored the Carboniferous rocks between Sassen Bay and Klaas Billen Bay and opened up an outcrop in the Culm measures, but in addition to disappearing below sea-level, the seam was only about three feet thick and very ashy; thicker, better coal was found further up the bay. The attempt to prove the Bruce City coalfield failed because the boreholes were not placed well, and there were problems with drilling in frozen ground. While the miners who opened up the Bridgeness seam did not manage to strike the best part, boreholes could have ascertained its nature. Yet, Brown had come home before the boreholes had located the coal. Tyrrell agreed that the party

---

[292] Hoel (1967a) p. 1075.
[293] Cadell, H. M. (1919) *Diary,* National Library of Scotland, Edinburgh.
[294] Cadell, H. M. (1919) *Diary,* National Library of Scotland, Edinburgh; *Extracts from H. M. Cadell's chronicle, Vol. 4 1910-1925*, Cadell of Grange papers, Acc. 5381, National Library of Scotland, Edinburgh.

could have stayed a fortnight longer and was certain that coal would have been reached in a week or two.[295] According to Cadell, the real value of the expedition lay in the map produced by Mathieson.

The money originally set aside for the expedition had been spent, and the remainder had been issued as shares, but the poor results had caused their value to drop. The unfinished work needed to be completed in 1920, for which new funding would have to be found. In Cadell's own words,

> The simplest solution of the whole matter will be for the Norwegian Government, now that Norway is the legalised owner, to buy out the British claim holders, as the Norwegians bought out the American mine owners at Advent Bay on Ice Fjord a few years ago. The Norwegians can well afford to do so as they have not suffered in the War and have no coalfields of their own. They are geographically in the best position to develop the place and the market would be mainly in Norway. Norwegian labour moreover can best work in the arctic climate all the year round.[296]

Cadell barely mentioned gypsum or any other natural resource in the Scottish areas. Perhaps the coalowner did not think their exploitation worthwhile, which sheds doubt on his ability as a director to suitably represent the syndicate's varied interests. Samples of gypsum had in any case been brought to Scotland, and an internal report was available in early December.[297] It concluded that the syndicate would be hard-pressed to find a market in Britain.

Cadell's private records offer additional insights into the achievements between 1920 and 1925. The director commonly wrote from the perspective of a shareholder with inside information. After his visit to Spitsbergen in 1920, Cadell determined that 'the territory of the Syndicate was not proved of much value and indeed apart from the coal it has nothing of value at all.'[298] Even the coal was high in ash and practically unsaleable at normal times. In summer 1920, it was worth £10 to £12 per ton in Norway, but by the end of the year it was down to £2 per ton

[295] Tyrrell, G. W. (1919) *Letter to H. M. Cadell, 30 September*, Cadell of Grange papers, Acc. 5381, National Library of Scotland, Edinburgh.
[296] *Extracts from H. M. Cadell's chronicle, Vol. 4 1910-1925*, Cadell of Grange papers, Acc. 5381, National Library of Scotland, Edinburgh.
[297] Wilson, W. (1919) *Gypsum report*, Scottish Spitsbergen Syndicate papers MS 311, Scott Polar Research Institute, Cambridge. One David Fisher, whose large Edinburgh business as a plasterer had won him contracts for the King at Windsor, was presented with two specimens of gypsum, which he thought excellent and equal to the best; the specimens of plaster he classed B, A, and O. Class B would cost 37s per ton free delivery, A 35s, and O 30s. In Fisher's opinion, the gypsum would produce high-quality plaster, but presently the consumption of plaster in Edinburgh was limited, with only 1,500 tons of class O being sold annually. Other towns shared the market, and competition from Carlisle was particularly stiff. Fisher provided the cautionary tale that a small firm had started up in Bo'ness, but Carlisle undercut the price by 10s per ton and put it out of business.
[298] *Extracts from H. M. Cadell's chronicle, Vol. 4 1910-1925*, Cadell of Grange papers, Acc. 5381, National Library of Scotland, Edinburgh.

and 'a drug on the market.'[299] To Cadell, it seemed unlikely that the syndicate would benefit from its Arctic possessions. Nonetheless, he judged the exploration of Gips Valley in the following year to have been quite successful. The three-foot seam of good coking coal about ten miles up the valley had given good results. It was the lower and better of the two seams of the Bruce City coalfield, but it lay largely below sea level. To work it without a shaft would require a ten-mile railway to the outcrops. An attempt to involve a subsidiary company was fruitless; in the financial depression, the £1 share had dropped to 15s. Cadell personally made a loss of £200. 1922 was again a bad year. The tractor worked well, but Cadell blamed the engineer in charge, the son of Prof. Dron, for the boreholes freezing up or getting jammed before coal could be reached. In fact, 'The expedition was a complete failure and wasted a lot of the Syndicate's vanishing resources.'[300] The shares had fallen to 6s, which meant a loss of £700 to Cadell. In 1923, he wrote off another £275 while the syndicate marked time, still unable to sell its coalfield. The depression continued throughout 1924, and the last and only comment regarding 1925 was a write-off of £25.

It is noteworthy that Cadell should regard a topographic map as the most valuable product of an expedition. Mathieson produced three such maps for the Scottish Spitsbergen Syndicate: *Klaas Billen Bay and Temple Bay* (1920), *Ebba Valley* (1921), and *Gips Valley* (1922).[301] His men peppered the polar landscape with countless Scottish place-names, which today constitute a lasting legacy of the syndicate. Whilst the toponyms aided effective occupation and meaningful communication during the expeditions, the maps had a more immediate and important purpose as reliable base maps for geological exploration. Correct contours assured the extrapolation of data from one outcrop to a whole seam on a different mountain, across a bay, or below sea level. A good base map could effectively present the results of geological fieldwork to the board and other stakeholders. It is surprising, therefore, that besides a few pencil marks, no draft or final geological map has been found, let alone published.

In 1924, a formal enquiry into its products and achievements reached the Scottish Spitsbergen Syndicate from another quarter. Adolf Hoel asked for information about the Scottish coalfield and provided a form to be filled in. The syndicate responded that the form was inappropriate because it was not a mining

[299] *Extracts from H. M. Cadell's chronicle, Vol. 4 1910-1925*, Cadell of Grange papers, Acc. 5381, National Library of Scotland, Edinburgh.
[300] *Extracts from H. M. Cadell's chronicle, Vol. 4 1910-1925*, Cadell of Grange papers, Acc. 5381, National Library of Scotland, Edinburgh.
[301] Mathieson, J., Cowan, G. M., and Campbell, A. F. (1920) *Map of Klaas Billen Bay and Temple Bay, 1:100,000;* Mathieson, J. and Pollock, C. M. (1921) *Plan of Ebba Valley, Klaas Billen Bay;* Mathieson, J. and Finlay, T. R. (1922) *Map of Gips Valley.*

company; it was an exploration company.[302] Hoel then requested any articles dealing with the Scottish work in any British periodicals for his personal library.[303] He received the reply that papers had been read at various learned societies but that the references included did not have the authority of the board.[304] Again, the Scots were keen to point out that they were an exploring company. Although over £100,000 had been spent on investigations and proving the coal, it was not in order to win any coal beyond samples for analysis, but with the goal to form a subsidiary company. There were several company reports, but they were deemed to be useless to Hoel because they largely dealt with finance.

A list of learned papers that resulted from the Scottish expeditions has been included in Appendix 3. Both Bruce and Brown began to publish on the subject of Spitsbergen before the First World War; Brown continued throughout and long after the conflict. The post-war boom that allowed a large scientific staff to attend the expeditions in 1919 and 1920 went hand in hand with a flurry of subsequent lectures and articles. Wordie was invited to speak before the Royal Geographical Society on March 21. In his opinion, thorough scientific exploration in unfamiliar regions preferably entailed continual work for at least three field seasons. 'Of these the first can then be devoted to reconnoitring the broad features of the country; in the second the main spade-work of detailed investigation can be carried out; the third can be usefully occupied in the filling up of gaps and in revision of the more difficult problems.'[305] Similarly, Tyrrell produced articles that suggested a focus on geological rather than commercial concerns. Nonetheless, the scientists were able to address select audiences in Britain's intellectual and financial hubs, and their work was distributed in publications ranging from the *Bo'ness Journal* to the *Colliery Guardian*.[306] Whether the scientists and the syndicate had an understanding can no longer be discerned. Wordie did, in any case, not make the same *faux pas* as Ponting, and any information given out will have been welcomed, underhand publicity for the company.

The majority of learned papers attracted comments of a scientific, commercial, or geopolitical nature. Occasionally, technical aspects were covered. Cadell's presentation on Arctic coal mining and Dron's account of boring in frozen strata generated genuine interest at the Institution of Mining Engineers. Cadell

[302] Scottish Spitsbergen Syndicate (1924) *Letter to Hoel, 22 January*, Norsk Polarinstitutt 239, Regional State Archive, Tromsø.
[303] Hoel, A. (1924) *Letter to the Scottish Spitsbergen Syndicate, 6 February*, Norsk Polarinstitutt 239, Regional State Archive, Tromsø.
[304] Scottish Spitsbergen Syndicate (1924) *Letter to A. Hoel, 11 February*, Norsk Polarinstitutt 239, Regional State Archive, Tromsø.
[305] Wordie, J. M. (1921) 'Present-day conditions in Spitsbergen', *Geographical Journal*, 58 (1), pp. 25-45.
[306] *Bo'ness Journal* (1921) 'Coal mining in Spitzbergen', 25 March; *Colliery Guardian* (1921) p. 962.

provided a long-awaited map of Spitsbergen's coalfields (Fig. 8.20) and a geological section showing the relation between the Tertiary, the Cretaceous, and the syndicate's Carboniferous coal at Klaas Billen Bay (Fig. 8.21).

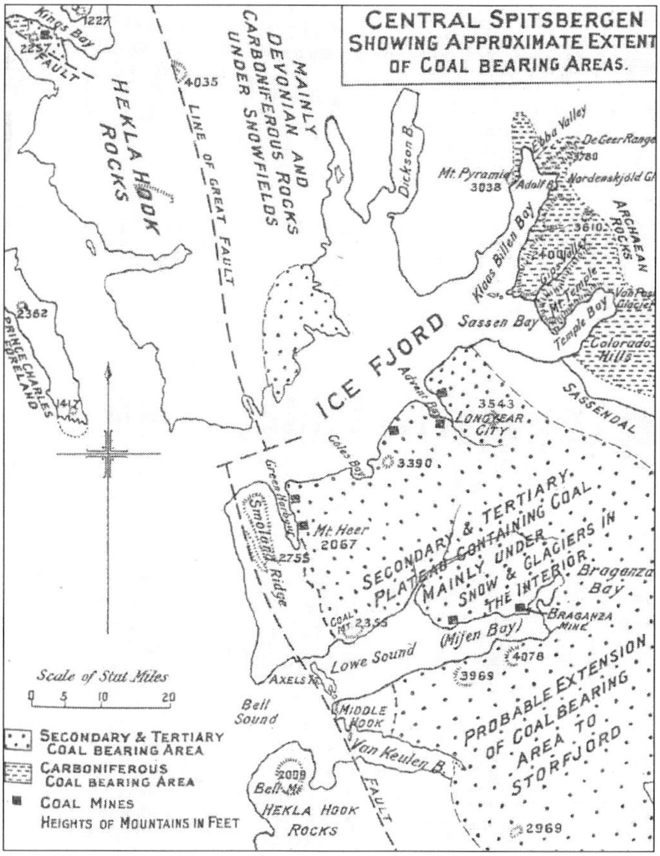

8.20 Map showing Spitsbergen's coalfields. Note that the working mines were confined to the Tertiary deposits. (Source: Cadell, H. M. (1921) 'Coal-mining in Spitsbergen', Transactions of the Institution of Mining Engineers, 60, 119-42.)

The discussion of Dron's paper was opened by Andrew Kyle, proprietor of the drilling firm in Galston. It is not known if he had visited Bruce City, but he knew details of the operations there. He made a series of suggestions for future work in frozen strata. Both papers undoubtedly bestowed some technical legitimacy on the Scottish expeditions as the work carried out addressed the little known engineering properties of permafrost and the challenges of working in an Arctic climate. Yet, without any output to speak of, the syndicate continued to be inconsequential to practical mining. Its shareholders may have appreciated the scientific achievements, but like Cadell, they would have made substantial losses on the financial market.

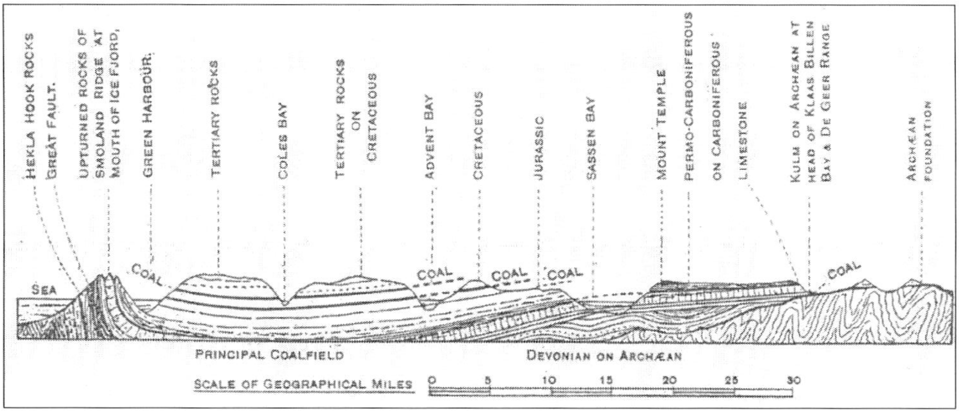

8.21 Geological section through Spitsbergen's coal-bearing horizons. (Source: Cadell, H. M. (1921) 'Coal-mining in Spitsbergen', Transactions of the Institution of Mining Engineers, 60, pp. 119-42.)

8.9 The Danish Commission

In April 1925, the Norwegian Government in the renamed capital of Oslo submitted a bill regarding Spitsbergen and Bear Island, proposing the joint official name of Svalbard.[307] Norway's sovereignty over Svalbard was proclaimed on August 14.[308] This went hand in hand with a public notice to all claimants of land on the archipelago to notify the Norwegian Government of the exact limits of their territory.[309] The arbitration of claims came at a cost; a penny per acre cleverly thwarted whole-sale claiming. Having formerly boasted 2,980 square miles, which would set it back £7,947, the Scottish Spitsbergen Syndicate put forward four greatly reduced areas.[310] These included:

1)	An area on Prince Charles Foreland	1,440	acres
2)	Ebba Valley and Wordie Crags	5,920	acres
3)	Bruce City, Gips Valley, and Anser Islands	53,248	acres
4)	The coast of Sassen and Temple Bays	3,712	acres
	Total	64,320	acres

These areas barely comprised four percent of the previous properties. On the one hand, they were a confession that the majority of claims had been valueless. On

[307] *The Times* (1925) 'The government of Spitsbergen', 24 April, p. 11.
[308] *The Times* (1925) 'The new status of Spitsbergen', 14 August, p. 9.
[309] *Edinburgh Gazette* (1925) 'Spitsbergen. Notice to claimants of land', 14 August, p. 940.
[310] *Afskrift. Claims to estates in Spitsbergen* (1925) Norsk Polarinstitutt 239, Regional State Archive, Tromsø.

the other hand, they expressed optimism that the remaining natural resources were worth the fee and future administrative costs.

In January 1926, it came to the attention of the *Colliery Guardian* that Brown would attend a conference in Oslo.[311] The meeting would decide on questions regarding mining laws and other rights. After his return, he composed a memorandum to the shareholders, which put renewed faith in the Danish Commissioner to adjudicate the claims swiftly and absolutely.[312] He presented several facts about Spitsbergen, which were supposedly well known to most people in other stakeholder countries, bar the British.[313] Notably, he pointed out what these countries were willing to do, which Britain was not, including 'working at a time when the cost of material was at its highest, when freights and wages were abnormal and with workers of little experience and methods of a primitive nature.'[314] He predicted that they would soon be independent of British coal and judged the syndicate's areas to be 'of immense value and possibility not only from the point of view of the Shareholders, but also as a national asset. [...] the capital invested [in their development] should not only give a satisfactory return to the Shareholders, but again from a national point of view, such work cannot fail to provide a great outlet for the unemployed in this country.'[315] His rhetoric was tailored to the prevailing situation in Britain and aimed at the well-to-do. Widespread dissatisfaction culminated in the General Strike in May 1926.

The conference in Oslo was probably where agents of the syndicate conversed with Arnold Christopher Ræstad, the legal scholar representing the interests of the Norwegian Government, which then became a subject between the British minister in Oslo and the Prime Minster Stanley Baldwin.[316] It transpired that Ræstad opposed part of the Scottish claims, which the Scots found entirely groundless. Without this interference, their claims might have been recognised

---

[311] 'Spitsbergen mines' (1926) *Colliery Guardian,* 131, p. 255.

[312] Scottish Spitsbergen Syndicate (1926) *Memorandum as to the company's properties in Spitsbergen, and their possible development*, Norsk Polarinstitutt 239, Regional State Archive, Tromsø.

[313] 'Comparatively few people in Britain know anything about Spitsbergen except that it lies in the Arctic regions. These regions seem to be associated in this country only with "Pole-hunting" and adventure among ice and snow and among walrus and whales. A few years ago there was an attempt to "boost" Spitsbergen as an El Dorado of gold and precious stones and metals, but the notoriety it earned in that respect in certain quarters was short lived. The pictures were so overdrawn and the reports of fabulous wealth so obviously imaginative that they were soon discredited by thinking persons. A revulsion of feeling set in, and when the actual facts as ascertained by scientific experts of the highest authority were published, little interest was taken in the publications. The harm was done in this country. Capitalists were afraid to venture, and it remained for other countries, driven by the necessity, to demonstrate the possibilities of Spitsbergen as a coal-producing source.' (Scottish Spitsbergen Syndicate, 1926).

[314] Scottish Spitsbergen Syndicate (1926).

[315] Scottish Spitsbergen Syndicate (1926).

[316] British Legation in Oslo (1926) *Letter to the Prime Minister, 10 February*, Norwegian Foreign Department Box 5374, National Archives of Norway, Oslo.

without difficulty, but now they felt that Norway resented a Scottish monopoly of the gypsum deposits and reluctantly agreed to meet a Norwegian delegate on Spitsbergen in summer 1926 to go over the areas and come to an agreement. Besides the unnecessary expenditure, the syndicate maintained that the Spitsbergen Treaty did not give Ræstad the right to oppose them since his Government had not put in a counterclaim and paid a penny per disputed acre. The British minister pointed out that Ræstad also opposed the claims of individual Norwegians and feared that if he did not tread more carefully, it would undermine his position as the governmental representative.

The matter had not been resolved by April 13, when the solicitors of the syndicate contacted the Norwegian foreign minister. They declared that 'it was of utmost importance that our clients get a title to the property claimed by them or meantime at least a letter from the Commissioner that there are no other claimants to the portions claimed by them.'[317] Only then would the board be able to take immediate steps for the floatation of a subsidiary company to open up the gypsum deposits. Although it was not stated explicitly, the agents presumably hoped that the Norwegian foreign minister would either reason with Ræstad to withdraw his objection or, less likely, prompt the Danish Commissioner to compose the desired letter. Each additional day in limbo incurred losses to its shareholders, which the syndicate wanted to curb.

The waiting resumed. At the end of May and the beginning of June 1926, meetings were held in Oslo, at which the Danish Commissioner mediated between Ræstad, representing the Norwegian Government, and the syndicate's director Bolton and Alexander Matthew, a lawyer. The Norwegians and the Scots arrived at an arrangement that made another voyage to Spitsbergen redundant.[318] At the same time, the Swedish counterclaim at Klaas Billen Bay was successfully repelled, the Scots having taken over the hut in Ebba Valley at a small cost.[319]

Despite the delay in a formal title, the directors routinely issued the report for the year to February 28, 1927. The balance sheet attested that all 100,000 shares had been issued and that the reserve account held a little over £20,130. Some £50 were owed to sundry creditors. Among the assets, the syndicate's auditors listed the estates on Spitsbergen at £15,000; the costs of the expeditions between 1919 and 1925 at approximately £89,180, to which the journeys made in 1926 and the depreciation on stock contributed another £680; as well as stores

[317] Hagart and Burn Murdoch (1926) *Letter to the Minister of Foreign Affairs of Norway, 13 April*, Norwegian Foreign Department Box 5374, National Archives of Norway, Oslo.
[318] Scottish Spitsbergen Syndicate (1927) *Proceedings of the eighth ordinary general meeting, 20 April*, Norsk Polarinstitutt 239, Regional State Archive, Tromsø.
[319] Scottish Spitsbergen Syndicate (1927) *Report and balance sheet, 11 April*, Scottish Spitsbergen Syndicate, Norwegian Polar Institute, Tromsø.

including huts, machines, instruments, and equipment worth about £620. Preliminary and managerial expenses amounted to roughly £13,660, and costs in connection with the Danish Commission had reached £680. Still, there was a cash balance of circa £370. Although the syndicate was debt-free and one document away from undisputed possession of its claims, the chairman lamented that the chance to attract a financial group to develop the areas had been missed.[320] Even if a formal title removed the primary obstacle to any negotiations, the coal industry and general trade experienced an abnormal slump (Fig. 7.28).

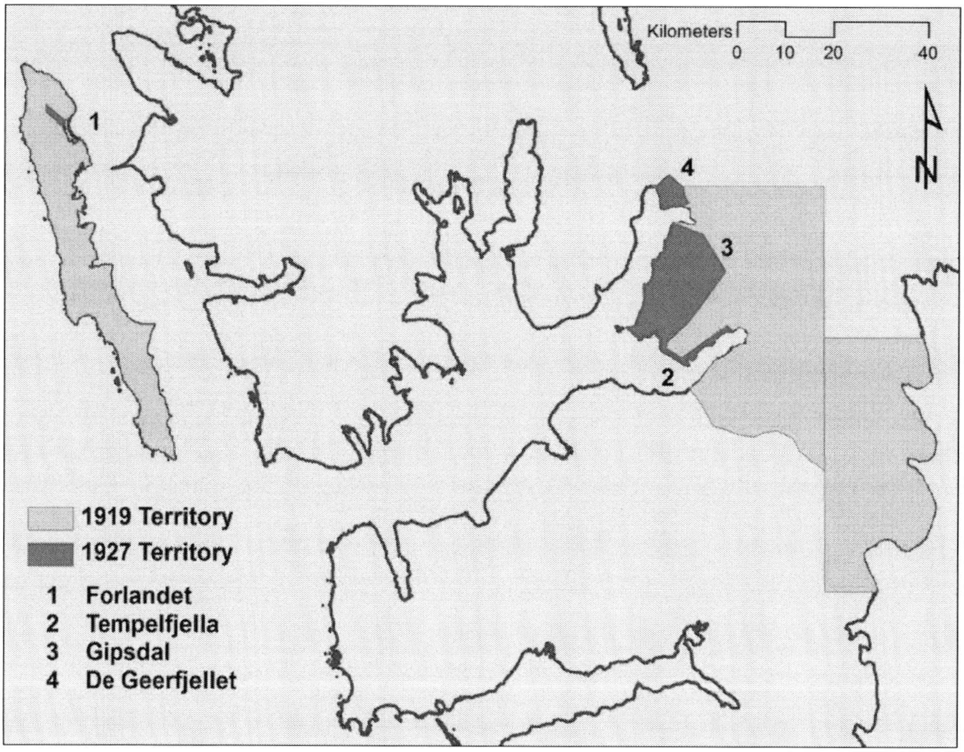

8.22 Map showing some of the syndicate's former claims and the areas awarded in 1927. (Source: Report of the Svalbard Commissioner concerning the claims to land in Svalbard, Pt 1, B. Maps (1927) Oslo: De Norske Svalbardekspedisjoner; Map: F. Kruse.)

On November 4, 1927, the Scottish Spitsbergen Syndicate circulated the longed-for news at last: the Danish Commissioner had completed and published his report,

[320] Scottish Spitsbergen Syndicate (1927) *Proceedings of the eighth ordinary general meeting, 20 April*, Norsk Polarinstitutt 239, Regional State Archive, Tromsø.

and the syndicate's title had thereby been assured.[321] In the process of arbitration, the claims shrunk considerably to less than 70,000 acres and received irrevocable Norwegian names (Fig. 8.22). Nonetheless, the Scottish were the unquestionable owners of these treaty properties for the next ten years.

8.10 Marking time

A subsidiary company had not materialised, and the market was unfavourable, yet the syndicate had 'not lost sight of the possibilities of other means of developing the company's estates.'[322] Its lawyer Matthews, for instance, corresponded with E. A. A. Thiis of Oslo, who had formerly been interested in the Swedish coal mines.[323] In November 1927, Thiis must have enquired into the suitability of the Scots' coal for distilling oil. Matthews initially provided the analysis results for coal samples from Gips Valley and asked if Thiis knew how these compared with the coal in Kings Bay and Braganza Bay. In a follow-up letter, he had subjected further samples to distillation tests and provided Thiis with the figures.[324] Other correspondence is not known. The example is presumably one of many attempts to salvage the situation, which led to nothing.

Sources for the ensuing period are scarce. They mainly comprise yearly directors' reports and proceedings of annual meetings for the benefit of the shareholders. In the year ending on February 28, 1928, the directors had found it inexpedient to take any action regarding the coal.[325] They had, however, advanced the negotiations concerning the gypsum. In the process, the chemists Woodcock and Mellersh of London sent the analyses of gypsum with the verdict of being excellent to the Mineral Development Syndicate in Edinburgh.[326] Further test results were provided by Gemmell and Thin in November 1928.[327] These were

---

[321] Aitken & Co. (1927) *Letter to the shareholders, 4 November*, Norsk Polarinstitutt 239, Regional State Archive, Tromsø.

[322] 'Scottish Spitzbergen Syndicate, Ltd.' (1927) *Colliery Guardian*, 135, p. 1053.

[323] Matthew, A. (1927) *Letter to E. A. A. Thiis, 17 November*, Norsk Polarinstitutt 239, Regional State Archive, Tromsø.

[324] Matthew, A. (1927) *Letter to E. A. A. Thiis, 26 November*, Norsk Polarinstitutt 239, Regional State Archive, Tromsø.

[325] Scottish Spitsbergen Syndicate (1928) *Report and balance sheet, 30 June*, Scottish Spitsbergen Syndicate, Norwegian Polar Institute, Tromsø; Scottish Spitsbergen Syndicate (1928) *Proceedings of the ninth ordinary general meeting, 10 July*, Scottish Spitsbergen Syndicate, Norwegian Polar Institute, Tromsø.

[326] Woodcock and Mellersh (1928) *Letter to the Mineral Development Syndicate, 4 July,* Norsk Polarinstitutt 239, Regional State Archive, Tromsø.

[327] Gemmell and Thin (1928) *Analyses of gypsum and anhydrite*, Norsk Polarinstitutt 239, Regional State Archive, Tromsø.

enclosed in a letter to Hoel, hoping that he could identify a market in Norway.[328] Bolton's name and the syndicate's telephone number had been crossed off the letterhead at this stage. Hoel replied that he thought it impossible to get the Norwegian Government or any private persons interested in the deposits.[329] Gypsum was a cheap mineral that required bulk purchases, and although there was a demand in Norway, the market was not large enough to make the exploitation of the resource profitable.

On June 24, 1929, the board reported that 'nothing of consequence to communicate to the Shareholders has taken place.'[330] Mathieson had travelled to Spitsbergen in 1928 in order to inspect the huts and other property, which he found to be in order, to secure samples of gypsum from various areas, and to ascertain conditions for shipping an experimental 500-ton cargo back to Scotland. While the analysis had again been favourable and the deposits could easily be worked, it had not been possible to get a good shipping rate, especially in view of high labour costs. Only bulk samples of over 5,000 tons were economical. The expenditure for the year had been about £130, which comprised £65 to cover Mathieson's journey and £45 for the agent in Norway. The directors personally provided these finances; they invited any willing shareholders to join them.

Urmston's death in June 1930 witnessed another change to the board, the vacant directorate being filled by Mathieson.[331] The coal market had not improved, and enquiries into the possibility of sending Spitsbergen coal to South America had been unproductive because freight rates were too high. The gypsum failed to attract interest for the same reason. Despite the inactivity, the Scots incurred losses in the form of the depreciation of stock and stores as well as managerial expenses. Although *The Times* briefly reported on the Swedish State Railways signing contracts with both Northumberland and Spitsbergen, suggesting some serious competition for England's coalfield from northern Europe, the economic situation remained unaffected in 1931.[332]

---

[328] Scottish Spitsbergen Syndicate (1929) *Letter to A. Hoel, 7 June,* Norsk Polarinstitutt 239, Regional State Archive, Tromsø.

[329] Hoel, A. (1929) *Letter to the Scottish Spitsbergen Syndicate*, *8 June*, Norsk Polarinstitutt 239, Regional State Archive, Tromsø.

[330] Scottish Spitsbergen Syndicate (1929) *Report and balance sheet, 24 June*, Scottish Spitsbergen Syndicate, Norwegian Polar Institute, Tromsø; Scottish Spitsbergen Syndicate (1928) *Proceedings of the tenth ordinary general meeting, 2 July*, Scottish Spitsbergen Syndicate, Norwegian Polar Institute, Tromsø.

[331] Scottish Spitsbergen Syndicate (1930) *Report and balance sheet, 30 June*, Norsk Polarinstitutt 239, Regional State Archive, Tromsø.

[332] *The Times* (1931) 'Northumberland coal for Sweden. Competition with Spitzbergen', 17 September, p. 9; *The Times* (1931) 'Spitsbergen coal', 18 September, p. 17; Scottish Spitsbergen Syndicate (1931) *Report and balance sheet, 4 July*, Scottish Spitsbergen Syndicate, Norwegian Polar Institute, Tromsø.

In mid-1932, Mathieson admitted to Hoel, 'Of course you know that the coal question is in the worst condition it has ever been in Great Britain and that is the reason we are not trying to develop our coal-measures in Spitsbergen. Indeed I think the Company would be only too pleased to dispose of their claims if a suitable offer was made.'[333] He wondered if the Norwegian Government would be interested. He did not directly refer to its purchase of the Northern Exploration Co.; perhaps he did not know about it. Throughout 1933 and 1934, the Scots were marking time.[334] Cadell had died after being on the board since incorporation 15 years ago. His seat remained vacant. The delayed 1935 report involved a subtle shift in language away from the former aim to float a subsidiary company to 'the settled policy of this Syndicate [...] to dispose of its properties in whole or in part to a purchaser or purchasers who would develop these properties'.[335] In recent months, there had been an improvement in trade, and a gentleman was currently visiting Spitsbergen in view of putting in an offer for the gypsum deposits. While the syndicate had not been able to travel north, it had permitted expeditions from Cambridge (1932) and Oxford (1933) to use their huts. They supplemented the syndicate's former investigations.

The gentleman in question was the Bergen ship owner Jacob A. M. Kjøde. He indeed put in an offer at the end of 1935. He proposed £10,000, but the syndicate envisaged £50,000 for its estates, including a copper vein on Prince Charles Foreland.[336] Kjøde replied, 'if you cannot believe my proposal to be serious I am afraid we shall not be able to make any headway.'[337] He pointed out that the Dutch had invested a million pounds at Green Harbour and had to sell Barentsburg for a fraction. Similarly, the Swedes had spent £700,000 in Braganza Bay, and when they had offered Sveagruvan to the Norwegian Government for only £50,000, it was not even accepted; Store Norske acquired the property for that price. Kjøde felt his proposal to be a fair one, 'if you will sell you must come very near my price, and if not we can both save ourselves the trouble.'[338] He invited the Scots to accompany him to Spitsbergen on one of his ships in summer 1936. At

---

[333] Mathieson, J. (1932) *Letter to A. Hoel, 30 June*, Norsk Polarinstitutt 239, Regional State Archive, Tromsø.

[334] Scottish Spitsbergen Syndicate (1933) *Report and balance sheet, 10 July*, Norsk Polarinstitutt 239, Regional State Archive, Tromsø; Scottish Spitsbergen Syndicate (1934) *Report and balance sheet, 14 July*, Norsk Polarinstitutt 239, Regional State Archive, Tromsø;

[335] Scottish Spitsbergen Syndicate (1935) *Report and balance sheet, 11 October*, Norsk Polarinstitutt 239, Regional State Archive, Tromsø;

[336] Mathieson, J. (1936) *Letter to A. Hoel, 30 October*, Norsk Polarinstitutt 239, Regional State Archives, Tromsø.

[337] Kjøde, J. (1936) *Letter to the Scottish Spitsbergen Syndicate, 24 January,* Norsk Polarinstitutt 236, Regional State Archive, Tromsø.

[338] Kjøde, J. (1936) *Letter to the Scottish Spitsbergen Syndicate, 24 January,* Norsk Polarinstitutt 236, Regional State Archive, Tromsø.

approximately the same time, in April 1936, the syndicate contacted the trade representative of the Soviet Republics in London; this they did not fail to mention to Kjøde.[339] To his resentment, the syndicate was stalling. He asserted, 'I have made you an offer, and had eventually expected to receive a counter offer. Under the present circumstances I have no interest in making you a new offer, and feel little inclined to plan a further inspection of your properties.'[340]

The syndicate was stalling because it was awaiting a second opinion as to the value of its properties. The agent Andr. Aagaard had previously mentioned the matter to Hoel, to whom £50,000 seemed too much. The board now desired Aagaard to ascertain what sum would be appropriate.[341] Hoel was disinclined to comment, but the Scots persisted.[342] Since the Norwegian frequently required information for his research, such as place-name evidence, he possibly offered to visit the Scottish properties in return for their cooperation.[343] In October, Mathieson conceded that they might be asking too high a price for the estates in light of the inopportune global market, thereby prompting Hoel to report from his site visit.[344] Only in December did he receive Hoel's opinion that the properties were worth £30,000 or proportionally less if parts of the deposit were found not to be gypsum but anhydrite.[345] At the end of yet another financial year, the board again reported that nothing had been achieved.[346] Certain standing charges such as the Norwegian Land Tax were once more met by the directors.

The trend continued until the outbreak of the Second World War.[347] Directors Usher and W. G. Burn Murdoch had died, and Maxtone-Graham and Mathieson were joined by John Edgar. By September 1939, any negotiations were stopped by the new international situation, the duration of which was impossible to predict. The syndicate's resources were exhausted, and what expenses there had

---

[339] Scottish Spitsbergen Syndicate (1936) *Letter to J. Kjøde, 22 April*, Norsk Polarinstitutt 239, Regional State Archives, Tromsø.

[340] Kjøde, J. (1936) *Letter to the Scottish Spitsbergen Syndicate, 7 May,* Norsk Polarinstitutt 239, Regional State Archive, Tromsø.

[341] Aitken & Co. (1936) *Letter to A. Aagaard, 7 February*, Norsk Polarinstitutt 239, Regional State Archive, Tromsø.

[342] Aitken & Co. (1936) *Letter to A. Aagaard, 5 June*, Norsk Polarinstitutt 239, Regional State Archive, Tromsø.

[343] Mathieson, J. (1936) *Letter to A. Hoel, 11 June*, Norsk Polarinstitutt 239, Regional State Archive, Tromsø; Hoel, A. (1936) *Letter to J. Mathieson, 15 June,* Norsk Polarinstitutt 239, Regional State Archive, Tromsø.

[344] Mathieson, J. (1936) *Letter to A. Hoel, 30 October*, Norsk Polarinstitutt 239, Regional State Archive, Tromsø.

[345] Hoel, A. (1936) *Letter to J. Mathieson, 3 December*, Norsk Polarinstitutt 239, Regional State Archive, Tromsø.

[346] Scottish Spitsbergen Syndicate (1936) *Report and balance sheet, 21 November,* Norsk Polarinstitutt 239, Regional State Archive, Tromsø.

[347] Scottish Spitsbergen Syndicate (1937) *Report and balance sheet, 23 November*, Norsk Polarinstitutt 239, Regional State Archive, Tromsø; Scottish Spitsbergen Syndicate (1938) *Report and balance sheet, 2 December*, Norsk Polarinstitutt 239, Regional State Archive, Tromsø.

been, were still met by the board. The existence of the syndicate needed to be otherwise ensured, and the directors proposed a voluntary contribution of a penny per shareholder to avoid liquidation. No one outwardly took note of Hoel's recent article, which marked the expiration of the ten-year period, during which owners of treaty properties had the sole right of establishing claims of ten square kilometres on their property.[348] Besides Norwegian and Russian owners, the Scots were the only ones, who still possessed two coal-bearing treaty properties. Yet, they had not staked out any claims to be worked, presumably on the grounds of being an exploration company. For the time being, no one asked what would happen to redundant treaty properties as the near future was overshadowed by war.

8.11 The expiry of treaty properties

During the Second World War, the War Office gave some consideration to Spitsbergen and the active mining companies but practically none to a Scottish exploration company. The syndicate, meanwhile, routinely held its annual general meetings. By 1942, the old guard had almost been replaced by a board comprising John Edgar, A. Burn Murdoch, and the Spitsbergen veteran John H. Kenneth.[349] The meetings were poorly attended, and besides reiterations that nothing could be done until the war was over, that the coal could find no market, that transport was problematic, and that something might yet be made of the gypsum in Norway or Sweden, they did not amount to much. In 1945, Mathieson died, which may have led to Brown becoming active on the board. Next, the secretary Aitken retired from Aitken, Methuen, and Aikman, and his duties were carried on by Munro and Forbes. The board that emerged from the war was almost unrecognisable.

Its immediate concern was money, of which it did not have much. It owed approximately £75 to the Norwegian Government and a further £44 to the agent Andr. Aagaard.[350] This exceeded the cash available, namely £95. Perhaps Aagaard could be persuaded to forgo compound interest since the board wished him to go on representing the syndicate. The directors encouraged the agent to also avoid the payment of interest on the tax for the years 1939 to 1945.[351] They felt that charging tax for the war years was in itself unjust because the Norwegian Government had been unable to provide any services during this time, let alone

[348] Hoel, A. (1938) 'Coal-mining in Svalbard', *Polar Record*, 16, pp. 74-85.
[349] Scottish Spitsbergen Syndicate (1942) *Proceedings of the twenty-third annual general meeting, 30 November*, Robert Neal Rudmose Brown SSS Correspondence, Scott Polar Research Institute, Cambridge.
[350] Scottish Spitsbergen Syndicate (1946) *Letter to R. N. R. Brown, 7 March*, Robert Neal Rudmose Brown SSS Correspondence, Scott Polar Research Institute, Cambridge.
[351] Scottish Spitsbergen Syndicate (1946) *Draft letter to A. Aagaard, 7 March*, Robert Neal Rudmose Brown SSS Correspondence, Scott Polar Research Institute, Cambridge.

protect the properties. To make a point, Aagaard might want to ask what services the Scots would have received if tax had been paid. Besides that, however, the syndicate had little financial bargaining power. Having remained debt-free meant that the syndicate was not dependent on the whim of any creditors, but lacking the necessary resources for decisive action, its hands were now tied.

On May 13, 1946, the syndicate received post from Gunnar Aagaard, *bergmesteren* on Spitsbergen.[352] He drew the board's attention to a notice in the Norwegian public gazette from March 21. It stated that the ship owner Kjøde made 12 notifications of gypsum and anhydrite. The notifications had initially been placed before the Norway's minister of commerce during the war, but the procedure had been postponed. An official survey of the claims was now being planned, and anyone who thought he had a better claim was asked to attend and look after his interests. *Bergmesteren* pointed out that most of Kjøde's claims lay on Scottish territory and that he had fixed a meeting to be held at Bjona Haven on July 30. The syndicate's agent subsequently explained to the Scots that Kjøde was entitled to carry out the survey on the their property, while they had omitted to do so themselves within the ten-year lease.[353] They might still take part in mining the claims, but they would have to pay their own expenses. Nonetheless, A. Aagaard thought that Kjøde was still willing to buy the property and seemed to remember a sum of £15,000 being mentioned.

Being confronted with a new claim dispute, the board promptly turned to the Foreign Office. It explained that Kjøde was claiming mineral resources on its territory, and although this could be opposed at a meeting on Spitsbergen in July, the costs involved were prohibitive to the syndicate.[354] The directors requested observation from the Foreign Office, whose Northern Department replied that it would instruct the British Embassy in Oslo to make enquiries and procure a copy of the mining regulations.[355] The British Embassy, in turn, found the regulations difficult to obtain even in Oslo, and although the ambassador was willing to send it to the syndicate, he would like to have it back. Meanwhile, he would look into Kjøde's claims and try to arrange the meeting to take place elsewhere than in Spitsbergen to facilitate international attendance.[356]

---

[352] Scottish Spitsbergen Syndicate (1946) *Letter to R. N. R. Brown, 13 May*, Robert Neal Rudmose Brown SSS Correspondence, Scott Polar Research Institute, Cambridge.
[353] Aagaard, A. (1946) *Letter to the Scottish Spitsbergen Syndicate, 23 May*, Robert Neal Rudmose Brown SSS Correspondence, Scott Polar Research Institute, Cambridge.
[354] Aitken and Co. (1946) *Letter to the Foreign Office, 17 May*, FO 371/56327 Affairs of the Scottish Spitsbergen Syndicate, Northern 1946 Norway File No. 6555, National Archives, Kew.
[355] Hankey, R. M. A. (1946) *Letter to the Scottish Spitsbergen Syndicate, 24 May*, FO 371/56327 Affairs of the Scottish Spitsbergen Syndicate, Northern 1946 Norway File No. 6555, National Archives, Kew.
[356] Collier, L. (1946) *Letter to the Foreign Office, 28 June,* FO 371/56327 Affairs of the Scottish Spitsbergen Syndicate, Northern 1946 Norway File No. 6555, National Archives, Kew; Hankey, R. M. A.

The Embassy's enquiries were not in the syndicate's favour. Ambassador Collier reported to the Foreign Office that *bergmesteren* had autonomy on Spitsbergen, and all communication regarding this matter had to go to him. G. Aagaard had been in touch with the syndicate before the war and warned that it would jeopardise its rights if it did not develop its property. According to the treaty, Kjøde had already definitely obtained the right to work the gypsum and anhydrite. The meeting taking place did not mean that the Scots were obliged to attend, especially in light of a representative being unable to do more than lodge a formal complaint. The only remaining question was whether the syndicate intended to take interest in a quarter of Kjøde's operations. The Embassy had become aware that Kjøde had previously offered to buy the property, and its commercial secretary was currently busy to ascertain the facts from Tromsø. The Embassy feared that none of it would be of use to the syndicate and felt that the Scots could have made more of an effort. They had plenty of warning, and the commercial secretary had wasted valuable time in chasing the Norwegian authorities.

The British vice-consul in Tromsø provided the additional detail that Kjøde had indeed offered £15,000.[357] The board was only interested in selling, not in a 25-per-cent share in Kjøde's operations, since this included a quarter of the costs of materials, labour, and the like. Agent Aagaard had not communicated with *bergmesteren* since before the war until he got the news that Kjøde proposed to commence a survey in July, which was in keeping with the law. The directors wanted Kjøde to make a new offer, while the Norwegian was still waiting for a counter offer from the Scots. These details from Tromsø found their way via the British Embassy in Oslo to the Northern Department at the Foreign Office. The Embassy concluded that 'it confirms our earlier impression that the Syndicate do not appear to have been attending very well to their business.'[358] The Foreign Office provided the syndicate with a summary of the enquiry, and seeing that this was not in the Scots' favour, they quickly dropped the matter.

In the meantime, the syndicate had explored other avenues still open to it. It thought it advisable to approach the British Government and composed a letter to the Ministry of Fuel and Power.[359] The ministry, however, was not interested in

(1946) *Letter to the Scottish Spitsbergen Syndicate, 9 July,* FO 371/56327 Affairs of the Scottish Spitsbergen Syndicate, Northern 1946 Norway File No. 6555, National Archives, Kew.

[357] Cumming, Vice-consul, Tromsø (1946) *Letter to Barber, Commercial secretary, Oslo, 18 July,* FO 371/56327 Affairs of the Scottish Spitsbergen Syndicate, Northern 1946 Norway File No. 6555, National Archives, Kew.

[358] Chancery, British Embassy, Oslo (1946) *Letter to the Northern Department, Foreign Office, 25 July,* FO 371/56327 Affairs of the Scottish Spitsbergen Syndicate, Northern 1946 Norway File No. 6555, National Archives, Kew.

[359] Scottish Spitsbergen Syndicate (1946) *Letter to R. N. R. Brown, 5 July,* Robert Neal Rudmose Brown SSS Correspondence, Scott Polar Research Institute, Cambridge; Scottish Spitsbergen Syndicate

acquiring the Scottish properties.[360] By January 1947, the board renewed talks with the Norwegian Government in view of affecting a sale.[361] Shortly afterwards, Group Captain C. W. B. Urmston entered the scene, but the directors were not greatly enthusiastic about his proposal and thought it well to temporise with the Norwegian Government.[362] In time, the Norwegian Government made a tentative offer of 100,000 kroner, but the directors found this offer insufficient and instead objected to Kjøde's claims.[363] After full consideration was given to Urmston's views, these objections were again withdrawn in January 1948.[364] Other interest came from the solicitors Farman, Daniell & Co., whose clients thought about taking over the syndicate.[365] This and other such interest were indicative of the post-war economic optimism that also existed after the Second World War.

8.12 A subsidiary company and voluntary liquidation

Group Captain Charles William Brabazon Urmston first wrote to the Scottish Spitsbergen Syndicate at the end of March 1947, introducing himself as the nephew of C. H. Urmston and cousin of W. G. Burn Murdoch.[366] The late directors had often spoken about the syndicate, and Urmston, who had been a shareholder for years, now requested information about its capital, the distribution of its shares, and its holdings. If he could obtain a controlling interest, he would propose a scheme to develop the properties with American and British support. Urmston was confident that if the development of Spitsbergen was practical, his wide contacts and extensive capital could achieve it, although time and money would have to be spent to get the best possible result.[367] By August, Urmston's negotiations were

---

(1946) *Letter to R. N. R. Brown, 8 July,* Robert Neal Rudmose Brown SSS Correspondence, Scott Polar Research Institute, Cambridge.

360 Scottish Spitsbergen Syndicate (1946) *Letter to R. N. R. Brown, 22 August,* Robert Neal Rudmose Brown SSS Correspondence, Scott Polar Research Institute, Cambridge.

361 Scottish Spitsbergen Syndicate (1947) *Letter to R. N. R. Brown, 22 January,* Robert Neal Rudmose Brown SSS Correspondence, Scott Polar Research Institute, Cambridge.

362 Urmston, C. W. B. (1947) *Letter to the Scottish Spitsbergen Syndicate, 26 March,* Robert Neal Rudmose Brown SSS Correspondence, Scott Polar Research Institute, Cambridge; Scottish Spitsbergen Syndicate (1947) *Letter to R. N. R. Brown, 3 April,* Robert Neal Rudmose Brown SSS Correspondence, Scott Polar Research Institute, Cambridge.

363 *Draft minutes of meeting of directors, 18 September* (1947) Robert Neal Rudmose Brown SSS Correspondence, Scott Polar Research Institute, Cambridge.

364 *Draft minutes of meeting of directors, 22 January* (1948) Robert Neal Rudmose Brown SSS Correspondence, Scott Polar Research Institute, Cambridge.

365 Farman, Daniell and Co. (1947) *Letter to R. N. R. Brown, 8 July,* Robert Neal Rudmose Brown SSS Correspondence, Scott Polar Research Institute, Cambridge.

366 Urmston, C. W. B. (1947) *Letter to the Scottish Spitsbergen Syndicate, 26 March,* Robert Neal Rudmose Brown SSS Correspondence, Scott Polar Research Institute, Cambridge.

367 Urmston, C. W. B. (1947) *Letter to Scottish Spitsbergen Syndicate,* Robert Neal Rudmose Brown SSS Correspondence, Scott Polar Research Institute, Cambridge.

nearing the necessity to form an Anglo-American cooperation.[368] The cooperation would only finance the preliminary exploration if some assurance of the progression to commercial development could be given. The effect of Kjøde's claims would need to be ascertained, but little could be done about the highly uncertain diplomatic situation of Spitsbergen in world strategy at the time.

Urmston reportedly obtained the support of the United States State Department and was sure of American capital once the production stage was reached.[369] He had also been in touch with the Brexim Overseas Corporation, a large company with world-wide offices. To go ahead, the security of the title had to be cleared up conclusively, and the firm had to be refinanced. Urmston envisaged taking complete control and entering into negotiations with the Norwegian Government prior to dispatching an expedition to Spitsbergen. When Brexim contacted the syndicate towards the end of the year, the company was happy to enter into an agreement with the Scots to purchase the whole share capital on condition that the Norwegian Government was sympathetic towards the investigations and development of the Scottish properties.[370]

It is difficult to reconstruct why the syndicate withheld crucial information from Urmston for as long as it did. By January 1948, Urmston himself had made intensive investigations into the position of the Scottish properties.[371] Like the Foreign Office before him, he discovered that the position was not altogether favourable. The ten-year period in which owners of treaty properties could file claims had expired in 1937, eleven years ago. Kjøde's claims had been delayed by the war, but these mitigating circumstances only applied to him, not to the Scots. Norway was not interested in purchasing the Scottish properties because they could be had for nothing if the Norwegian Government filed its own claims. This the Norwegians apparently already knew when they made their 100.000-kroner offer. Urmston had spoken to people, who did not think the syndicate's former coal properties particularly attractive, and the Scots could in any case only exploit them if they filed new claims. In spite of all of this, Urmston suggested to go ahead with

---

[368] Urmston, C. W. B. (1947) *Letter to the Scottish Spitsbergen Syndicate, 13 August,* Robert Neal Rudmose Brown SSS Correspondence, Scott Polar Research Institute, Cambridge.

[369] Urmston, C. W. B. (1947) *Letter to the Scottish Spitsbergen Syndicate, 20 October,* Robert Neal Rudmose Brown SSS Correspondence, Scott Polar Research Institute, Cambridge.

[370] Brexim (1947) *Letter to Scottish Spitsbergen Syndicate, 19 November,* Robert Neal Rudmose Brown SSS Correspondence, Scott Polar Research Institute, Cambridge; Scottish Spitsbergen Syndicate (1947) *Letter to Brexim, 21 November*, Robert Neal Rudmose Brown SSS Correspondence, Scott Polar Research Institute, Cambridge; Brexim (1947) *Letter to Scottish Spitsbergen Syndicate*, Robert Neal Rudmose Brown SSS Correspondence, Scott Polar Research Institute, Cambridge.

[371] Urmston, C. W. B. (1948) *Letter to Scottish Spitsbergen Syndicate, 9 January,* Robert Neal Rudmose Brown SSS Correspondence, Scott Polar Research Institute, Cambridge.

the scheme, although Brexim was suddenly no longer interested.[372] Urmston felt that everybody would be let down unless he gambled in Brexim's stead. He was willing to pay their contribution of £5,000 himself.

This money probably revived an expedition to Spitsbergen between August 10 and September 2, 1948.[373] Led by Captain Ian R. H. Black, it aimed to file new mineral claims. The party enjoyed the assistance of many Norwegians both on the mainland and on the archipelago. The nephew of Fridtjof Nansen, for example, was 'almost fanatically pro-British.' From Longyearbyen (diarist John Elbo almost exclusively used the Norwegian place-names), *sysselmannen* conveyed them to Gipsbukta. At the head of Gipsdalen, the men found the old hut in good order, and the first claims were staked out to the north of it. On each new claim, a cairn was erected next to a marked rock. The marks eventually numbered S.S.1 to S.S.27.[374] Over the next few days, Elbo observed the coal outcrops in the valley as well as the signs of old workings, collapsed adits, and the tractor tracks from 1922 on the east side of the lake. The hut in Gipsbukta was in fairly poor condition, and the tractor had been left outside. On August 19, the party arrived at Brucebyen, where the men cleaned out the main hut before claiming commenced in the area, focussing on coal as well as gypsum. The reconnaissance in Ebbadalen on August 24 lasted less than a day, and in the afternoon, the men were back in Longyearbyen. Black typed out the witness declarations pertaining to the new claims on the spot. Another couple of days were spent on various trips and visits before the men returned home.

Black submitted 27 witness declarations and mineral samples from Gipsdalen, Brucebyen, and Ebbadalen to *bergmesteren* while still on Spitsbergen.[375] In October, he additionally sent reports in duplicate on the discoveries of 27 mineral occurrences and enclosed maps of the areas claimed, the discovery points, and the positions of marks in the field as well as geological notes and diagrams.[376] *Bergmesteren* replied that the official inspection to verify and approve the claims could take place early in summer 1949 and suggested a date in mid-July.[377] To cover the charges for dealing with the application for the claims, the syndicate had to pay 6,000 kroner. The time for the survey was notified in the

---

[372] *Notes of a Telephone conversation between Urmston and the Syndicate, 19 March* (1948) Robert Neal Rudmose Brown SSS Correspondence, Scott Polar Research Institute, Cambridge.
[373] Elbo, J. G. (1948) *Diary, 10 August to 2 September*, MS 270/1, John Elbo Collection, Scott Polar Research Institute, Cambridge.
[374] This proves that the engraving 'SS19' recorded by LASHIPA 3 in 2006 dated from 1948.
[375] Black, I. R. H. (1948) *Letter to bergmesteren, 25 August,* Bergmesteren 54, Regional State Archive, Tromsø.
[376] Black, I. R. H. (1948) *Letter to bergmesteren, 10 October*, Bergmesteren 54, Regional State Archive, Tromsø.
[377] Aagaard, G. (1948) *Letter to I. R. H. Black, 15 November*, Bergmesteren 54, Regional State Archive, Tromsø.

Norwegian gazette at the end of March. From July 27, 1949, Aagaard fixed and described each claim. He asked for the photos, which Gee had taken of all marking cairns in 1948.[378] If he had known about them earlier, he would have requested them before the survey. As it was, he could not discern six of the marks, despite a party of men having searched for them in a particularly monotonous district. He hoped the images would assist him to describe these. On February 28, 1950, *bergmesteren* announced the publication of the 27 claims granted to the syndicate. Just how meticulous he had been in the process is evident in the resultant claim map (Fig. 8.23).

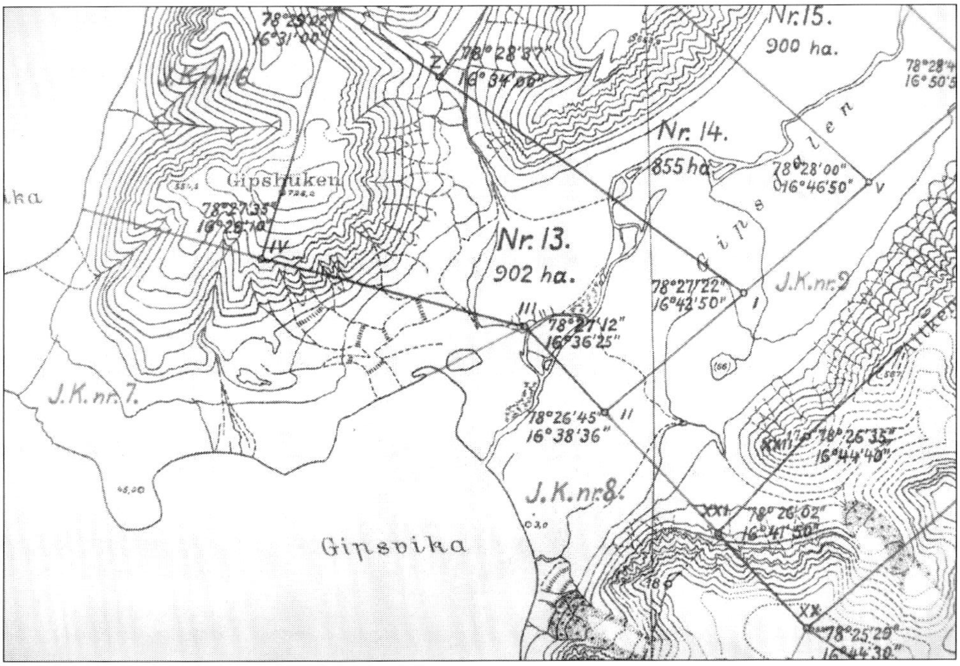

8.23 Details of a map showing both Kjøde's claims of 1946 ('J.K.nr.x.') and the syndicate's claims of 1948 ('Nr.x.'). (Source. Aagaard, G. (1950) 'Utmålskart for the Scottish Spitsbergen Syndicate', Bergmesteren R99 II, Regional State Archive, Tromsø.)

Securing these 27 claims may have been the last official act of the Scottish Spitsbergen Syndicate as agreed with Urmston. Shortly after the expedition, its solicitors were instructed by the financial agents of Urmston's new venture to prepare the necessary documents and to advise on the proper procedure in

[378] Aagaard, G. (1950) *Letter to C. W. B. Urmston, 28 January,* Bergmesteren 54, Regional State Archive, Tromsø; Aagaard, G. (1950) *Letter to E. R. Gee, 30 January*, Bergmesteren 54, Regional State Archive, Tromsø.

connection with the liquidation of the firm and the acquisition of its assets by their clients.[379] Neither party was, however, in any hurry to rush the liquidation until Aagaard's survey had been completed.[380] Following the survey, Urmston's side got the impression that the syndicate was attempting to withdraw from the arrangements, and if not withdraw, to alter them substantially.[381] They hoped to be mistaken. The matter would otherwise undermine the confidence and trust between the stakeholders.

Several months later, all suspicions were resolved, and the drawn-out negotiations between the syndicate and Urmston resulted in agreements dated May 10 and 11, 1950.[382] Subject to the approval of the shareholders, the board would sell the assets for £5,000, and the shareholders would be allotted founders shares in the new company. At the ensuing extraordinary general meeting, the chairman explained that the financial situation was such that it would not be able to continue without borrowing money.[383] For many years, the syndicate had tried to sell its assets, and the offer that had now been made was the only firm offer it had received. The directors thought it advisable to accept it rather than prolong life only to go into liquidation on less favourable terms. Subsequently, the *Edinburgh Gazette* published the notification that a special resolution had been passed to wind up the syndicate and to register a new company.[384]

Scottish Spitsbergen (Development) Ltd. was incorporated on October 29, 1950.[385] Besides the assets and several shareholders, it had little in common with the former Scottish Spitsbergen Syndicate and will not be treated here. Suffice to say that it soon fell to Urmston to try and dispose of the properties in the early days of the Cold War.[386] The liquidation of the syndicate took place at a leisurely pace. By April 1951, the liquidator had received the sum of £5,000 and was in the

---

379 Bell and Acroyd (1948) *Letter to the Scottish Spitsbergen Syndicate, 15 September*, Robert Neal Rudmose Brown SSS Correspondence, Scott Polar Research Institute, Cambridge; Scottish Spitsbergen Syndicate (1948) *Letter to R. N. R. Brown, 17 September,* Robert Neal Rudmose Brown SSS Correspondence, Scott Polar Research Institute, Cambridge.

380 Urmston, C. W. B. (1949) *Letter to C. L. Forbes, 18 March,* Robert Neal Rudmose Brown SSS Correspondence, Scott Polar Research Institute, Cambridge.

381 Swirsky (1949) *Letter to C. L. Forbes, 12 August,* Robert Neal Rudmose Brown SSS Correspondence, Scott Polar Research Institute, Cambridge.

382 Aitken & Co. (1950) *Letter to members of the Syndicate, 6 June*, Norsk Polarinstitutt 239, Regional State Archive, Tromsø.

383 *Chairman's remarks to the extraordinary general meeting, 29 June* (1950) Robert Neal Rudmose Brown SSS Correspondence, Scott Polar Research Institute, Cambridge.

384 *Edinburgh Gazette* (1950) 'The Scottish Spitsbergen Syndicate Limited', 4 July, p. 812.

385 *Certificate of incorporation* (1950) Scottish Spitsbergen (Development) BT31/36510/487321, National Archives, Kew.

386 Scottish Spitsbergen (Development) BT31/36510/487321, National Archives, Kew; Scottish Spitsbergen Syndicate, Northern 1951 Norway File No. 1461, FO 371/94695, National Archives, Kew.

process of paying the creditors.[387] At the end of November, the liquidator was able to report that after paying the syndicate's debts and the expenses of the liquidation, there remained a cash balance of £2,500.[388] This sum permitted the first and only dividend in 33 years. The shareholders needed to sign a declaration to claim 6d for each £1 share they held. Brown, for instance, sent a declaration for 680 shares.[389] A final general meeting was held on June 16, 1953.[390] Shortly afterwards, the Spitsbergen veteran, aged 73, received a cheque – the grand total of £2 9s 9d being due to him.[391]

8.13 Summary and conclusion

The archaeological record of the syndicate's sites in Chapter 4 produced an informative but limited actor-network (Fig. 4.87). Since several former members and employees bequeathed their private documents to public archives, these files closed some crucial gaps (Fig. 8.24). It thereby became possible to closely correlate the little known archaeology of exploration with the intricate policies and strategies of an exploration company. The syndicate existed for 44 turbulent years. As with the Northern Exploration Co., the most defining global occurrence was the First World War. Unlike the English, the Scots were not overwhelmed by later events, but they were forced into inactivity during which they failed to respond adequately to the expiry of the treaty properties in 1937. They were fortunate to find a buyer in 1950, which led to the only British dividend on record.

The Scottish Spitsbergen Syndicate had a promoter deeply rooted in science. Bruce had ample experience raising funds for scientific expeditions, which did not require the formation of a company. His explicit suggestion to incorporate a syndicate to make something of the mineral resources on Spitsbergen is therefore paramount. A private limited syndicate formed solely for the purpose of exploration was deemed to be the best means by which to facilitate commercial exploitation, thereby making a profit. The syndicate never had an active mining policy. On the board were several prominent Scots, of whom Davidson resigned very soon, demonstrating little trust in the undertaking. After the war, the Edinburgh management seemed unsuitable, the difference of a 'Glasgow sprinter' probably

[387] Scottish Spitsbergen Syndicate (1951) *Letter to R. N. R. Brown, 24 April,* Robert Neal Rudmose Brown SSS Correspondence, Scott Polar Research Institute, Cambridge.

[388] Lamb (1951) *Scottish Spitsbergen Syndicate (in voluntary liquidation), 29 November*, Robert Neal Rudmose Brown SSS Correspondence, Scott Polar Research Institute, Cambridge.

[389] Scottish Spitsbergen Syndicate (1951) *Letter to R. N. R. Brown, 3 December,* Robert Neal Rudmose Brown SSS Correspondence, Scott Polar Research Institute, Cambridge.

[390] *Edinburgh Gazette* (1953) 'The Scottish Spitsbergen Syndicate Limited', 15 May, p. 271.

[391] Ryland (1953) *Letter to R. N. R. Brown*, Robert Neal Rudmose Brown SSS Correspondence, Scott Polar Research Institute, Cambridge.

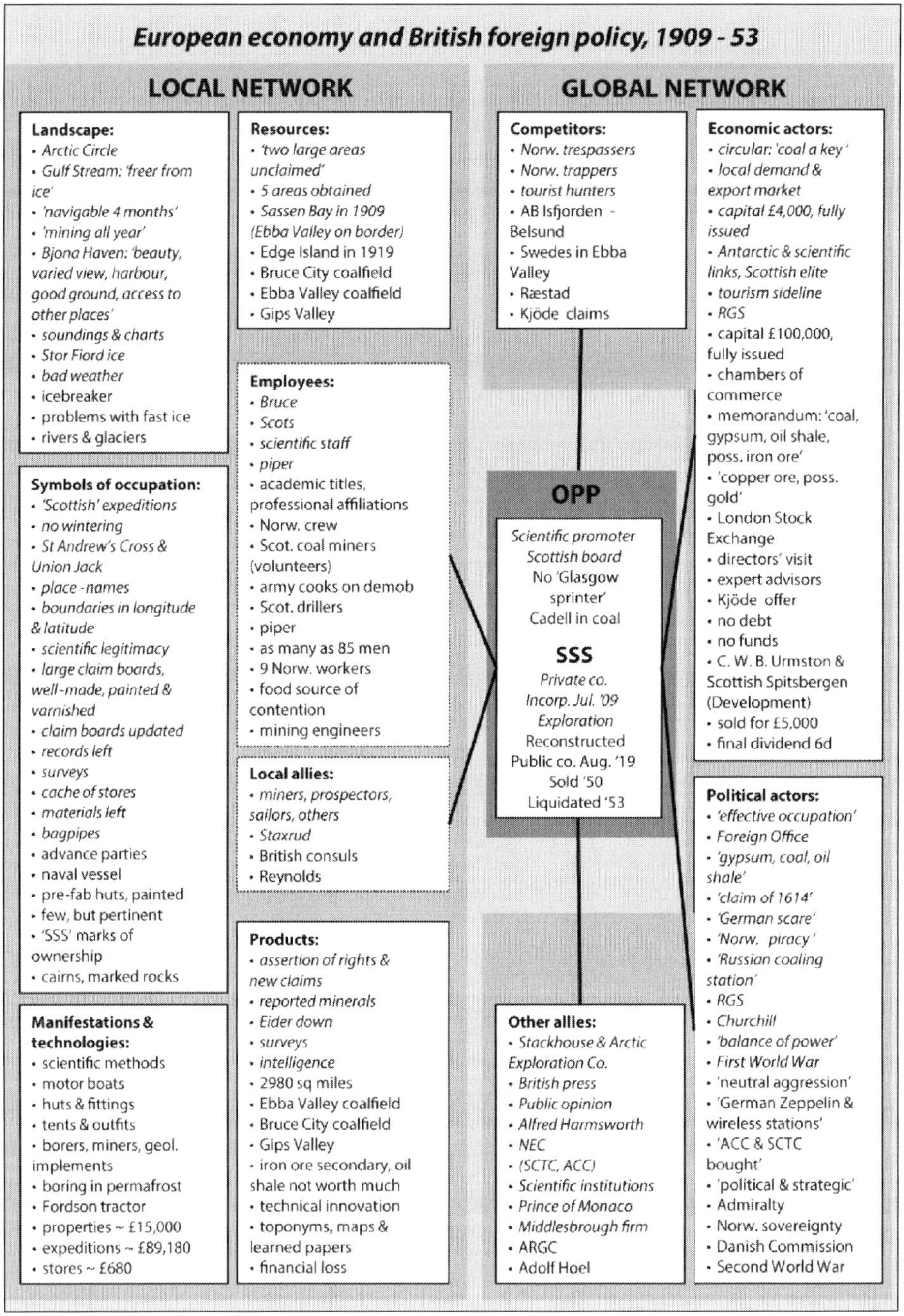

8.24 The company's history is characterised by an actor-network before the First World War (in italics) and one after. (Chart: F. Kruse.)

being industrial know-how. Only Cadell had a background in coal but had himself moved into earth sciences. The limitations on finance were overcome by restructuring the old syndicate into a new one. It is noteworthy that it was *not* converted into a public company at the same time; the benefits of appealing to the public were only recognised a few months later.

The promoters issued a first circular to encourage a syndicate. They identified coal to be the key resource and outlined local demand as well as an export market for the benefit of potential investors. They addressed the problems of title and climate. The original capital of £4,000 was divided into expensive £10-shares and subscribed to by members of the Scottish elite only. This funded the 1909 expedition, which appears to have been an uninspiring one-off. There is no evidence as to what transpired in the boardroom afterwards, but the economic actors no longer held much interest. It is a tell-tale sign that Brown thought a follow-up expedition useless. What prompted the expeditions in 1912 and 1914 is not known. If the latter had concluded unsuccessfully and had not been interrupted by war, it is questionable if a post-war renewal would have occurred. As it was, the situation called for a return to Spitsbergen and justified new capital. The associated memorandum was optimistic about coal and a host of other economic minerals. Nonetheless, Britain headed into the post-war depression and the syndicate with it. It was inactive until a Norwegian made an offer to buy its properties in 1935. The syndicate was under no pressure to accept. Rather than decline politely, it stalled, and Kjøde angrily withdrew. Although the syndicate remained largely debt-free, its funds were consumed by charges and depreciation. After the Second World War, it had no choice but to arrange the very fortunate sale to Urmston. For every £1 the shareholders had invested, they ultimately only received 6d in return.

The syndicate, presumably through Bruce's experiences in the Antarctic, knew about effective occupation before contacting the Foreign Office to obtain the official titles to its claims. The assurance that the ministry would safeguard its interests was simply not enough to pacify its stakeholders. To hurry the proceedings along, the Scots utilised a range of potent rhetorical tools such as 'the claim of 1614', 'a German scare', 'Norwegian piracy', and 'the balance of power'. Neither geopolitical construct swayed the Foreign Office, the Royal Geographical Society, nor Winston Churchill. The First World War gave rise to yet more powerful actants such as 'neutral aggression', 'German Zeppelin and wireless stations', and 'the Norwegian acquisition of former American and English properties'. Despite Bruce's assumption that the Admiralty would recognise the 'political and strategic' importance of Spitsbergen, the Government never intended to annex the islands. Before long, the Scots accepted Norwegian sovereignty, which eventually secured their properties for a period of ten years. Failure to respond in 1937 meant that the syndicate lost its rights, which necessitated the staking out of new claims according to the mining code.

The syndicate must have contacted countless other actors of economic or political potential, of whom only a tiny fraction is visible in the sources. The majority of contacts did not bear fruit. The involvement of Stackhouse looked like it could lead to a subsidiary tourism company, but he was not able to keep his promises. The Scots failed to interest the British press and public opinion in their course. The Northern Exploration Co. could not be trusted. The Arctic Coal Co. and the Spitzbergen Coal & Trading Co. were discontinued before any alliances could be achieved. If some Scottish institutions and the Prince of Monaco continued to be involved, it was probably due to scientific interest. After the war, it looked like the Middlesbrough firm or the Anglo-Russian Grumant Co. could be enrolled to mine the Scottish properties, but the negotiations led to nothing. Surprisingly, Adolf Hoel became an ally once the Spitsbergen Treaty had been ratified.

Norwegian and Swedish firms featured among the competitors of the syndicate. It is noteworthy that the Scots also repeatedly censured Norwegian trappers, who were said to deplete an important food source. The criticism was rooted in the desire to preserve Arctic wildlife and aimed at a particular class of investors in Britain. After the war, the Scots experienced problems with the Norwegian representative Ræstad, who did not find mention to the same extent in the sources of the Northern Exploration Co. In later years, Kjøde could have been an ally, but he resented his treatment at the hands of the Scots. He patiently waited his turn to legally stake out new minerals claims on the former Scottish territory. It has not been ascertained if his claims proved profitable and if the syndicate lost out on an opportunity.

Until 1920, the syndicate's staffing was strongly influenced by Bruce. It was of a Scottish and scientific character, the most pertinent question being whether a geologist who can recognise a mineral resource can also correctly judge its economic potential. Scientific credentials continued to play a key role after the war. Mining engineers probably had a secondary part and supervised the miners and drillers. The only time the syndicate employed Norwegians, it appears to have caused problems of an unknown nature. Food seems to have been a source of contention, but it never amounted to outright resistance, i.e. strikes.

The syndicate appears to have acted largely in isolation. Its expeditions seem not to have relied on the assistance of any local allies, although Bruce and Brown admitted that other miners, prospectors, and sailors were a valuable source of local knowledge. Bruce needed Staxrud to translate a contract for him, but it is not known how the Scots further handled the contact with Norwegians. Perhaps British consuls arranged more of their affairs than can now be traced. Nothing came of the acquaintance with the oil geologist Reynolds.

From the outset, the Arctic landscape was depicted realistically, probably because Bruce had experienced the Antarctic extremes. The syndicate's investors and especially potential subsidiary companies needed to know how long the shipping season was and when mining could be carried out, which were four months and all year, respectively. Bruce also stressed the Arctic merits as a tourist attraction. Scottish attempts to improve the Spitsbergen landscape are evident in the soundings and charts prepared by the expeditions. Nevertheless, ice in Stor Fjord and bad weather could considerably hinder a summer programme. The use of an icebreaker to extend the shipping season found mention, but none was utilised by the syndicate.

Bruce originally promoted two coal-bearing areas, although five changed hands by means of an agreement. In 1909, Sassen Bay was additionally occupied. Ebba Valley lay along its northern margin, which raises the question if it was ever properly taken possession of before the war. Despite the wholesale claim of Edge Island in 1919, further investigations and drilling showed that the Bruce City coalfield and coal and gypsum in Gips Valley had the greatest economic potential. Since the Ebba Valley coalfield and reported copper ore on Prince Charles Foreland could yet prove lucrative, they were included in the treaty properties of 1927. The LASHIPA project visited Bruce City, Gips Valley, and the Foreland: whether the natural resources here were valuable or not, the claims were shown never to have been worked by the Scots or anyone else.

The syndicate's expeditions were unmistakably Scottish. Before the war, they were sporadic, but immediately after they occurred annually, always flying St Andrew's Cross and the Union Jack. Yet, the Scots never thought it necessary to winter. The scientific focus led to widespread surveys and the peppering of the landscape with Scottish place-names. The claims were staked out according to natural coastlines as well as coordinates that gave the undertaking potent scientific legitimacy, which became undeniable once scientific papers and maps were published. These assured that the place-names are still in use today.

To defend its mineral claims further, the syndicate erected claim signs, but nowhere near as many as the Northern Exploration Co. The main ones were well-built, painted, and varnished. They occupied very visible locations, although it could not be ascertained if the syndicate was guilty of referring back to the acquisition date in 1909 on later properties. The boards were regularly updated by follow-up expeditions. The Scots also mentioned the deposition of records, possibly parchment buried under cairns, and they left stores and equipment in strategic places. They used the distinctive sound of bagpipes to announce Scottish occupation on the spot. Pre-fabricated huts were only used after the war, and there were only a handful of them. Individual parts were stencilled with clear marks of ownership. The practice of building cairns was continued in 1948, when the Scots also used rock carvings to number their newly staked out claims. Besides these highly visual aids in the field, the most effective way to defend a claim against trespassers was to notify the Foreign Office. The foreign officials used the earliest dates of notification in attempts to settle disputes.

The costs of the syndicate's local network were small compared to the Northern Exploration Co. The properties had cost £15,000, and the expeditions approached £90,000, probably because staffing and transportation had taken their toll. The amount for stores, which included the pre-fabricated houses that still stand today, was far from extravagant. Although the syndicate appreciated the need for symbols to mark its claims, it chose not to be unnecessarily showy. The strategy to sink boreholes to prove the coal and other resources at depth was the most cost-effective means of mineral exploration. All in all, the syndicate was financially very conservative.

The syndicate generated the customary products of an exploration company. It asserted its rights on its acquired properties and took over more claims, on which it reported the occurrence of mineral deposits. Alternative ventures into tourism and eider down were attempted but amounted to nothing. The Scots carried out surveys and gathered intelligence. Shortly after the war, they held almost 3,000 square miles and had proven the Ebba Valley coalfield, the Bruce City coalfield, and resources in Gips Valley. On the scientific stage, they had surveyed large parts of Spitsbergen and published learned papers and topographic maps. They had advanced the drilling techniques in permafrost. To meet their goal and be truly successful though, they needed to attract a subsidiary company to develop the properties. This did not happen. While the syndicate waited, it incurred large financial losses. The fact that C. W. B. Urmston eventually came forward was not the syndicate's accomplishment. It was lucky not to have lost him during the negotiations of the sale of the properties.

In conclusion, the Scottish Spitsbergen Syndicate was founded with purely commercial aims in mind. Some members were fervently nationalistic, but the only time they formulated political goals was to bring about British sovereignty after the First World War. Sovereignty would have secured the Scottish properties and reduced the risks for investors. The syndicate was not a financial company. It produced the products expected of true exploration. The fact that it remained debt-free assured its longevity, and reasons for the failure to attract a subsidiary mining company must be sought externally. Few sources concerning negotiations with potential firms have been found that clarify their reluctance to become involved. It was most likely a combination of factors. The status of Spitsbergen was initially unsettled; the British had been accused of territorial grabbing; gold-diggers had damaged the islands' reputation; and the syndicate's coal and other mineral deposits were not thought profitable during the ongoing depression. It is surprising, therefore, that the aging directors so haughtily declined Kjøde's offer. Perhaps they were unable to see past the traditional Anglo-Norwegian rivalry. They were lucky that Urmston did not simply stake out new claims himself. Perhaps strong family and social ties with members of the syndicate prevented him from acting too shrewdly. He bought the properties for a third of Kjøde's price. The Scots had ventured into the risky business of Arctic exploration, and their conservative strategy ensured their long survival against the odds, which led to the only British dividend being paid on Spitsbergen.

PART IV

THE COMBINED IMAGE

9 Four British companies on Spitsbergen: discussion and conclusion

9.1 Introduction

The objective of the research underlying this book is to explain the involvement of four British companies in the industrialisation of Spitsbergen in the European Arctic in the first half of the twentieth century as well as their impact on the geopolitical situation and the natural environment. A background chapter places this involvement in the historical context of Britain's coal industry, global empire, and polar exploration. Six empirical chapters then consider the archaeological and written evidence for the companies in search of the driving forces behind their development. This discussion is structured according to the four associated sub-questions. The first question asks why the British companies were started. The second enquires how they chose to operate and why. The third looks at why the companies were discontinued. Lastly, the consequences of their involvement are evaluated in terms of economic output, political influence, and environmental impact. The actor-network theory (ANT) is used to generate answers and provide a plausible explanation for the rise and fall of the British companies. To conclude, the chapter revisits the central research question and tests the related hypothesis in addition to offering some final remarks regarding the methodology of historical archaeology and opportunities for future research.

9.2 Why were the British companies started?

According to ANT, project leaders are able to start a project because there is a favourable historical context for doing so. As network builders, they use aspects of this context in order to interest and enrol actors to a global network to support the project. In turn, this support creates the negotiation space in which the network builders construct the local network consisting of resources and the facilities to exploit them.

The background chapter has shown the historical context at the turn of the twentieth century to include Britain's long tradition in mining. After a peak in 1900, coal prices were on the rise again in 1904. In view of increasing industrial development at home and abroad, it was a good time to invest in this resource. The seafaring nation had also acquired a vast empire and played a leading role in polar exploration. It was therefore susceptible to the various, commonly prosperous opportunities in distant and unfamiliar lands. Knowledge of the Arctic archipelago of Spitsbergen and the first commercial shipment of coal from the uninhabited no man's land to Norway spread rapidly in Britain and elsewhere. It fuelled an international hype for the polar region, not least for financial gain.

This book distinguishes between the promoters of company formation and the British companies themselves. The charismatic promoters had personal goals that a successful business would help to achieve: Fangen was in search of greater prestige; Mansfield was a gold digger in search of new adventures; Bruce was a polar scientists in search of funding; and Salisbury-Jones, given the Rhodesian example, probably envisaged himself in a central role in the colonial administration of Spitsbergen. It is unlikely that these companies would have been founded or revived in the men's absence. Other firms might perhaps have filled the niche.

For the purposes of this study, the British companies, i.e. the company boards, were the project leaders and network builders. After initial contact with a Norwegian predecessor, the Spitzbergen Coal & Trading Co. identified an Arctic coal mine in Advent Bay and the extraction of coal to meet local and Norwegian demand as a project worth pursuing. Based on British pioneering, the focus of the Spitzbergen Mining & Exploration Syndicate was twofold: it secretly prospected for gold while openly promoting the possibilities of coal. Unwittingly, the Northern Exploration Co. inherited the residual interest in gold and coal while additionally turning its attention to marble, iron ore, and a host of other potential resources on the islands. Rooted in Bruce's previous surveys, the Scottish Spitsbergen Syndicate was convinced that the presence of coal on its claims was the key to the exploitation of any other minerals. Each company therefore had a crucial economic motive for becoming involved on Spitsbergen.

Each company enrolled economic as well as political actors to its global network, who could provide the capital to develop the natural resources and exert political pressure to protect the claims. The companies were only partially successful in enrolling these actors. The network builders behind the Spitzbergen Coal & Trading Co. founded it as a private limited mining company in the belief that it would need no more than 50 investors to subscribe to its nominal capital of £2,500. These investors included professional and personal acquaintances, friends and family. Their money enabled the acquisition of the claim in Advent Bay, but only a mortgage provided the operational capital to send an expedition. Political actors were not yet essential. The Foreign Office was only approached after the inspection of the mine.

The promoters of the Spitzbergen Mining & Exploration Syndicate had already attracted influential backers like Earl Morton to their Arctic proposal prior to the formation of the public limited mining company worth £5,250. The company issued a prospectus in order to enrol investors to its global network that could fund an expedition. The expedition sailed, although not all shares had been taken up.

Similarly, the first discoverers had previously tried to enrol the Foreign Office to their global network because any investments on Spitsbergen were in danger unless the ministry could grant mining concessions and protect the claims. Their success was limited due to the unclear legal status of the islands. The Foreign Office stated that any mining operations were undertaken at their own risk. The British Government had no intention to take responsibility for the no man's land, but it would not confiscate any mining output either. The application to other powers showed that no one assumed control over the islands. The Foreign Office reluctantly accepted the syndicate's claim maps in recognition of its claims.

Because of his continued self-interest in gold, Mansfield and his associate network builders formed the Northern Exploration Co. The private limited company, and this is crucial, was not a mining company; it was an exploration company. It envisaged a role in the employment of capital worldwide, but it started out with only £100 until the prospector could produce the title deeds to his claims. Thereafter, the company leadership increased the nominal capital to a considerable £125,000. Who subscribed could not be ascertained. Although the successful Exploration Co. was probably not an actor, it was most likely the ideal behind the first expedition. At the time, a change in the historical context made it possible for the company to enrol the Foreign Office, who readily registered the company's claims. When the firm objected to trespassers on its properties, the department also attempted to mediate between the parties.

Bruce set out to build a global network by distributing a circular promoting the formation of a private limited syndicate to probable subscribers. The Scottish Spitsbergen Syndicate, too, was an exploration company. It created a local network by purchasing existent rights and claiming additional areas in view of mineral exploration and the flotation of subsidiary mining companies. The idea that it marketed appealed to Bruce's Antarctic connections and his scientific network. The Scottish elite, friends, and family also invested. They raised a capital of £4,000, half of which enabled the fitting out of the first expedition. Afterwards, the network builders were able to also enrol the Foreign Office, who assured the syndicate that it would safeguard the Scottish interests.

It emerges that each company's primary concern was to find the capital for a maiden voyage to Spitsbergen that would decide how to proceed. There were differences in the type of company, three being private and one public. There were also differences in the amounts deemed necessary. The Spitzbergen Coal & Trading Co. already took out a mortgage of £3,000; the Scots embarked on the voyage with a conservative £2,000; and the Northern Exploration Co. had £50,000 at its disposal. These amounts probably reflected different operational strategies. Initially, political support was secondary. Personal risks were willingly taken to journey into the unknown. Additional investments, however, would be jeopardised unless the claims enjoyed official protection. When the companies were founded, British foreign policy had no intention to assert control in the European Arctic. The Foreign Office subtly adjusted its position, however, in view of growing international attention on the islands.

9.3 How did the companies choose to operate and why?

ANT proposes that the network builders maintain their actor network by ensuring that the actors in the global network keep supplying the local network with the means to sustain it and that the local network, in turn, delivers the natural resources which the actors in the global network desire. For each company, the original financial outlay created the negotiation space in which the local network took shape. The LASHIPA expeditions proved that neither company had been fraudulent. All had been on Spitsbergen. All had given rise to an archaeological record that, substantiated by archival evidence, revealed the existence and character of their local networks. They had constructed facilities in an attempt to live and work in the Arctic and exploit its mineral deposits.

The Spitzbergen Coal & Trading Co. was not interested in wholesale claiming and focused instead on Advent Bay. The settlement and the mine were its most prominent symbols of occupation, supported by two hunting huts and a number of claim signs. Advent City was well-built and differentiated. The pre-fabricated houses could accommodate a large workforce and were fit for wintering. The staff and some key personnel were English, while cheap labour was recruited in Scandinavia. The presence of women, a post box, and livestock perfected the image of a civilised Arctic colony. Little is known about the mine, which was worked all year around. Rudimentary mining technology was quickly replaced with state-of-the-art equipment. An engine house prophesied the electrification of the site and promised rapid development of the resource. A pier was said to enable the tipping of coal into lighters that could be got away at any tide to supply local steamers. Other output was shipped to Norway. The board gained knowledge of Advent City in the form of claim maps, buoyant reports, and wonderful photographs. The directors also visited the site to inspect the development. Although it is not known how much of their knowledge was disseminated in the global network, the shareholders were satisfied with the progress and agreed to increase the capital to £25,000. When this was not fully issued, they also consented to a series of mortgages and debentures until the company was deeply in debt. In other words, the local network was initially able to deliver enough resources to keep the actors in the global network satisfied.

Although the physical remains of the Spitzbergen Mining & Exploration Syndicate were masked by later activities, archaeological fieldwork gave the impression that the company's claims were confined to the northern shore of Bell Sound and Lowe Sound in order to prove a coalfield and prospect for gold. Camp Morton was a simple mining settlement named after its most illustrious backer. English miners and Norwegian workers opened two seams in three headings, and the coal was reported to be splendid. Camp Bell was a prospecting camp and claim hut, in the vicinity of which Mansfield had made some gold finds. Archival research, however, revealed a whole series of claims, also on the southern shore of Lowe Sound, which were marked by claim boards in each corner. The syndicate merely owned one of these. It is difficult to establish if the others had been made for powerful shareholders or were entirely separate. If the latter, Camp Bell was not a syndicate asset. Either way, the directors again obtained claim maps, reports, and photographs that proudly sported the Union Jack. By using these actants, the network builders managed to increase the capital to £15,350, which enabled them to finance two additional expeditions. Yet, the 1908 expedition was the final one.

The local network of the Northern Exploration Co., its business being exploration, was characterised by wholesale claiming. The company occupied vast areas anew or purchased them from former Norwegian owners. The territory reached its maximum extent shortly after the First World War. It was guarded by a host of material actants such as mining settlements, prospecting camps, claim huts, and claim signs. Most but not all were associated with workings. Some sites were not suitable for wintering. The war breathed new life into the company and its installations. Mining settlements such as Marble Island and Camp Morton were arguably advanced prospecting camps, while Calypso Beach functioned as an administrative centre. Common were camps where a simple hut or tents were linked to a mineral outcrop. Pure claim huts were erected by allied trappers to claim ownership of an area. The workings included coal and 'gold' mines, marble and asbestos quarries, trial trenches in iron ore and copper ore, and shafts into zinc blende. The cutting-edge technology at the marble quarries, which witnessed boring and channeling, and the wireless stations was offset against simple prospecting and shallow workings elsewhere. Large workforces were divided into advance parties, main expeditions, and winterers. British staff coordinated the efforts of specialised miners and quarrymen among general workers. The company phased out increasingly expensive Scandinavian labour in favour of British men. Norwegian trappers, however, persisted.

It was paramount to the exploration company to demonstrate the wealth and workability of Spitsbergen in order to attract subsidiary mining firms to work the mineral deposits. In order to enrol such actors to its global network, the board of directors produced a host of agents that would present their local network in the most advantageous and credible way. Claim maps, eyewitness testimonies of employees as well as visiting stakeholders, expert statements, photographs, samples, test results, and inferences were probably available for public viewing at the London office. The network builders disseminated selected information in seven elaborate brochures, later switching to company reports. They most likely withheld detrimental details. On the back of this strategy, the Northern Exploration Co. succeeded in raising its operational funds to £150,000. It then used the greatly politicised context of the war to boost its nominal capital to £500,000, and, after converting to a public company, to £1,000,000. The time was right to lobby for the British annexation of Spitsbergen among actual and potential economic and political actors. Sovereignty would secure the company's claims on the most favourable terms and therefore aid its primary economic goal. The Foreign Office, however, continued to arbitrate in claim disputes and to assure the safeguarding of British interests.

Archaeologically, the local network of the Scottish Spitsbergen Syndicate before the war is practically non-existent. Documents, however, showed that the Scots had claimed widely. The post-war representations of effective occupation were few but well-built and strategically placed. Pre-fabricated huts were recorded at Bruce City, on Prince Charles Foreland, at Gips Bay, and in Gips Valley. The pre-fabricated hut at Bjona Haven and the Swedish hut purchased in Ebba Valley were known of, but they were not surveyed in this project. The syndicate further made extensive use of tent camps, which barely left a trace, and erected few claim markers, which were nonetheless tactically positioned. The syndicate routinely carried out topographic surveys, which bestowed the landscape with Scottish place-names, generated detailed base maps and geological knowledge, and gave the undertaking scientific legitimacy. The surveys aided the positioning of trial trenches and boreholes. LASHIPA recorded the remains of drilling rigs near Bruce City and in Gips Valley, which indicated a coordinated drilling programme to prove the coal (and gypsum) at depth. In Gips Valley, large distances were overcome using a tractor. The syndicate made wide use of scientific staff on its expeditions immediately after the war but eventually reduced it to mining engineers. The group hired Norwegian workers on one occasion only. Before the war, the syndicate's products of exploration and a new circular warranted additional funds of £1,000. As with the Northern Exploration Co., the Scots used the optimistic post-war climate for a reconstruction with a nominal capital of £100,000 and shortly afterwards converted to a public company. They also appealed to likely actors for British sovereignty. This, too, served an overall economic purpose, but foreign policy did not change in its favour.

It transpires that each company laboured to construct a local network with the greatest possible appeal to actual and potential economic and political actors in the global network. These actors expected a return on their investments and support. Since exploration and mining could not generate immediate revenue, the preliminary products were reports, samples, test results, and inferences, which continually promised profitability in the near future. There was a difference in the scale of operations. The mining company at Advent City operated on a small claim with a focus on extracting coal. The Spitzbergen Mining & Exploration Syndicate is best viewed as a hybrid with a coal mine at Camp Morton surrounded by large properties on which to prospect for gold. Unfortunately, these companies did not last long enough to observe how they would have accommodated the changes during and after the First World War. It was in the nature of exploration companies to claim widely in the hope that somewhere on the vast territory lay mineral deposits attractive to subsidiary firms. These companies had a greater interest in the settled legal status of Spitsbergen in order to secure their properties. The Northern Exploration Co. opted to be highly speculative and very showy, while the Scottish Spitsbergen Syndicate based itself on scientific methods and was conservative. Both, in effect, tried to sell an idea (accompanied by workable claims). For a time, they appealed to their respective target audiences, but they were ultimately unsuccessful.

9.4 Why were the companies discontinued?

The failure of an actor network can be explained by two different but related processes. Either actors in the global network lose their interest and withhold their support with the consequence that the local network collapses and the project is over; or the local network fails to deliver the natural resources that the global network has come to expect.

Archaeological and archival sources have revealed serious deficiencies in the local network of the Spitzbergen Coal & Trading Co. Despite Emerson Muschamp Bainbridge, one of Britain's leading coalowners, having been the company's founder, it lacked his knowledge, experience, and competence. Bainbridge seemingly never assumed active control of the operations; if he had done, Advent City may have been a moderately successful mine yet. As it was, the company had barely mastered detrimental environmental aspects such as exposure, a poor harbour, and difficult landing and loading conditions. It had overspent on the settlement and the mining technology, some unsuitable equipment remaining on site till this day, while the coalfield had not been proven. The debt could have been repaid once the mine produced a payable output. However, it repeatedly failed to meet its targets due to the incompetence of the local managers and the coal being of low quality. In addition, the strained relation between the leadership and the workforce resulted in a prolonged strike and a temporary worker take-over of the mining town. The conflict exposed the lack of political support from both the British Government and the Norwegian authorities. Where the deficiencies of the local network became known to the global network, they discouraged the stakeholders. Advent City had not met their expectations. It had not given a return on the money so far invested, and there was no prospect of it doing so in the future. The shareholders did not re-invest, and the debenture holders demanded their money back. The company was under pressure to turn its property into cash but managed to do so very badly. After a period of inactivity, the First World War was arguably a fortunate circumstance in which a Norwegian buyer came forward. The sale occurred at a loss. The Cretaceous coal was never worked again. In ANT terms, despite the favourable pre-war context, the company was discontinued because its local network had for a combination of factors failed to deliver any coal.

The problems of the Spitzbergen Mining & Exploration Syndicate began in the boardroom, although it is questionable whether its directors ever met. A small London office juggled its dual objectives. Despite being a public company, it was very secretive about the search for gold and stated the exploitation of coal to be its main objective. This led to a poorly defined operational strategy. The local network did not produce gold in lucrative amounts, while a professional report concluded that the coal at Camp Morton was uneconomical. The shares of the syndicate had not been popular from the start. Disillusioned with the prospecting and discouraged by the mine, the shareholders withdrew, which spelt the end of the development in Bell Sound. Disagreements within the syndicate are evident in the disintegration of the management and the refusal to wind up properly. The group was unceremoniously crossed off the register. The fact that another attempt was later made to work the coal at Camp Morton suggests that the damaging report had been concealed from the Northern Exploration Co. In this case, too, the company failed due to the inability of its local network to produce a marketable output.

The operational strategy of the Northern Exploration Co. raises the question if hasty claiming still lay within the company's comfort zone or whether the wide spread had reached its geographical, financial, and practical limits. The company had occupied more barren ground than mineral occurrences, and the need to prove effective occupation coupled with the struggle against the elements, mismanagement, and the pressure of transportation and communication drew relentlessly on the company's resources. Although it is difficult to assess when an exploration company was guilty of overspending, the appropriateness of many installations is questioned in view of costly equipment remaining on the sites today. The company was reportedly bankrupt before the war and prior to global events taking their toll. Yet, it appears never to have considered relinquishing any of its claims. However, the deficiencies of the pre-war local network are of surprisingly little interest to this study because the greatly changed post-war context allowed the company to reinvent itself.

The Northern Exploration Co. apparently had not learned from earlier shortcomings. If anything, it adhered to territorial grabbing and speculation more vehemently. Yet, its unfounded company propaganda appealed to the contemporary masses. Importantly, the board with only limited experience of exploration overseas was chasing the success stories of other parts of the Empire, and although the directors attracted the capital to do so, they failed to emulate the crucial mining expertise. The company adopted professional reports after the war, which by and large concluded that the claims were uneconomical, but the directors chose to ignore this, stuck to a policy of floating a subsidiary company, and retained all claims until the arbitration by the Danish Commission. The post-war slump meant that capitalists on the whole were no longer interested in Spitsbergen. The British coal industry was in decline, the markets were bad, and oil was the centre of future energy visions.[1] In addition, sovereignty, which could yet have swayed the company's fortunes, was not in the interest of British foreign policy due to priorities elsewhere and a reluctance to provoke the Scandinavian countries.[2] The company's inflated promises to economic and political actors during the good years had the repercussion that it was under pressure to succeed, but there was no subsidiary company. Its conversion to active mining came too late; the limited output could not recover the investments. When the last candidate for a subsidiary firm dropped out, the company sold its properties to the Norwegian Government at a large loss to the stakeholders. The company may once have had legitimate rights to Kings Bay and Braganza Bay, but it had lost out on the opportunity of extracting marketable coal there. As it was, none of its claims were ever worked commercially. To sum up, the local network was not entirely to blame while the reluctance of subsidiary firms to come forward is as of yet poorly understood. Perhaps speculation as the company had undertaken it belonged to a bygone era.

Evidence for the pre-war failure of the Scottish Spitsbergen Syndicate is sparse. Unlike the Northern Exploration Co., it was possibly not flamboyant enough to ignite far-reaching enthusiasm and the matching investments. Yet, it survived long enough to also benefit from the altered economic and political climate after the war. That its reconstruction was quickly followed by the conversion to a public

---

[1] Notz, W. (1918) 'The world's coal situation during the war', *Journal of Political Economy*, 26 (6), pp. 567-611; Supple, B. (1989) 'The British coal industry between the wars', *Refresh*, 9, pp. 5-8; Engdahl (1993).

[2] Goldstein, E. and McKercher, B. J. C. (2003) 'Power and stability in British foreign policy, 1865-1965', *Diplomacy and statecraft*, 14 (2), pp. 1-22; Steiner and Dockrill (2003); Salmon, P. (1997) *Scandinavia and the great powers 1890-1940,* Cambridge: Cambridge University Press.

company puts into question if the Scots had thought their approach through. Still, in the peak years, the public was the most likely investor. The local network was not unnecessarily crowded and produced the genuine products of exploration. By the mid-1920s, the syndicate had undertaken what exploration it could without a return on the original outlay. Its shareholders now expected successful negotiations with a subsidiary mining firm. There may have been a time when the syndicate's properties could have provided a cautious firm with saleable output. The changing historical context, however, had several effects. The ongoing depression made the coalfields and other resources uneconomical. The syndicate left machinery in anticipation of a return to the islands, but it never exploited its claims. It presumable contacted countless capitalists, but they were reluctant to come forward. This reluctance was probably rooted in Spitsbergen's unsettled status, Britain being accused to territorial aggrandisement, the bad reputation of other companies, and the coal not being lucrative under the prevailing circumstances. The syndicate could not influence this. Although it did not revert to active mining to recover the investments, its inability to float another company did not spell its immediate end either. The promise that the Spitsbergen Treaty would secure the claims and the investments kept the hope alive that the deposits were profitable yet. Being largely debt-free, the syndicate could await the ratification and thereafter adopt a waiting policy. The decision not to sell to Kjøde when the chance arose was a mistake probably embedded in the traditional Anglo-Norwegian animosity. Thereafter, the syndicate was indeed guilty of managing its affairs abysmally. Its increasing inability to meet the standing costs eventually forced it to sell to the only buyer available. It was luck rather than a shrewd sense for business that allowed for a small dividend being paid. Yet, the Scottish Spitsbergen Syndicate may be seen as the only successful British company: it achieved its goal of selling its proven claims to a mining company, albeit for a miserable fortieth of the original share price.

It is safe to conclude that the explanation for pre-war failure lies in the local networks' inability to satisfy the expectations of the global networks. If the context had remained unchanged, none of the economic actors would have re-invested, and the political support from the British Government would have remained at a bare minimum. A general explanation for the post-war failure of the exploration companies is more difficult to arrive at. The Northern Exploration Co. was not ideally managed, but the wave of post-war buoyancy and rhetoric swept it along. The Scottish Spitsbergen Syndicate implemented a better strategy, but both companies missed the short window of opportunity at peak times to attract subsidiary firms. During the depression, continued survival hinged on the ability to keep the promise of returns from the claims alive until a better market would facilitate their exploitation. The Northern Exploration Co. folded before the Scottish syndicate because it failed to meet the costs involved sooner. Although these reasons are anchored in the historical context and the global network, they were exacerbated by the failure of the local networks to deliver at key moments.

The four British companies that were active on Spitsbergen thus support Bone's notion that the industrial development of peripheral regions is subject to external forces.[3] They were highly dependent on global demand, but they never enjoyed governmental intervention to assure their survival against the odds. If only using the core-periphery model, however, there is a danger of disregarding the industrial operations at the resource frontier. The model's one-sided focus on global demand is therefore best complemented by ANT, which throughout this book has placed emphasis on the importance on how well the local networks functioned.

The companies further supported Wilkins' trend for British free-standing companies, which were probably the most typical mode of British foreign direct investment.[4] The purpose of free-standing companies was to interest British investors in potentially profitable operations oversees and to mobilise capital for the projects. Since the costs of promotion, legal fees, and other dues could be as high as 33 per cent, Wilkins asks why such companies were formed and reasons that the British would not invest in the totally unknown nor trust foreigners with their money. The companies provided the institutional framework to put investors at ease. A crucial feature was the limited size of a typical head office, which normally comprised a corporate secretary and several directors. The limitation is best shown in comparison with America, where the vastness of the country aided companies to grow into large multi-functional, multi-regional businesses that developed appropriate managerial talents. The compact domestic market of Britain was ill-suited to develop similar skills. According to Wilkins, 'the need to manage the business overseas [...] provided a formidable challenge, and one that the free-standing companies often failed to meet.'[5]

Wilkins recognises the directorial role as network builders, but she states that 'in most cases the connections [...] were too partial and too weak to be designated.'[6] She unwittingly identifies the main criticism of using ANT throughout this study: the British companies on Spitsbergen barely established lasting let alone constructive bonds with their actors. The negative nature of a bond commonly had the most bearing on policy and strategy. Generally, profitable free-standing companies were those that either developed successful in-house management or those under the management of a mining engineering firm or a

[3] Bone (2003).
[4] Wilkins, M. (1988) 'The free-standing company, 1870-1914: an important type of British foreign direct investment', *Economic History Review*, 41(2), pp. 259-82.
[5] Wilkins (1988) p. 264.
[6] Wilkins (1988) p. 265.

trading company. Such companies represented a very small percentage of the total number registered. Yet, British investors never learned their lesson. According to Wilkins, 'as some enterprises were profitable, the potential for success existed; free-standing companies were presented in such a way as to generate confidence, even though the prospectus might cloak the reality; nominal, projected returns always looked attractive; and few investors appreciated the managerial difficulties inherent in these ventures.'[7] The bottom line is that while the British free-standing companies were as a whole highly successful in raising capital, they were less apt at providing management, technology, and business know-how.

Despite the logic behind these reasons, which also apply to the British companies on Spitsbergen, Burt argues that they do not sufficiently explain the failure of hundreds such companies across the world.[8] He points out that while properly qualified, informed, and experienced mining promoters would have identified any pitfalls, the promoters and directors behind these firms were 'innocents abroad'[9]. They were unable to judge the value of the properties they obtained and how to effectively exploit them. Any sensible learning process was regularly disrupted by what Burt calls foreign 'mining manias'. He concludes that, 'the great majority of British overseas mining promoters and promotions were poorly informed, undercapitalized, and badly managed. Most of the ventures never progressed as far as the production phase and very few indeed ever produced a respectable return on capital [...]. The explanations [...] reflect an essentially British problem, probably located in the London market, where the great majority of overseas enterprises were promoted.'[10] In his mind, significantly greater experience was needed to make the fine judgements necessary.

9.5 What were the economic, political, and environmental consequences?

After he witnessed Klondike, Jack London contemplated who had profited from the gold rush and who had lost, how much gold was taken out of the ground and how much went back in.[11] He saw the event as a great transfer of energy and argued

[7] Wilkins (1988) p. 271.

[8] For exemplary studies see Woodland, J. (2011) 'Experiences of the first gold mining companies in Australia', *Mining Perspective: the 8th International Mining History Congress 2009.* Penventon Park Hotel, Redruth 12-15 June. Truro: Cornwall and West Devon Mining Landscape World Heritage Site, Cornwall Council; Harvey, C. and Taylor, P. (1987) 'Mineral wealth and economic development: foreign direct investment in Spain, 1851-1913', *Economic History Review*, 40 (2), pp. 185-207.

[9] Burt, R. (1997) 'British investment in the American mining frontier', *Business and Economic History*, 26 (2), p. 517.

[10] Burt (1997) pp. 523-4.

[11] London, J. (1900) 'The economics of Klondike', *American monthly review of reviews*, pp. 70-4.

that the capital and labour must have gone somewhere. He concluded that the Yukon returned only a tenth of what was put into it. Yet, 'while this sudden and immense application of energy had proved disastrous to those involved, it has been of inestimable benefit to the Yukon country, to those who will remain in it, and to those yet to come.'[12]

The British involvement on Spitsbergen can be subjected to the same contemplation. British investments and accrued debts can be added to know how much went in. A third was said to be taken up during company registration and administration. This invariably stimulated the British economy as did the costs for British staff, equipment, and shipping requirements. How much money benefitted the Scandinavian market, especially Norwegian outfitting towns, in the form of workers' wages and the payments to stores and shipping firms has not been ascertained. There was little chance of 'dropping money on the trail', as London called it. Spitsbergen initially gained materially, and once the treaty was ratified, British dues swelled the Spitsbergen Fund. Ultimately, a very small percentage of the original outlay was returned to the companies. A tiny dividend was paid once. The economic consequences of the involvement, therefore, boiled down to benefitting Britain's and Norway's markets through the spending of cash.

The network builders failed to fully enrol and mobilise influential political actors in order to pursue their interests. The Foreign Office undertook the diplomatic minimum to safeguard British interests while maintaining the balance of power in the European Arctic. Unlike other stakeholder powers, the British Government never subsidised the firms. Despite this seeming lack of support, its influence on the legal status of Spitsbergen was profound. The British companies voiced their concerns before and during the negotiations of the settlement, and the British Government supported the limitations imposed on Norwegian sovereignty in the Spitsbergen Treaty. To the dismay of the companies, British annexation was not attempted and therefore never achieved, which eventually contributed to their demise. However, the treaty enabled the archipelago to become the sphere of international understanding and cooperation that Svalbard is today.

The environmental impact is best assessed using the archaeological record. None of the British sites reached a sustainable producing stage, which fortunately also prevented the disasters often associated with extractive industries today. The Arctic landscape nonetheless indicates enduring effects in terms of the localised reduction of rock, changes in topography, and the disturbance of vegetation. Some of these are permanent, while others will gradually go back to normal. The developments and related hunting activities may once have removed all wildlife in the vicinity, but the LASHIPA expeditions observed the return of polar

[12] London (1900).

bear, Arctic fox, reindeer, and other animals. Some people may regard the physical remains of the British installations as unsightly in what they imagine to be a pristine environment. However, the archipelago gained from the benefits of industrial archaeology as outlined in Chapter 1. Moreover, the remnants of British mining and exploration frequently enhance their surroundings, are a source of income from tourism, and communicate not only the experiences of early Arctic industry to our post-industrial society but also of geopolitical conflicts in unclaimed territories.

9.6 Conclusion

The conclusion returns to the central research question as to the driving forces behind the development of the British mining industry on Spitsbergen between 1904 and 1953. The hypothesis proposes that while economic driving forces initially determined the British presence on the archipelago, political motivations amplified at specific moments in time and their repercussions sustained the companies beyond economic feasibility. This study has shown that financial gain either through active mining or through exploration coupled with the floatation of a subsidiary firm was the primary goal of each of the four companies. The two mining companies had failed before the specific moment, i.e. the heavily politicised context of the First World War, allowed the two exploration companies to reinvent themselves. The repercussion of their struggle for British sovereignty in the hope of better business was to continually conjure the potential of success, which proved to be another of the world's mining manias. It prolonged the companies' existence beyond economic feasibility. Some shareholders willingly waited more than four decades for a negligible return on their investments.

Despite the core-periphery model being an integral part of the theoretical framework, these conclusions have been arrived at primarily with the assistance of the actor-network theory. ANT provided the structural building blocks on which each of the empirical chapters is founded and allowed for a concise graphic representation to sum up the actors and actants involved. However, Wilkins pinpointed a major flaw in the theory, which also applies to this study: the connections were often too partial or too weak to be designated.[13] The problem was especially inherent during prolonged episodes of passively waiting for a favourable change in the historical context. ANT needs to be improved for a greater relevance to the feeble and ailing networks of free-standing companies and to mining and exploration finance.

If this research had not been done using a combination of archaeology and history, the archaeological landscape of Arctic mining and in particular Arctic

[13] Wilkins (1988).

exploration would either remain poorly understood, or the historical image of the British companies would be incomplete and difficult to interpret. Owing to the fact that LASHIPA has been in the field, as opposed to the majority of previous studies, the combined image of the British involvement on Spitsbergen emphasises that the companies did what their documents said they would do or had done. Archaeological fieldwork provided another, more reliable perspective of past events. Seen with one's own eyes, the companies' interactions with the Arctic, especially their impacts on the natural environment, become more clear. Not to have done the surveys would have doomed this work to only offer a partial, unreliable impression.

The research has given rise to some recommendations for further work. In the first instance, the remainder of British sites on Spitsbergen should be visited and surveyed in an attempt to complete the industrial archaeological landscape, especially the landscape of exploration, and formulate a conservation strategy that highlights important scientific and educational aspects. Marble Island and other marble outcrops in Kings Bay and Cross Bay deserve additional attention to investigate why the quarries failed. Reasons may need to be sought in local fault patterns as well as changes in the global marble market over time. This study could be embedded in the work on British free-standing companies and global mining finance. It would also be instructive to know why the countless capitalists who must have been approached during company promotion did *not* invest in Spitsbergen. Such work may need to be based on in-depth biographical research.

Acknowledgements

I am grateful to Prof. Dr. Louwrens Hacquebord for his vision and endurance to get the LASHIPA project off the ground and for enabling me to play a part in it. He has spent countless hours reading my work and providing the necessary guidance and constructive criticism to see it completed and published.

I also thank my co-promoter Dr. Dag Avango. Dag tirelessly led the fieldwork on Spitsbergen that I have been lucky to participate in twice. He repeatedly read my chapters and made invaluable comments. Imagine the thickness of this book if both Louwrens and Dag hadn't asked me to keep it 'short'!

In light of 'short' not being my forte, a special thanks to my manuscript committee, too: Prof. Dr. Maarten Duijvendak, Prof. Dr. Peter Jordan, Prof. Dr. Luis Lobo-Guerrero, and Dr. Miles Oglethorpe.

Others who have helped me in one noteworthy way or other are Ypie Aalders, Ann Kristin Balto, Susan Barr, Nienke Boschman, Caroline Botman, Mark Carlyle, Peter Darrington, Hidde de Haas, Seth DePascal, Peter Falding, Heinz Fromke, Mark Gilbert, Ulf Gustafsson, Jan Hardy, Cameron Hartnell, Stig Henningsen, Andreas Höhne, Britta Jung, Gesa Kruse, Harm Kruse, Niels Kruse, Waltraud Kruse, Julia Lajus, Pia Leminen, Herdis Lien, Maarten Loonen, Janina Lux, Grant Mangham, Ruth Mangham, Patrick Martin, Margot McCuaig, Nicole Miedema, Larry Mishkar, David Munro, Michael Muschamp, Jennifer Kelly, Richard Pilley, Terje Planke, Jouke Prop, Per Kyrre Reymert, Kathryn Rohweder-Lux, Cecilia Sandström, Annette Scheepstra, Vadim Starkov, Frits Steenhuisen, Ellie Swinbank, Geoff Swinney, Tekke Terpstra, Aileen Thompson, Elske Tielens, Marion Tielens, Marlies van Kruining, Marion van Rijssel, Ronald Visser, Sheila Watt, Ian West, John Willis, Inger Woltinge, John Woodland, and Frank Zibull.

Tu, was jeder loben müsste, wenn die ganze Welt es wüsste;
Tu es, dass es niemand weiß und gedoppelt sei sein Preis.

- Wilhelm Busch

References

Aalders, Y. I. and Hacquebord, L. (2008) 'Europese walvisvaarders en Russische jagers in Green Harbour, Spitsbergen', *Paleoaktueel,* 19, pp. 210-5.

Allen, J. B. (1966) *The company town in the American West*, Norman: University of Oklahoma Press.

Arisholm, T., Devold, E. M., Hoem, S., and Rossnes, G. (2000) *Gruve 2b, Longyearbyen, Svalbard. Rapport – dokumentasjon 2000.* Oslo: DAR-gruppen.

Arlov, T. B. (1989) *A short history of Svalbard*, Oslo: Norwegian Polar Institute.

Arlov, T. B. (1991) *Store Norske 75 år*, Longyearbyen: Store Norske Spitsbergen Kulkompani A/S.

Arlov, T. B. (1996) *Svalbards historie 1596-1996*, Oslo: Aschehoug.

Arlov, T. B. (2005) 'The discovery and early exploitation of Svalbard. Some historical notes', *Acta Borealia*, 22 (1), pp. 3-20;

Avango, D. (2003) 'Aktanter i ingenmanslandet', in Avango, D. and Lundström, B. (eds.) *Industrins avtryck*, Stockholm: Brutus Östlings Förlag Symposium, pp. 173-206.

Avango, D. (2005) *Sveagruvan – Svensk gruvhantering mellan industry, diplomati och geovetenskap*, Stockholm: Jernkantoret.

Avango, D., Bekooy, B., Gustafsson, U., Hacquebord, L., and Hartnell, C. (2006) *LASHIPA 2, Archaeological Expedition on Svalbard, August 8-22, 2005.* Field report. Groningen: Arctic Centre.

Avango, D., Gustafsson, U., Hacquebord, L., and Hartnell, C. (2008a) *LASHIPA 3, Archaeological expedition on Spitsbergen, August 7-24, 2006.* Field report. Groningen: Arctic Centre.

Avango, D., Aalders, Y., Gustafsson, U., Haas, H. R. de, Hacquebord, L., and Hartnell, C. (2008b) *LASHIPA 4, Archaeological expedition on Spitsbergen, August 2-25, 2007.* Field report. Groningen: Arctic Centre.

Avango, D. and Hacquebord, L. (2008) 'The history and heritage of natural resource exploitation in the Arctic and Antarctic: the LASHIPA project', *Industrial Patrimony*, 19, pp. 7-16.

Avango, D., Aalders, Y., Gustafsson, U., Haas, H. R. de, Hacquebord, L., Hartnell, C., Kruse, F., and DePasqual, S. (2009) *LASHIPA 5, Archaeological expedition on Spitsbergen, 27 July – 17 August 2008.* Field report. Groningen: Arctic Centre.

Avango, D., Haas, H. R. de, and Kruse F. (2010) *LASHIPA 9, Archaeological expedition to Spitsbergen, 31 July – 15 August 2010.* Field report. Groningen: Arctic Centre.

Avango, D., Hacquebord, L., Aalders, Y., Haas, H. R. de, Gustafsson, U., and Kruse, F. (2011a) 'Between markets and geo-politics: natural resource exploitation on Spitsbergen from 1600 to the present day', *Polar Record*, 47 (240), pp. 29-39.

Avango, D., Basberg, B. L., Gustafsson, U., and Rossnes, G. (2011b) *LASHIPA 6, Archaeological expedition on South Georgia, 3 March – 12 April 2009.* Field report. Groningen: Arctic Centre.

Avango, D., Gustafsson, U., and Hacquebord, L. (forthcoming) *LASHIPA 8, Archaeological expedition to Deception Island and South Shetland Islands, 2010.* Field report. Groningen: Arctic Centre.

Baigent, E. (2004) 'Scoresby, William, junior (1789-1857)', *Oxford Dictionary of National Biography*, Oxford: Oxford University Press.

Baldwin, R. C. D. (2004) 'Poole, Jonas (bap. 1566, d. 1612)', *Oxford Dictionary of National Biography*, Oxford: Oxford University Press.

Barr, S. (1985) 'Ernest Mansfield, drømmer, svindler, gentleman og eventyrer' in Amundsen, B. (ed.) *Svalbardboka 1985-1986*, Tromsø: Ursus Forlag, pp. 159-73;

Barr, S., Newman, D., and Nesteroff, G. (2012) *Ernest Mansfield (1862-1924). "Gold, or I'm a Dutchman."*, Trondheim: Akademika Publishing.

Barr, W. (2008) 'The Arctic cruises of Prince Albert I of Monaco', *Polar Record*, 44 (228), pp. 1-14.

Bernstein, R. E. (1985) 'The Scottish National Antarctic Expedition 1902-04', *Polar Record*, 22 (139), pp. 379-92.

Bolhuis, E., Haas, H. R. de, Hacquebord, L., and Steenhuisen, F. (2009) *LASHIPA 7, Archaeological expedition on Spitsbergen, 2009.* Field report. Groningen: Arctic Centre.

Bone, R. M. (2003) *The geography of the Canadian North*, Oxford: Oxford University Press.

Bravo, M. (2006) 'Geographies of exploration and improvement: William Scoresby and Arctic whaling, 1782-1822', *Journal of Historical Geography*, 32, pp. 512-38.

Brown, R. N. R. (1943) 'Review: Place-names of Svalbard', *Geographical Journal*, 102 (4), pp. 180-4.

Bruce, W. S. (1908) 'The Exploration of Prince Charles Foreland, 1906-1907', *The Geographical Journal*, 32 (2), pp. 139-48.

Buchanan, A. (2005) 'Industrial archaeology: past, present and prospective', *Industrial Archaeology Review* 27 (1), pp.19-21.

Burnett, D. G. (2000) *Masters of all they surveyed. Exploration, geography, and a British El Dorado,* Chicago: The University of Chicago Press.

Burt, R. (1997) 'British investment in the American mining frontier', *Business and*

Economic History, 26 (2), pp. 515-25.

Callon, M. (1986) 'Some elements of a sociology of translation: domestication of the scallops and the fishermen of St Brieuc Bay', in Law, J. (ed.) *Power, action and belief: a new sociology of knowledge*, London: Routledge & Kegan Paul.

Cassell, M. S. (2005) 'The landscape of Iñupiat eskimo industrial labor', *Historical Archaeology* 39 (3), pp. 132-51.

Cassell, M. S. and Stachiw, M. O. (2005) 'Perspectives on landscapes of industrial labour', *Historical Archaeology,* 39 (3), pp. 1-7.

Catford, K. (2002) 'The industrial archaeology of Spitsbergen', *Industrial Archaeology Review,* 24 (1), pp. 23-36.

Catford, K. (2004) 'Svalbard coal – a hundred years'. [Online]. Available at: http://website.lineone.net/~polar.publishing/svalbardcoal.htm (Accessed: 7 November 2011).

Church, R. (1986) *The history of the British coal industry: Vol. 3 1830-1913 Victorian pre-eminence,* Oxford: Clarendon Press.

Coleman, E. C. (2006) *The Royal Navy and polar exploration from Frobischer to Ross*, Stroud: Tempus.

Coleman, E. C. (2007) *The Royal Navy and polar exploration Vol 2: from Franklin to Scott*, Stroud: Tempus.

Conway, M. (1906) *No man's land*, Cambridge: Cambridge University Press.

Cullen, B. T. and Pretes, M. (2000) 'The meaning of marginality: interpretations and perceptions in social science', *Social Science Journal*, 37 (2), pp. 215-29.

Debenham, F. (1942) 'Place-names in the polar regions', *Polar Record*, 3, pp. 541-552.

DePasqual, S. (2009) *Winning coal at 78° north: mining, contingency and the Chaîne Opératoire in Old Longyear City.* Unpublished MSc thesis. Michigan Technological University.

Dinius, O. J. and Vergara, A. (eds.) (2011) *Company towns in the Americas: landscape, power, and working-class communities,* Athens: University of Georgia Press.

Dodds, K. (2002) *Pink Ice: Britain and the South Atlantic empire*, London: I. B. Tauris.

Dodds, K. (2007) *Geopolitics – A very short introduction,* Oxford: Oxford University Press.

Dole, H. (1922a) *America in Spitsbergen. The Romance of an Arctic coal-mine. Part 1.* Boston: Marshall Jones Company.

Dole, H. (1922b) *America in Spitsbergen. The Romance of an Arctic coal-mine. Part 2.* Boston: Marshall Jones Company.

Engdahl, F. W. (1993) *A century of war. Anglo-American oil politics and the new world order,* Wiesbaden: Dr Bottinger Verlags-GmbH.

Erskine, A. B. (1994) 'Victor Campbell and Michael Barne in Svalbard: the 1914 voyage of *Willem Barents*', *Polar record*, 30 (173), pp. 117-22.

Evjen, B. (1996) *Longyearbyen 1916-1975. Fra arktisk arbeidsplass til etablert industrisamfunn?,* PhD thesis, Tromsø;

Foucault, M. (1977) *Discipline and punishment: the birth of the prison*, New York/NY: Pantheon Books.

Friedmann, J. (1966) *Regional development policy: a case study of Venezuela.* Cambridge/MA: MIT Press.

Given, M. (2005) 'Mining landscapes and colonial rule in early-twentieth-century Cyprus', *Historical Archaeology* 39 (3), pp. 49-60.

Goldstein, E. and McKercher, B. J. C. (2003) 'Power and stability in British foreign policy, 1865-1965', *Diplomacy and statecraft*, 14 (2), pp. 1-22.

Good, G. A. (2004) 'Sabine, Sir Edward (1788-1883)', *Oxford Dictionary of National Biography*, Oxford: Oxford University Press.

Gustafsson, U. I. (2008) 'Modern whaling industry in Spitsbergen as a tool for territorial claiming and national sovereignty strives', *Industrial Patrimony,* 19, pp. 17-24.

Gustafsson, U. I. (2010) 'Early whaling in the north. Industrialising the Arctic: settlement design of modern whaling stations in Spitsbergen and Bear Island', *The 3rd Symposium on Whaling and History.* Commander Chr. Christensen Whaling Museum, Sandefjord, June 2009. Sandefjord: Commander Chr. Christensen Museum, pp. 45-58.

Haas, H. R. de (2008a) 'Spitsbergen's resources, Dutch entrepreneurs and geo-politics', *Industrial Patrimony*, 19, pp. 25-36.

Haas, H. R. de (2008b) 'Nederlandse ondernemers op poolavontuur. De Nederlandsche Spitsbergen Compagnie, 1920-1932', in Veerman, D. (ed.) *Tegenpolen. Op expeditie naar contrastrijke poolgebieden,* Zutphen: Walburg Pers.

Haas, H. R. de (2009) 'NV Nederlandsche Spitsbergen Compagnie. The forgotten history of a Dutch coal mining company in the High Arctic', *Natural Resource*, 0 (3), pp. 26-8.

Haas, H. R. de (2011) 'In search of profit in the High Arctic', in Hacquebord, L. (ed.) *LASHIPA – History of large-scale resource exploitation in polar areas*, Groningen: Arctic Centre and Barkhuis, pp. 47-60.

Hacquebord, L. (1984) *Smeerenburg. Het verblijf van Nederlandse walvisvaarders op de westkust van Spitsbergen in the 17e eeuw*, Unpublished PhD thesis, University of Amsterdam;

Hacquebord, L. (1996) 'Whaling stations as bridgeheads for exploration of the

Arctic regions in the sixteenth and seventeenth century', *International Conference on Shipping, Factories and Colonization 1994*. Brussels 24-26 November, Brussels: Koninklijke Academie van België, pp. 289-97.

Hacquebord, L. (ed.) (2012) *LASHIPA. History of large scale exploitation in polar areas,* Groningen: Barkhuis.

Hacquebord, L. and Avango, D. (2009) 'Settlements in an Arctic Resource Frontier Region', *Arctic Anthropology*, 46 (1-2), pp. 25-39.

Hacquebord, L., Steenhuisen, F., and Waterbolk, H. J. (2003) 'English and Dutch whaling trade and whaling stations in Spitsbergen (Svalbard) before 1660', *International Journal of Maritime History*, 2 (15), pp. 117-34.

Hanoa, R. (1993) *Kings Bay Kull Comp. A/S 1917-1992 – fra gruvedrift til forskningsservice på Svalbard*, Oslo: Schibsted.

Hardesty, D. L. (2010) *Mining archaeology in the American West: a view from the Silver State,* Lincoln: University of Nebraska Press and the Society for Historical Archaeology.

Hartnell, C. C. (2009) *Arctic network builders: the Arctic Coal Company's operations on Spitsbergen and its relationship with the environment.* PhD thesis. Michigan Technological University.

Harvey, C. and Taylor, P. (1987) 'Mineral wealth and economic development: foreign direct investment in Spain, 1851-1913', *Economic History Review*, 40 (2), pp. 185-207.

Hayter, R., Barnes, T. J., and Bradshaw, M. J. (2003) 'Relocating resource peripheries to the core of economic geography's theorizing: rationale and agenda', *Area*, 35 (1), pp. 15-23.

Heldal, T. and Neeb, P. R. (2000) 'Natural stone in Norway: production, deposits and developments', *NGU-Bulletin*, 436.

Hoel, A. (1942) *The place-names of Svalbard*, Oslo: Norges Svalbard- og Ishavs-Undersøkelser.

Hoel, A. (1966) *Svalbard: Svalbards historie 1596-1965, Part 1*, Oslo: Sverre Kildahls Boktrykkeri.

Hoel, A. (1967a) *Svalbard: Svalbards historie 1596-1965, Part 2,* Oslo: Sverre Kildahls Boktrykkeri.

Hoel, A. (1967b) *Svalbard: Svalbards historie 1596-1965, Part 3,* Oslo: Sverre Kildahls Boktrykkeri.

Hoem, S. (2005) Rapport istandsetting 2005. Brucebyen, Billefjorden – diverse istandsetting av flere hytter, Longyearbyen: Sysselmannen på Svalbard.

Hoem, S. (2007a) Befaringsrapport 2007 Prins Karls Forland – Fredete hytter mm., Longyearbyen: Sysselmannen på Svalbard.

Hoem, S. (2007b) Rapport istandsetting 2007. Brucebyen, Billefjorden – diverse istandsetting, Longyearbyen: Sysselmannen på Svalbard.

Huntford, R. (1985) *Shackleton*, London: Hodder and Stoughton.

Johannessen, L. J. (1997) *Hiorthhamn. Coal mining under difficult conditions*, Longyearbyen: Governor of Svalbard.

Johansen, B. F. (ed.) (2011) Cruise handbook for Svalbard, Tromsø: Norwegian Polar Institute.

Knudson, J. W. (1993) 'Late to the feast: newspapers as historical sources', *Perspectives*, 31 (7).

Krupnik, I. *et al.* (eds) (2011) *Understanding earth's polar challenges: International Polar Year 2007-2008,* Edmonton: CCI Press in collaboration with the University of the Arctic and ICSU/WMO Join Committee for International Polar Year 2007-2008.

Kruse, F. (2011) 'Four former British mining settlements on Spitsbergen', *Mining Perspective: the 8th International Mining History Congress 2009.* Penventon Park Hotel, Redruth 12-15 June. Truro: Cornwall and West Devon Mining Landscape World Heritage Site, Cornwall Council, pp. 117-24.

Lajus, J. (2004) 'From fishing to mining: the change of priorities in the development of the North and Russian expeditions to Spitsbergen in the early 20th century', in Wråkberg, U. (ed.) *Arktisk gruvdrift II. Teknik, vetenskap och historia i nor*, Stockholm, Jernkantoret, pp. 93-106.

Latour, B. (1987) *Science in action: how to follow scientists and engineers through society*, Milton Keynes: Open University Press.

Latour, B. (2005) *Reassembling the social: an introduction to actor-network-theory*, Oxford: Oxford University Press.

Laughton, J. K. (2004) 'Parry, Sir (William) Edward (1790-1855), *Oxford Dictionary of National Biography*, Oxford: Oxford University Press.

Law, J. (1987) 'Technology and heterogeneous engineering: the case of Portuguese expansion' in Bijker, W. E., Hughes, T. P., and Pinch. T. J. (eds.) *The social construction of technological systems: new directions in the sociology and history of technology,* Cambridge/MA: MIT Press.

Law, J. and Callon, M. (1992) 'The life and death of an aircraft: a network analysis of technical change', in Bijker, W. E. and Law, J. (eds) *Shaping technology/building society, studies in sociotechnical change.* London: MIT Press, pp. 21-52.

Law, J. and Hassard, J. (eds.) (1987) *Actor network theory and after*, Oxford & Keele: Blackwell & Sociological Review.

London, J. (1900) 'The economics of Klondike', *American monthly review of reviews*, pp. 70-4.

Lund-Mathiesen, I. (1974) *Grubesamfunnene på Spitsbergen 1905-1917 – syndikaliststreiken og militærintervensjonen sommeren* 1917, Oslo.

Mansfield, E. (1910) *Astria: the Ice Maiden*, London: Lonsdale Press.

Marks, S. and Trapido, S. (2004) 'Rhodes, Cecil John (1853-1902)', *Oxford Dictionary of National Biography*, Oxford: Oxford University Press.

Martin, P., Avango, D., Hartnell, C., Martin, S. R., Mishkar, L., Oglethorpe, M., Rosness, G., Tennant, E., and West, I. (2006) *[LASHIPA 1] Industrial heritage in the arctic: research and training on Svalbard, August 2004*, Michigan: MTU.

Mathisen, T. (1954) *Svalbard in international politics*, Oslo: Norwegian Polar Institute.

Michigan Technological University (2005) *Svalbard archaeology.* [Online]. Available at: http://www.svalbardarchaeology.org/ (Accessed 8 November 2011).

Mikalsen, H. (1995) *The Scottish Spitsbergen Syndicate og the Northern Exploration Company på Svalbard 1909-1920*, unpublished thesis, Oslo;

Mitchell, B. R. (1984) *Economic development of the British coal industry 1800-1914*, Cambridge: Cambridge University Press.

Morris, J. (1968) *Pax Britannica: the climax of an empire*, London: Penguin Books.

Notz, W. (1918) 'The world's coal situation during the war', *Journal of Political Economy*, 26 (6), pp. 567-611.

Oglethorpe, M. and Nisser, M. (2004) 'Industrial archaeology field course', *TICCIH Bulletin*, 27, p. 5.

Oglethorpe, M. (2005) 'Industrial heritage and national identity – sharing data, the importance of context and strategic priorities', *Industrial Archaeology Review* 27 (1), p. 27-31.

Pimlott Baker, A. (2004) 'Bainbridge, Emerson Muschamp (1845-1911)', *Oxford Dictionary of National Biography*, Oxford: Oxford University Press.

Planke, T. and Wammen, M. (2008) *Dokumentasjon av fangsthytter på Prins Karls Forland.* [Online]. Available at: http://www.sysselmannen.no/hoved.aspx?m=44267&amid=2380979 (Accessed: 8 November 2011).

Porteous, J. D. (1970) 'The nature of the company town', *Transactions of the Institute of British Geographers*, 51, pp. 127-42.

Reilly, J. T. (2009) *Greetings from Spitsbergen*, Trondheim: Tapir Academic Press.

Reymert, P. K. (2004) 'Den förste kulldrift på Svalbard. Søren Zakariassen og Kulkompagniet Isefjord Spitsbergen', in Wråkberg, U. (ed.) *Arktisk gruvdrift II. Teknik, vetenskap och historia i nor*, Stockholm, Jernkantoret, pp. 107-18.

Rice, T. (1986) *British oceanographic vessels 1800-1950*, London: The Ray Society.

Salmon, P. (1997) *Scandinavia and the great powers 1890-1940*, Cambridge: Cambridge University Press.

Singh, E. (1980) *The Spitsbergen (Svalbard) Question: United States Foreign Policy, 1907-1935*, Oslo, Universitetsforlaget.

Sörlin, S. (2002) 'Rituals and resources of natural history: the North and the Arctic in Swedish scientific nationalism', in Sörlin, S. and Bravo, M. (eds.), *Narrating the Arctic. A cultural History of Nordic scientific practices*, Canton/MA: Science History Publications, pp. 73-122.

Speak, P. (1992) 'William Speirs Bruce: Scottish nationalist and polar explorer', *Polar Record*, 28 (167), pp. 285-92.

Speak, P. (2003) *William Speirs Bruce. Polar explorer and Scottish nationalist,* Edinburgh: National Museums of Scotland Publishing.

Speak, P. (2004) 'Bruce, William Speirs (1867-1921)', *Oxford Dictionary of National Biography*, Oxford: Oxford University Press.

Spufford, F. (1996) *I may be some time,* London: Faber and Faber.

Starkov, V. F., Hacquebord, L., and Avango, D. (2010) 'Archaeological studies at Svalbard archipelago according to IPY Programme', in Kotlyakov, V. M. (ed.) *Changes of natural environment and climate; natural and possible consequent human-induced catastrophes. Vol. 3 Part 2: Natural processes in Polar Regions.* Moscow: Institute of Geography RAS, Institute of Physics of the Earth RAS, pp. 301-7.

Steiner, Z. and Dockrill, M. L. (2003) 'The Foreign Office reforms, 1919-21, *Historical Journal,* 17 (1), pp. 131-56.

Stoker, D. (1984) *The history of the British coal industry Vol. 2 1700-1830 The Industrial Revolution*, Oxford: Clarendon Press.

Strahan, A. (1919) 'The political status of Spitsbergen: discussion', *Geographical Journal*, 53 (2), pp. 91-6.

Sugden, D. E. (1982) *Arctic and Antarctic: a modern geographical synthesis*, Oxford: Basil Blackwell.

Supple, B. (1987) *The history of the British coal industry: Vol. 4 1913-1946 The political economy of decline*, Oxford: Clarendon Press.

Supple, B. (1989) 'The British coal industry between the wars', *Refresh*, 9, pp. 5-8.

Svendsen, H., Beszczynska-Møller, A., Hagen, J. O., Lefauconnier, B., Tverberg, V., Gerland, S., Ørbæk, J. B., Bischof K., Papucci, C., Zajaczkowski, M., Azzolini, R., Bruland, O., Wiencke, C., Winther, J.-G., and Dallmann, W. (2002) 'The physical environment of Kongsfjorden-Krossfjorden, an Arctic fjord system in Svalbard', *Polar Research,* 21 (1), pp. 133-66.

Swinney, G. N. (2001) 'Some new perspective on the life of William Speirs Bruce (1867-1921), with a preliminary catalogue of the Bruce collection of manuscripts in the University of Edinburgh', *Archives of Natural History*, 28 (3), pp. 285-311.

Swinney, G. N. (2002a) 'The training of a polar scientist: Patrick Geddes and the student career of William Speirs Bruce', *Archive of Natural History*, 29 (3), pp. 287-301.

Swinney, G. N. (2002b) 'William Speirs Bruce, the Ben Nevis Observatory, and Antarctic Meteorology', *Scottish Geographical Journal*, 118 (4), pp. 263-82.

Swinney, G. N. (2003) 'From the Arctic and Antarctic to 'the back parts of Mull': The life and career of William Gordon Burn Murdoch (1862-1939)', *Scottish Geographical Journal*, 119 (2), pp. 121-52.

Swinney, G. N. (2007) 'The Scottish National Antarctic Expedition (1902-04) and the founding of Base Orcadas', *Scottish Geographical Journal*, 123 (1), pp. 48-67.

Teisch, J. B. (2011) *Engineering nature: water, development, and the global spread of American environmental expertise*, Chapel Hill: University of North Carolina Press.

Thomson, J. O. (ed.) (1933) *Dr Salter of Tolleshunt D'Arcy: diary and reminiscences: 1849-1932*, London: The Bodley Head.

Turrell, R. V. and Van Helten, J.-J. (1986) 'The Rothschilds, the Exploration Company and mining finance', *Business History*, 28 (2), pp. 181-205.

Ulfstein, G. (1995) *The Svalbard treaty*, Oslo: Scandinavian University Press.

Vattens, O. (1980) *Longyearbyen 1905-1935: et norsk samfunns oppbygging og utvikling*, unpublished thesis, Trondheim.

Warman, R. M. (2003) 'The erosion of Foreign Office influence in the making of foreign policy, 1916-1918', Historical *Journal,* 15 (1), pp. 133-59.

Wilkins, M. (1988) 'The free-standing company, 1870-1914: an important type of British foreign direct investment', *Economic History Review*, 41(2), pp. 259-82.

Wilmot, D. (2005) 'Emerson Bainbridge of Newcastle & Sheffield, an overlooked entrepreneur', *Yorkshire Archaeological Journal*, 77, pp. 241-52.

Woodland, J. (2011) 'Experiences of the first gold mining companies in Australia', *Mining Perspective: the 8th International Mining History Congress 2009.* Penventon Park Hotel, Redruth 12-15 June. Truro: Cornwall and West Devon Mining Landscape World Heritage Site, Cornwall Council.

Wråkberg, U. (2002) 'The politics of naming: contested observations and the shaping of geographical knowledge', in Bravo, M. and Sörlin, S. (eds.) *Narrating the Arctic. A cultural history of Nordic scientific practices,* Canton, MA: Watson Publishing International.

Østreng, W. (1978) *Politics in high latitudes - The Svalbard archipelago,* Montreal: McGill-Queen's University.

Appendix 1

Place-names used in the text

Contemporary British place-names	Current Norwegian place-names
Adolf Bay	Adolfbukta
Advent Bay	Adventfjorden
Advent City	Advent City
Advent Valley	Adventdalen
Agardh Bay	Agardhbukta
Aitken Valley	Aitkendalen
Anser Islands	Gåsøyane
Antarctic Bay	Antarcticbogen
Axel Island	Akseløya
Barents Island	Barentsøya
Bear Island	Bjørnøya
Beinn Dhubh ('Black Hill')	Svarthaugen
Bell Sound	Bellsund
Berzelius Valley	Berzeliusdalen
Bjona Haven	Bjonahamna
Braganza Bay	Braganzavågen
Breccia Island	Juttaholmen
Bruce City	Brucebyen
Bünsow Land	Bünsow Land
Calypso Bay	Calypsostranda
Calypso Beach	Calypsostranda
Camp Asbestos	Asbestodden
Camp Bell	Camp Bell
Campbell Range	Campbellryggen
Camp Campbell	Lægerneset
Camp Claus Andersen	
Camp Franklin	
Camp Gilson	
Camp Jacobsen	Camp Jacobsen

Contemporary British place-names	Current Norwegian place-names
Camp Jacobsen	(Probably) Iron Mountain Camp
Camp Margaret	
Camp Marie Killengreen	
Camp Millar	Camp Millar
Camp Morton	Camp Morton
Camp Parry	Hi-Fix
Camp Petter Trondsen	
Camp Sabine	
Camp Scoresby	
Camp Smith	
Camp Svendsen	
Camp Violet	Camp Violet
Camp Williamson	Camp Williamson
Camp Zoe	Camp Zoe
Cape Ahlstrand	Ahlstrandhaløya
Cape Borthen	Kapp Borthen
Cape Ekholm	Kapp Ekholm
Cape Lee	Kapp Lee
Cape Lyell	Kapp Lyell
Cape Müller	Müllerneset
Cape Napier	Kapp Napier
Cape Thordsen	Kapp Thordsen
Carron River	Carronelva
Changing Point	Mistakodden
Coal Mountain	Kolfjellet
Coles Bay	Colesbukta
Copper Camp	Copper Camp
Cross Bay	Krossfjorden
Davis City	Camp Morton
Davis Harbour	Davishamna
Davis Island	Observasjonsholmen
De Geer Valley	De Geerfjellet
Dead Bear Bay	Daudbjørnfjellet

Contemporary British place-names	Current Norwegian place-names
Duck Island	Bellsundholmen (Reiniusøyne)
Dunder Bay	Dunderbukta
Dun Islands	Dunøyane
East Island	Fjørholmen (Dunøyane)
Ebba Valley	Ebbadalen
Ebeltoft Haven	Ebeltofthamna
Edge Island	Edgeøya
Eider Island	Eholmen
Ekholm River	
Farm Harbour	Farmhamna
Foreland Sound	Forlandsundet
Forland Laichs	Hyttevatna
Gerrit River	Gerritelva
Gerrit Valley	
Gips Bay	Gipsvika
Gips River	Gipsdalselva
Gips Valley	Gipsdalen
Goose Harbour	Gåshamna
Green Harbour	Grønfjorden
Grumant City	Grumantbyen
Hamburg Bay	Hamburgbukta
Hayes Glacier	Hayesbreen
Horn Sound	Hornsund
Ice Fjord	Isfjorden
Icefiord	Isfjorden
Icefiord Lead and Zinc Mine	
Icefiord Zinc and Lead Mine	
Inchcolm Hut	Inchcolmhytta
Inglefield Glacier	Inglefieldbreen
Iron Mountain	Jarnfjellet
Iron Mountain Camp	
Isbjorn Haven	Isbjørnhamna
Ivory Gate Valley	

Contemporary British place-names	Current Norwegian place-names
Kenmore	Kenmore
Kings Bay	Kongsfjorden
Klaas Billen Bay	Billefjorden
Lady Davis Harbour	Davishamna
Lesser Islands	Løvenøyane
Lizet Rocks	Lisettholmane
Longyear City	Longyearbyen
Longyear Valley	Longyeardalen
Lowe Sound	Van Mijenfjorden
Magdalena Bay	Magdalenabukta
Maples Island	Storholmen
Marble Island	Blomstrandhaløya
Martin Range	Martinfjella
Michiel Rinder's Bay	Rindersbukta
Middle Hook	Midterhuken
Miller Glacier	Millerbreen
Mimer Bay	Mimerbukta
Möller Bay	Möllerhamna
Mohn Bay	Mohnbukta
Mount Cadell	Cadellfjellet
Mount Hedgehog	Hedgehogfjellet
Mount Keilhau	Keilhaufjellet
Mount Tjosaas	Tjosaasfjellet
Mount Tyrrell	Tyrrellfjellet
No. 2 Marble Island	Storholmen
Nordenskiöld Glacier	Nordenskiöldbreen
Osborne Glacier	Osbornebreen
Oscar II Land	Oscar II Land
Phantom Bay	Phantomvika
Point McVitie	McVitiepynten
Point Napier	Dawespynten
Port Peirson	Peirsonhamna
Post Glacier	Von Postbreen

Contemporary British place-names	Current Norwegian place-names
Prince Charles Foreland	Prins Karls Forland
Rabot Glacier	Rabotbreen
Recherche Bay	Recherchefjorden
Richard Lagoon	Richardlaguna
River Carron	Carronelva
River Gerrit	Gerritelva
Safe Harbour	Trygghamna
Sassen Bay	Sassendalen
Sassendal River	Sassenelva
South Cape	Sørkapp
South Cape Islands	Sørkappholmane
St. Johns Bay	St. Jonsfjorden
Stor Fiord	Storfjorden
Stor Fjord	Storfjorden
Temple Bay	Tempelfjorden
Thousand Islands	Tusenøyane
Tinayre Bay	Tinayrebukta
Tinayre Glacier	Tinayrebreen
Usher Glacier	Usherbreen
Van Keulen Bay	Van Keulenfjorden
Vogel Hook	Fuglehuken
Volage Glacier	Antoniabreen
West Spitsbergen	Spitsbergen
Whales Bay	Kvalvågan
Wiche Bay	Wichebukta
Windy Point	Vindodden
Wordie Crags	Wordiekammen
Zinc Island	Sinkholmen

Appendix 2

Staff of the Scottish Spitsbergen Syndicate

1909

Brown, Robert Neal Rudmose
Botanist
DSc
University of Sheffield

Bruce, William Speirs
Leader
LLD
Scottish Oceanographical Laboratory

Bryce, F.

Burn Murdoch, James Victor
Geologist
FRGS, FGS, AIMM
Cousin of W. G. Burn Murdoch

Geddes, Alastair C. B.
Geologist
Son of Prof. Patrick Geddes

Hannay, Harry
Geologist, mining engineer
RE, MC

Kerr, Gilbert
Piper, handy man
Scotia expedition 1902-4

Mathieson, John
Surveyor
FRSGS
HM Ordnance Survey

Miller, Ernest Alexander
Meteorologist
Spitsbergen expedition 1906

Napier, Frank B.
Master 'Conqueror'

1909 cont.

Peach, Angus MacEwen
Geologist
Son of Sir Benjamin Peach

Sword, James
Mate 'Conqueror'

1912

Brown, Robert Neal Rudmose

Bruce, William Speirs

1914

Brown, Robert Neal Rudmose

Bruce, William Speirs
Leader

Burn Murdoch, James Victor

Craig, Robert M.
Geologist
University of St Andrews

Koeppern, John Henry
Naturalist
Father: friend of Rottenburg
Uncle: German minister in Kristiania
Changed his name to Kenneth in 1918

'Four others'

1919

Allan, Douglas Alexander
Geologist

Andrew Kyle & Co.,
Drillers
Galston

Bell, Richard
Cook

Brown, Robert Neal Rudmose
Second in command

Brown, Robley
Medical officer

Bruce, William Speirs
Leader

Burn Murdoch, James Victor
Senior geologist, mining engineer

Campbell, A. Fleming
Mining engineer

Carr, George
Practical miner

Charlesworth, John Kaye
Senior geologist

Cowan, George M.
Surveyor

Ferrier, Alexander
Cook

Ferrier, John
Fitter

Grant, Annan
Practical miner

1919 cont.

Grant, John
Practical miner
Halcrow, M.
Second engineer
Halcrow, W. J.
Second officer
Horne, John
Consultant
Kenneth, John Henry
Naturalist
Laffey, Thomas
Practical miner
Leitch, Colin M.
Surveyor
Mathieson, John
Surveyor
Miller, George W.
Leader of mining party
Napier, Frank B.
Master, 'Petunia'
Patterson, I.
Third engineer
Raeburn, C.
Junior mining engineer
Ross, George
Junior geologist
Ross, J.
Chief engineer
Sandys, H.
Representative, Middlesbrough
Scott, Chester M.
Organising secretary
Scott, James
Mining engineer
Shanks, James
General worker
Stanners, Thomas
Practical miner

1919 cont.

Stanners, Walter
Practical miner
Stevens, Alexander O.
Geologist
Thomson, Andrew
Practical miner
Thomson, G.
First officer
Tjosaas
Master, 'Phantom'
Tyrrell, George Walter
Chief geologist
Uncles, James
Piper, handy man
Wordie, James Mann
Senior geologist
'About 40 officers and crew'
'A party of up to 100'

1920

Allan, Douglas Alexander
Campbell, A. Fleming
Chief mining engineer
Campbell, Donald
Assistant to Tyrrell
Campbell, Robert
Geologist
Dron, R. L. A.
Head of miners
Fairweather, G. W. C.
Medical officer
Flett, W. G.
Master 'Lady of Avenal'
Johannesen, Kristian
Master 'Autumn'
Mason, David L.
Head of drillers

1920 cont.

Mathieson, John
Leader
McDonald
Moir, J.
Assistant to A. F. Campbell
Napier, Frank B.
Master 'Easonian'
Pollock, C. M.
Assistant to Mathieson
Scott, Chester M.
Organising secretary
Tyrrell, George Walter
Geologist
Watson, Hugh
Assistant to Mathieson
Wordie, James Mann
Geologist
'9 Norwegians'
'Staff larger than 1919'
'In all about 50'

1920 Directors' Visit

Aitken, Alfred Niven Gillies
Secretary
Allan, Douglas Alexander
Brown, Robert Neal Rudmose
Leader of party
Bruce, William Speirs
Cadell, Henry Moubray
Director
Cadell, John
Medical officer
Maxtone Graham, James
Director

1920 Directors' Visit cont.

Meakin, J.
Medical officer
Morton, A.
Valet
Skelton, A. N.
Counsel
Urmston, Charles Hanson
Director
Usher, Thomas Leslie
Director

1921

Allan, Douglas Alexander
Mining engineer
Cooper, James
Mining engineer
Dron, Robert Wilson
Mining engineer
Gypsum expert
Finlay, T. R.
Geologist, assistant to Allan
Maclean, D.
Assistant
Mathieson, John
Leader
Scott, Chester M.
Organising secretary
'3 miners'

1922

Alexander, James
Engineer
Blaik, Hugh
Engineer

1922 cont.

Dron, R. L. A.
Mining engineer
Mason, David
Head driller
Mathieson, John
Leader
Mitchell, Andr.
Sands, Thomas
Scott, Chester M.
Organising secretary
Tait, James
Assistant surveyor
'15 participants in all'

1923

Brown, Robert Neal Rudmose
Cowan, George M.
Mathieson, John

1924

Burn Murdoch, James Victor
Campbell, Ivor
Mathieson, John

1925

Brown, Robert Neal Rudmose
Mathieson, John
Made two trips

1928

Mathieson, John

1948

Black, I. R. H.
Captain
Elbo, John Gregers
Gee, E. R.
Glen, A. R.
Prytz, T.
Urmston, C. W.

Appendix 3

Learned papers of the Scottish Spitsbergen Syndicate

Brown, R. N. R. (1908) 'The flora of Prince Charles Foreland', *Transactions of the Botanical Society of Edinburgh*, 23, 8 pages.

Brown, R. N. R. (1911) 'British work in Spitsbergen: some historical notes', *Scottish Geographical Magazine*, 27, pp. 180-7.

Brown, R. N. R. (1912) 'The commercial development of Spitsbergen', *Scottish Geographical Magazine*, 28, pp. 561-71.

Brown, R. N. R. (1915) 'Spitsbergen in 1914', *Geographical Journal*, 46, pp. 10-23. *(The paper was meant to be given by Bruce who had already left for the Seychelles.)*

Brown, R. N. R. (1919) 'Spitsbergen, terra nullius', *Geographical Review*, 7 (5), pp. 311-21.

Brown, R. N. R. (1919) 'The present state of Spitsbergen', *Scottish Geographical Magazine*, 35, pp. 201-12.

Brown, R. N. R. (1920a) 'Recent developments in Spitsbergen', *Scottish Geographical Magazine*, 36, pp. 111-6.

Brown, R. N. R. (1920b) *Spitsbergen: an account of exploration, hunting, the mineral riches & future potentialities of an Arctic archipelago*, London: Seeley, Service & Co.

Brown, R. N. R., Trevor-Battye, A., Yate, C., Wordie, J. M. (1921) 'Discussion: Present-day conditions in Spitsbergen', *Geographical Journal*, 58 (1), pp. 45-9.

Brown, R. N. R. (1922) 'Obituary W. S. Bruce, LL.D.', *Scottish Geographical Magazine*, 38, pp. 46-8.

Brown, R. N. R. (1922) 'Mining development in Spitsbergen', *Scottish Geographical Magazine*, 38, pp. 115-7.

Brown, R. N. R. (1923) *A naturalist at the poles: The life, work, and voyages of Dr. W. S. Bruce, the polar explorer*, London: Seeley, Service.

Brown, R. N. R. (1927) *The polar regions: A physical and economic geography of the Arctic and Antarctic*, Methuen.

Brown, R. N. R. (1927) 'Some problems of polar geography', *Scottish Geographical Magazine*, 43 (5), pp. 257-81.

Brown, R. N. R. (1933) 'Recent polar work – some criticisms', *Polar Record* 1 (5), pp. 62-6.

Brown, R. N. R. (1944) 'Svalbard place-names', *Scottish Geographical Magazine*, 60 (3), pp. 67-70.

Brown, R. N. R. (1950) 'Svalbard of to-day', *Scottish Geographical Magazine*, 66(3-4), pp. 173-7.

Brown, R. N. R. (1955) 'Spitsbergen', *Encyclopædia Britannica*, 21, pp. 248–249.

Brown, R. N. R. (1960) 'Spitsbergen characters', *Polar Record*, 10 (66).

Bruce, W. S. (1900) 'Spitsbergen 1898 and 1899; voyages with H.S.H. the Prince of Monaco', *Scottish Geographical Magazine*, 16 (9), pp. 534-50.

Bruce, W. S. (1907) 'Prince Charles Foreland', *Scottish Geographical Magazine*, 23 (3), pp. 141-56.

Bruce, W. S. (1908) 'The exploration of Prince Charles foreland 1906-1907, *Geographical Journal*, 32 (2), pp. 139-48.

Bruce, W. S. (1911) 'The Prince of Monaco and oceanographical research', *Scottish Geographical Magazine*, 27 (3), pp. 142-6.

Bruce, W. S. (1913) 'The completion of the map of Prince Charles Foreland, Spitsbergen', *Scottish Geographical Magazine*, 29 (11), pp. 598-9.

Bruce, W. S. (1914) *Spitsbergen: past and present*, London: Sherratt & Hughes. *(Reprinted from the Journal of the Manchester Geographical Society, Parts II and III, 1913)*

Cadell, H. M. (1920) 'Spitsbergen in 1919', *Scottish Geographical Magazine*, 36, pp. 1-10.

Cadell, H. M. (1921) 'Coal-mining in Spitsbergen', *Transactions of the Institution of Mining Engineers*, 60, pp. 119-42.

Cadell, H. M. (1925) *The rocks of West Lothian*, Edinburgh and London: Oliver and Boyd.

Craig, R. M. (1916) 'The geology of Prince Charles Foreland', *Transactions of the Edinburgh Geological Society*, 10, p. 276 ff.

Dron, R. L. A. (1924) 'Boring in frozen strata', *Transactions of the Institution of Mining Engineers*, 67, pp. 186-91. (Discussion incl. Andrew Kyle (Galston), James Cooper (Edinburgh) pp. 333-6)

Mathieson, J. (1932) 'The story of Antarctic exploration, 1716-1931', *Scottish Geographical Magazine*, 48 (6), pp. 321-9.

Mathieson, J. (1934) 'The story of Arctic voyages and exploration', *Scottish Geographical Magazine*, 50 (5), pp. 281-308.

Peach, A. M. (1916) 'The peri-glacial platform and raised beaches of Prince Charles Foreland', *Transactions of the Edinburgh Geological Society*, 10, pp. 289-307.

Tyrrell, G. W. (1921) 'Geographical observations in Spitsbergen', *Scottish Geographical Magazine*, 37, pp. 227-42.

Tyrrell, G. W. (1922) 'The pre-Devonian basement complex of Central Spitsbergen', *Transactions of the Royal Society of Edinburgh*, 53 (1), pp.

209-29.

Tyrrell, G. W. (1924) 'The geology of Prince Charles Foreland, Spitsbergen', *Transactions of the Royal Society of Edinburgh*, 53(2), pp. 443-78.

Wordie, J. M. (1921) 'Present-day conditions in Spitsbergen', *Geographical Journal*, 58 (1), pp. 25-45

English summary

This study concerns the involvement of four British companies in the industrialisation of Spitsbergen in the European Arctic in the first half of the twentieth century. The companies were the Spitzbergen Coal & Trading Co., the Spitzbergen Mining & Exploration Syndicate, the Northern Exploration Co., and the Scottish Spitsbergen Syndicate. The combined historical context of Britain's leading coal industry, its global empire, and its contributions to polar exploration and science bore down on them. Against this multifaceted background, it is particularly interesting to ask, 'What were the driving forces behind the development of the British mining industry on Spitsbergen between 1904 and 1953?'

Rooted in the complementary fields of industrial archaeology and mining history, the study is embedded in the LASHIPA (Large-scale Historical Exploitation of Polar Areas) Project of the recent International Polar Year (IPY) 2007-2008. It employs a theoretical framework based on the core-periphery model and the actor-network theory (ANT). Archaeological fieldwork and extensive archival research gave rise to six empirical chapters, each of which addresses the four sub-questions: why were the companies started; what were their operational choices; why were they discontinued; and what were the economic, political, and environmental consequences?

It transpires that the archaeological landscape of exploration is on the whole poorly recognised, let alone understood, and therefore deserving of further fieldwork and standardisation. The companies were founded for primary economic reasons. Two were mining companies that were terminated prior to the First World War because their local networks failed to produce the natural resources their global network was expecting. Two were exploration companies that made use of the greatly politicised climate after the international conflict to prolong their existence until it was no longer commercially viable. Although the Scottish syndicate was able to sell its properties to a development company, it is doubtful if the negligible dividend that was paid can be called a success.

In conclusion, the research confirmed the hypothesis: while economic driving forces primarily determined the British presence on Spitsbergen, political motivations amplified at specific moments in time, i.e. after the war, and the repercussions of these motivations sustained the companies beyond economic feasibility. Although ANT in particular helped to generate these answers, the study nonetheless revealed weaknesses in the theory that should be addressed in future work.

Nederlandse samenvatting

Dit onderzoek behandelt de betrokkenheid van vier Britse bedrijven in de industrialisatie van Spitsbergen in de eerste helft van de twintigste eeuw. De bedrijven waren de Spitzbergen Coal & Trading Co., Spitzbergen Mining & Exploration Syndicate, Northern Exploration Co., en Scottish Spitsbergen Syndicate. Deze bedrijven kunnen niet los gezien worden van het wereldomspannende Britse imperium, de dominantie van de Britse mijnbouwindustrie en de Britse bijdragen aan de ontdekking van het Arctisch gebied. In deze context is interessant om de vraag te stellen: 'wat waren de drijvende krachten achter de ontwikkeling van de Britse mijnbouwindustrie op Spitsbergen tussen 1904 en 1953?'

Deze gecombineerd archeologische en historische studie is ingebed in het LASHIPA (*Large-scale Historical Exploitation of Polar Areas*) Project dat weer deel uitmaakt van het laatste Internationale Pooljaar (2007-2008). Het theoretische uitgangspunt van dit onderzoek is een combinatie van het *core-periphery model* en de *actor-network theory* (ANT). Archeologisch veldwerk en uitgebreid archiefonderzoek vormen de basis voor zes empirische hoofdstukken, welke elk vier deelvragen behandelen. De deelvragen zijn: waarom warden deze bedrijven opgericht; welke operationele keuzes maakten deze bedrijven; waarom warden ze uiteindelijk opgeheven; en wat was de invloed van de activiteiten van deze bedrijven op de natuurlijke omgeving en de politieke en economisch context?

Omdat de materiële cultuur van deze periode op Spitsbergen slecht herkend en begrepen bleek, was veldwerk erg belangrijk. De bedrijven zelf bleken vooral vanwege economische motieven opgericht. Twee mijnbouwbedrijven werden al voor de Eerste Wereldoorlog opgeheven omdat hun lokale netwerk er niet in slaagde het mondiale netwerk te voorzien van de beoogde grondstoffen. Twee bedrijven gericht op exploratie maakten gebruik van het politieke klimaat na de oorlog om hun activiteiten voort te zetten tot deze niet meer rendabel waren. Hoewel het Schotse syndicaat zijn eigendommen verkocht aan een ontwikkelingsmaatschappij, is het zeer de vraag of het bedrijf door het uitkeren van een minimaal dividend een succes genoemd kan worden.

Uiteindelijk bevestigde het onderzoek de hypothese dat hoewel economische belangen de belangrijkste drijfveer waren voor de Britse aanwezigheid op Spitsbergen, ook politieke motieven een rol speelden direct na de Eerste Wereldoorlog. Deze politieke motieven zorgden er bovendien voor dat de bedrijven actief bleven nadat het economische belang was weggevallen. Hoewel de toepassing van ANT bijzonder geschikt bleek bij het beantwoorden van de vragen, toont dit onderzoek wel enige zwakke punten in de theorie die in toekomstig onderzoek aangepakt dienen te worden.

14W83161/ T1/ 9789491431333